Building Materials and Construction

Building Materials and Construction

Edited by **Seth Royal**

New York

Published by Willford Press,
118-35 Queens Blvd., Suite 400,
Forest Hills, NY 11375, USA
www.willfordpress.com

Building Materials and Construction
Edited by Seth Royal

International Standard Book Number: 978-1-68285-082-4 (Hardback)

Printed in the United States of America.

Contents

Preface

Building materials can be natural substances or man-made products. Different properties of building materials are studied under various disciplines of engineering. This book attempts to understand the multiple branches that fall under the vast field of building materials and construction. It is a compilation of chapters that strive to provide an in-depth knowledge of this area by discussing topics such as shear moments and forces, surveying, environmental engineering, soil mechanics, etc. The researches and case studies incorporated in this book will prove to be beneficial to students, academicians as well as professionals engaged in this field.

The researches compiled throughout the book are authentic and of high quality, combining several disciplines and from very diverse regions from around the world. Drawing on the contributions of many researchers from diverse countries, the book's objective is to provide the readers with the latest achievements in the area of research. This book will surely be a source of knowledge to all interested and researching the field.

In the end, I would like to express my deep sense of gratitude to all the authors for meeting the set deadlines in completing and submitting their research chapters. I would also like to thank the publisher for the support offered to us throughout the course of the book. Finally, I extend my sincere thanks to my family for being a constant source of inspiration and encouragement.

Editor

Magneto–thermo-viscoelastic material with a spherical cavity

M. A. Ezzat* and H. M. Atef

Department of Mathematics, Faculty of Education, Alexandria University, Alexandria, Egypt.

The present paper is concerned with a homogeneous isotropic perfect conducting viscoelastic body with a spherical cavity subjected to both ramp-type heating and external constant magnetic field. The model of the equations in the context of generalized thermoelasticity with one relaxation time is introduced. The closed form solution for distributions of displacement; temperature, strain, and stress are obtained by using the Laplace transform and the state-space approach. Numerical results applicable to a material - like copper are presented graphically.

Key words: Thermoelasticity, magneto-thermo-viscoelasticity, Lord-Shulman theory, state space approach.

INTRODUCTION

The classical uncoupled theory of thermoelasticity predicts two phenomena not compatible with physical observations. First, the equation of heat conduction of this theory does not contain any elastic terms; second, the heat equation is of a parabolic type, predicting infinite speeds of propagation for heat waves.

Biot (1956) introduced the theory of coupled thermoelasticity to overcome the first shortcoming. The governing equations for this theory are coupled, eliminating the first paradox of the classical theory. However, both theories share the second shortcoming since the heat equation for the coupled theory is of a mixed parabolic-hyperbolic type.

Three generalizations to the coupled theory are considered here. The first is due to Lord and Shulman (1967), who obtained a wave-type heat equation by postulating a new law of heat conduction (the Maxwell–Cattaneo equation) to replace the classical Fourier law. Because the heat equation of this theory is of the wave-type, it automatically ensures finite speeds of propagation for heat and elastic waves. The remaining governing equations for this theory, namely, the equations of motion and constitutive relations, remain the same as those for the coupled and the uncoupled theories. Joseph and Preziosi (1989, 1990) state that the Maxwell–Cattaneo equation is the most obvious and simple generalization of the Fourier law that gives rise to a finite propagation speed.

The second generalization to the coupled theory of elasticity is what is known as the theory of thermoelasticity with two relaxation times or the theory of temperature-rate-dependent thermoelasticity. Müller (1971), in a review of the thermodynamics of thermoelastic solids, proposed an entropy production inequality, with the help of which he considered restrictions on a class of constitutive equations. A generalization of this inequality was proposed by (Green and Laws (1972). Green and Lindsay (1975) obtained another version of the constitutive equations. These equations were also obtained independently and more explicitly by Şuhubi (1975). This theory contains two constants that act as relaxation times and modify all the equations of the coupled theory, not only the heat equation. The classical Fourier law of heat conduction is not violated if the medium under consideration has a center of symmetry. One can refer to Ignaczak (1991), for a review, presentation of the two theories, and some important results obtained in this field.

The third generalization to the coupled theory is known as the dual-phase-lag thermoelasticity, proposed by Chandrasekhraiah (1998), in which the Fourier law is

*Corresponding author. E-mail: maezzat2000@yahoo.com.

replaced by an approximation to a modification of the Fourier law with two different translations for the heat flux and the temperature gradient. One can refer to Hetnarski and Ignaczak in their survey article (1999) in which they examined five generalizations to the coupled theory and obtained a number of important analytical results.

Viscoelastic materials are those for which the relationship between stress and strain depends on time. All materials exhibit some viscoelastic response. In common metals such as steel, aluminum, copper etc. at room temperature and small strain, the behavior does not deviate much from linear elasticity. Synthetic polymer, wood as well as metals at high temperature display significant viscoelastic effects. With the rapid development of polymer science and plastic industry, as well as the wide use of materials under high temperature in modern technology and application of biology and geology in engineering, the theoretical study and applications in viscoelastic materials has become an important task for solid mechanics.

Linear viscoelastic materials are rheological materials that exhibit time temperature rate-of-loading dependence. When their response is not only a function of the current input, but also of the current and past input history, the characterization of the viscoelastic response can be expressed using the convolution (hereditary) integral. A general overview of time-dependent material properties has been presented by Tschoegl (1997). Additionally, a detailed description of the physical response of linear voiscoelastic materials has been explained by Lee and Knauss (2000), based on ramp tests to determine the relaxation modulus, which is a time-domain linear viscoelastic response function. The mechanical-model representation of linear viscoelastic behavior results was investigated by Gross (1953), Staverman and Schwarzl (1956), Alfery and Gurnee (1956) and Ferry (1977). One can refer to Atkinson and Craster (1995) for a review of fracture mechanics and generalizations to the viscoelastic materials.

The theory of coupled thermo-viscoelasticity, and the solutions of some boundary value problems of thermo-viscoelasticity were investigated by Biot (1954), Morland and Lee (1960), Ilioushin and Pobedria (1970) and Gurtin (1972).

The theory of magneto-thermo-elasticity (MTE) was developed with the possibilities of their extensive practical applications in diverse fields such as geophysics, optics and acoustics and so on. A survey of relevant magneto-thermo–elasticity theories were studied by Wilson (1963) and Paria (1967) in the second half of the last century. Using generalized theory of heat conduction of Lord–Shulman, a large number of research workers made valuable contributions in magneto-thermo-elasticity during the last three decades. Öncü and Moodie (1989, 1990) made an analysis of the thermal transient generated by non-uniform sources applied to circular cavities and circular hole in inhomogeneous conductor. (Sherief and Ezzat, 1996) solved a thermal shock half-space problem using asymptotic expansions. Lately, Sherief and Ezzat (1998) solved a problem for an infinitely long annular cylinder, while Ezzat (1997) and Ezzat and Youssef (2005) solved a two-dimensional problem for perfectly conducting media.

The theory of magneto-thermo-viscoelasticity (MTVE) has aroused much interest in many industrial appliances, particularly in nuclear devices, where there exists a primary magnetic field. Various investigations have been carried out by considering the interaction between magnetic, thermal and strain fields. Analyses of such problems also influence various applications in biomedical engineering as well as in different geomagnetic studies. Misra et al. (1992), has studied a one-dimensional uncoupled magnetic-thermoelastic problem in a viscoelastic medium using Maclaurin's approximation method valid for only a specific range of parameters.

The generalized thermo-viscoelasticity models ignoring the relaxation effects of the volume are established by El-Karamany and Ezzat (2004a, b). Among the theoretical contributions to the subject are the proofs of uniqueness theorems under different conditions by Ezzat and El-Karamany (2002) and the boundary element formulation were presented by (El-Karamany and Ezzat, 2002). A state-space method for the calculation of dynamic response of systems made of viscoelastic materials with exponential type relaxation kernels was introduced by (Menon and Tang, 2004). Extensions of thermo-viscoelastic and magneto-thermo-viscoelastic problems in generalized theory are found to be present in the works of many researchers amongst whom are Mukhopadhyay and Bera (1992) and Rakshit and Mukhopadhyay (2005).

Youssef (2005), studied the problem of generalized thermoelasticity with one relaxation time with variable modulus of elasticity and the thermal conductivity were used to solve a problem of an infinite material with a spherical cavity. The inner surface of the cavity was taken to be traction free and acted upon by a thermal shock to the surface and Youssef and Al-Harby (2007) solved the previous problem in two-temperature theory when elastic parameters are taken as constant values.

This paper introduces a model for generalized magneto- thermo- viscoelasticity with one relaxation time. For this model, we shall formulate the state space approach developed in Ezzat (2008) to problems of magneto-thermoviscoelasticity. The resulting formulation is applied to a thermal stresses problem of an electrically perfect conducting infinite solid with a spherical cavity subjected to ramp-type heating in the presence of a constant magnetic field. Laplace transform technique is used throughout. The inversion of the transforms is carried out using a numerical inversion technique (Honig and Hirdes, 1984).

Formulation of the problem

We shall consider a homogeneous isotropic thermo-viscoelastic medium occupying the region R ≤ r < ∞ of a perfect electrically conductivity permeated by an initial constant magnetic field H_o, where R is the radius of the shell. Due to the effect of this magnetic field there arises in the conducting medium an induced magnetic field *h* and induced electric field *E*. Also, there arises a force F (the Lorentz Force). Due to the effect of this force, points of the medium undergo a displacement u, which gives rise to a temperature.

The linearized equations of electromagnetism for slowly moving media are as described by Ezzat (1997):

$$\text{Curl } \boldsymbol{h} = \boldsymbol{J}, \tag{1}$$

$$\text{curl}\,\mathbf{E} = -\mu_0 \frac{\partial \mathbf{h}}{\partial t} \tag{2}$$

$$\boldsymbol{B} = \mu_o \boldsymbol{H}, \tag{3}$$

$$div\,\boldsymbol{B} = 0, \tag{4}$$

where, $\boldsymbol{H} = \boldsymbol{H_o} + \boldsymbol{h}$ is the total magnetic field vector, $\boldsymbol{J}$ is current density vector, $\boldsymbol{B}$ is the magnetic induce vector, μ_0 is the magnetic permeability.

The above field equations are supplemented by constitutive equations which consist first of Ohm's law

$$\boldsymbol{E} = -\mu_0 \frac{\partial \boldsymbol{u}}{\partial t} \times \boldsymbol{H}_0. \tag{5}$$

The second constitutive equation is the one for the Lorenz force which is

$$\boldsymbol{F}=\boldsymbol{J}\times\boldsymbol{B}. \tag{6}$$

The third constitutive equation is the stress – displacement – temperature relation for viscoelastic medium of Kelvin – Voigt type.

$$\tau_{ij} = 2\left(\mu_e + \mu_v \frac{\partial}{\partial t}\right) e_{ij} + \left(\lambda_e + \lambda_v \frac{\partial}{\partial t}\right) e\,\delta_{ij} - \gamma\theta\,\delta_{ij}, \tag{7}$$

where, u = (u_r, u_ψ, u_ϕ) are displacement vector, τ_{ij} are the components of the stress tensor, e_{ij} are the components of the strain tensor, $e = e_{ii}$ is the dilatation, λ_e and μ_e are Lame's elastic constants, λ_v and μ_v are Lame's viscoelastic constants for the viscoelastic solid, $\gamma = (2\mu_e + 3\lambda_e)\alpha_t$ is a constant material, α_t being the coefficient of linear thermal expansion, $\theta = T - T_o$ is the temperature increment such that $|\theta / T_o|$ <<1 and δ_{ij} is the Kronecker's delta.

The equation of motion is given by

$$\rho\frac{\partial^2 u_i}{\partial t^2} = \left[(\lambda_e+\mu_e)+(\lambda_v+\mu_v)\frac{\partial}{\partial t}\right]u_{j,ij} + \left[\mu_e+\mu_v\frac{\partial}{\partial t}\right]u_{j,ij} - \gamma\theta_{,i} + \mu_0(\boldsymbol{J}\times\boldsymbol{H_o})_i \tag{8}$$

where, ρ is the density, the comma denotes material derivatives and the summation convention is used.

The generalized heat conduction equation is given by

$$K\theta_{,ii} = \rho C_E(\dot{\theta} + \tau_0 \ddot{\theta}) + \gamma T_0(\dot{e} + \tau_0 \ddot{e}), \tag{9}$$

where, *K* is the thermal conductivity, C_E is the specific heat at constant strain, τ_0 is the relaxation time.

The strain displacement relation is given by

$$e_{ij} = \frac{1}{2}(u_{i,j} + u_{j,i}) \tag{10}$$

Together with the previous equations, constitute a complete system of generalized magneto–thermo–viscoelasticity equations for a medium with a perfect electric conductivity.

Let (*r*, ψ, ϕ) denote the radial coordinates, the co- latitude, and the longitude of a spherical coordinates system, respectively. Due to spherical symmetry, all the considered function will be functions of r and t only.

The components of the displacement vector will be taken the form

$$u_r = u(r,t) \quad , \quad u_\psi = u_\phi = 0. \tag{11}$$

The strain tensor components are thus given by

$$e_{rr} = \frac{\partial u}{\partial r} \quad , \quad e_{\psi\psi} = e_{\phi\phi} = \frac{u}{r} \quad , \quad e_{r\phi} = e_{\phi\psi} = 0. \tag{12}$$

It follow that the cubical dilatation is of the form

$$e = \frac{\partial u}{\partial r} + \frac{2u}{r} = \frac{1}{r^2}\frac{\partial(r^2 u)}{\partial r}. \tag{13}$$

From Equation (7) we obtain the components of the stress tensor as

$$\tau_{rr} = 2\left(\mu_e + \mu_v \frac{\partial}{\partial t}\right)\frac{\partial u}{\partial r} + \left(\lambda_e + \lambda_v \frac{\partial}{\partial t}\right) e - \gamma\theta, \tag{14}$$

$$\tau_{\phi\phi} = \tau_{rr} = 2\left(\mu_e + \mu_v \frac{\partial}{\partial t}\right)\frac{u}{r} + \left(\lambda_e + \lambda_v \frac{\partial}{\partial t}\right)e - \gamma\theta, \quad (15)$$

$$\tau_{r\phi} = \tau_{r\psi} = \tau_{\psi\phi} = 0 \quad (16)$$

Assume now that the initial magnetic field acts in the ϕ–direction and has the components (0, 0, H_0).

The induced magnetic field h will have one component h in the Φ–direction, while the induced electric field ***E*** will have one component *E* in the ψ – direction.

Then, equations (1), (2) and (5) yield

$$J = H_0 \frac{\partial e}{\partial r}, \quad (17)$$

$$h = -H_0\left(\frac{\partial u}{\partial r} + \frac{u}{r}\right), \quad (18)$$

$$E = \mu_0 H_0 \frac{\partial u}{\partial t}. \quad (19)$$

From Equations (17) and (6), we get that the Lorentz force has only one component F_r in the *r* – direction:

$$F_r = \mu_0 H_0^2 \frac{\partial e}{\partial r}. \quad (20)$$

Also, we arrived at

$$\rho\frac{\partial^2 u}{\partial t^2} = \left[(\lambda_e + 2\mu_e) + (\lambda_v + 2\mu_v)\frac{\partial}{\partial t} + \mu_0 H_0^2\right]\frac{\partial e}{\partial r} - \gamma\frac{\partial \theta}{\partial r}. \quad (21)$$

Equation (21) is to be supplemented by the constitutive Equation (13) and the heat conduction equation

$$K\nabla^2\theta = \left(\frac{\partial}{\partial t} + \tau_0\frac{\partial^2}{\partial t^2}\right)(\rho C_E\theta + \gamma T_0 e), \quad (22)$$

where ∇^2 is Laplace's operator in spherical coordinates which is given by

$$\nabla^2 = \frac{1}{r^2}\frac{\partial}{\partial r}(r^2\frac{\partial}{\partial r}) + \frac{1}{r^2 sin\psi}\frac{\partial}{\partial \psi}(sin\psi\frac{\partial}{\partial \psi}) + \frac{1}{r^2 sin^2\psi}\frac{\partial^2}{\partial \varphi^2}$$

In case of dependence on r only, this reduce to

$$\nabla^2 = \frac{1}{r^2}\frac{\partial}{\partial r}\left(r^2\frac{\partial}{\partial r}\right)$$

Now, we shall use the following non dimensional variables

$$r' = C_1\eta r,\ u' = C_1\eta u,\ t' = C_1^2\eta t,\ \tau_0^1 = C_1^2\eta\tau_0,$$

$$\tau'_{ij} = \frac{\tau_{ij}}{\mu_e},\ \lambda'_e = \lambda_e,\ \mu'_e = \mu_e,\ \lambda_v = C_1^2\eta\lambda_v,$$

$$\mu'_v = C_1^2\eta\mu_v,\ e' = e,\ \theta = \frac{\theta}{T_0},\ h' = \frac{h}{H_0},$$

$$E' = \frac{E}{\mu_0 H_0 C_1},\ J' = \frac{J}{\eta H_0 C_1}.$$

Equations (14)–(19), (21) and (22) take the following form (droping the primes for convenience).

$$J = \frac{\partial e}{\partial r}, \quad (23)$$

$$h = -\left(\frac{\partial u}{\partial r} + \frac{u}{r}\right), \quad (24)$$

$$E = \frac{\partial u}{\partial t}, \quad (25)$$

$$\tau_{rr} = \left(\beta_e^2 + \beta_v^2\frac{\partial}{\partial t}\right)e - \left(4 + 2a\frac{\partial}{\partial t}\right)\frac{u}{r} - b\,\theta, \quad (26)$$

$$\tau_{\psi\psi} = \left(\beta_e^2 - 2 + a_1\frac{\partial}{\partial t}\right)e + \left(2 + a\frac{\partial}{\partial t}\right)\frac{u}{r} - b\theta, \quad (27)$$

$$\tau_{r\phi} = \tau_{r\psi} = \tau_{\phi\psi} = 0, \quad (28)$$

$$\frac{\partial^2 u}{\partial t^2} = \left(1 + a_2\frac{\partial}{\partial t} + R_H\right)\frac{\partial e}{\partial r} - b_1\frac{\partial \theta}{\partial r}, \quad (29)$$

$$\nabla^2\theta = \left(\frac{\partial}{\partial t} + \tau_0\frac{\partial^2}{\partial t^2}\right)(\theta + g\,e), \quad (30)$$

where,

$$\eta = \frac{\rho C_E}{K},\ C_1^2 = \frac{\lambda_e + 2\mu_e}{\rho},\ \beta_e^2 = \frac{\lambda_e + 2\mu_e}{\mu_e},\ \beta_v^2 = a + a_1,$$

$$a = \frac{2\mu_v}{\mu_e},\ a_1 = \frac{\lambda_v}{\mu_e},\ a_2 = \frac{\lambda_v + 2\mu_v}{\lambda_e + 2\mu_e},$$

$$b_1 = \frac{b}{\beta^2},\ b = \frac{\gamma T_0}{\mu_e},\ g = \frac{\gamma T_0}{\rho C_E},\ R_H = \frac{\mu_0 H_0^2}{\lambda_e + 2\mu_e},$$

where the coefficient R_H represent the effect of the applied magnetic field on the thermoelastic process proceeding in the body. Equation (29) could be written in the form

$$\frac{1}{r^2}\frac{\partial}{\partial r}(r^2\frac{\partial^2 u}{\partial t^2})=\left(1+a_2\frac{\partial}{\partial t}+R_H\right)\frac{1}{r^2}\frac{\partial}{\partial r}\left(r^2\frac{\partial e}{\partial r}\right)-b_1\frac{1}{r^2}\frac{\partial}{\partial r}\left(r^2\frac{\partial \theta}{\partial r}\right)$$

Using Equation (13), we obtain

$$\left(1+a_2\frac{\partial}{\partial t}+R_H\right)\nabla^2 e-b_1\nabla^2\theta=\frac{\partial^2 e}{\partial t^2}. \quad (31)$$

We shall now define the Laplace transform with respect to a function $f\,(r,t)$ by the relation

$$L[f(r,t)]=\bar{f}(r,p)=\int_0^\infty e^{-pt}f(r,t)\,dt\ ,\ \mathrm{Re}(p)>0$$

Applying the Laplace transform to both sides of Equations (23)–(28), (30) and (31), we get

$$\bar{\tau}_{rr}=\left(\beta_e^2+\beta_v^2\,p\right)\bar{e}-(4+2a\,p)\frac{\bar{u}}{r}-b\,\bar{\theta}\,, \quad (32)$$

$$\bar{\tau}_{\varphi\varphi}=\bar{\tau}_{\psi\psi}=\left(\beta_e^2-2+a_1p\right)\bar{e}+(2+a_1p)\frac{\bar{u}}{r}-b\bar{\theta}\,, \quad (33)$$

$$\nabla^2\bar{\theta}=\left(p+\tau_0p^2\right)\left(\bar{\theta}+g\bar{e}\right), \quad (34)$$

$$(1+a_2p+R_H)\nabla^2\bar{e}-b_1\nabla^2\bar{\theta}=p^2\bar{e}, \quad (35)$$

$$\bar{J}=\frac{\partial\bar{e}}{\partial r}, \quad (36)$$

$$\bar{h}=-\left(\frac{\partial\bar{u}}{\partial r}+\frac{\bar{u}}{r}\right), \quad (37)$$

$$\bar{E}=P\bar{u}\,. \quad (38)$$

State space formulation

Equations (34) and (35) can be written in the form

$$\nabla^2\bar{e}=\frac{p^2+b_1g\,(p+\tau_0p^2)}{1+a_2p+R_H}\bar{e}+\frac{b_1(p+\tau_0p^2)}{1+a_2p+R_H}, \quad (39)$$

$$\nabla^2\bar{\theta}=g\,(p+\tau_0p^2)\bar{e}+(p+\tau_0p^2)\bar{\theta}\,. \quad (40)$$

Choosing as state variable the temperature increment and the strain component, Equations (39) and (49) can be written in the matrix from

$$\nabla^2\bar{V}(r,p)=A(P)\bar{V}(r,p), \quad (41)$$

where,

$$\bar{V}(r,p)=\begin{bmatrix}\bar{e}(r,p)\\ \bar{\theta}(r,p)\end{bmatrix}\text{, and}$$

$$A\,(p)=\begin{bmatrix}\dfrac{p^2+b_1g\,(p+\tau_0p^2)}{1+a_2p+R_H} & \dfrac{b_1(p+\tau_0p^2)}{1+a_2P+R_H}\\ g\,(p+\tau_0p^2) & (p+\tau_0p^2)\end{bmatrix}$$

The formal solution of system (Equation 41) can be written in the form

$$\bar{V}(r,p)=\frac{e^{-\sqrt{A(P)}\,r}}{r}C_1+\frac{e^{\sqrt{A(p)}\,r}}{r}C_2, \quad (42)$$

where, C_1 and C_2 are constants.

For a bounded solution as $r\rightarrow\infty$, we have to choose $C_2=0$, hence we have

$$\bar{V}(r,p)=\frac{e^{-\sqrt{A(P)}\,r}}{r}C_1\,. \quad (43)$$

Since $r=R$ must satisfy the last equation. We can get the constant C_1 in the form

$$C_1=R\,e^{\sqrt{A(p)}R}\,\bar{V}(R,p)\,. \quad (44)$$

Hence, Equation (43) will take the form

$$\bar{V}(r,p)=\frac{R}{r}e^{-\sqrt{A(p)}(r-R)}\,\bar{V}(R,p)\,,\ r\geq R\,, \quad (45)$$

where,

$$\bar{V}(R,p)=\begin{bmatrix}\bar{e}(R,p)\\ \bar{\theta}(R,p)\end{bmatrix}. \quad (46)$$

We will use the well - known Cayley – Hamilton theorem to find the form of the matrix of exp $\left(-\sqrt{A(p)}(r-R)\right)$.

The characteristic equation of the matrix A (p) can be written as

$$\kappa^2 - m_1\kappa + m_2 = 0, \tag{47}$$

where,

$$m_1 = \frac{p^2+(p+\tau_0 p^2)(1+\varepsilon+a_2 p+R_H)}{1+a_2 p+R_H},$$

$$m_2 = \frac{p^2+(p+\tau_0 p^2)}{1+a_2 p+R_H},$$

where, $\varepsilon = b_1 g$.

The roots of Equation (47) namely, κ_1 and κ_2 , satisfy the relations

$$\kappa_1+\kappa_2 = \frac{p^2+(p+\tau_0 p^2)(1+\varepsilon+a_2 p+R_H)}{1+a_2 p+R_H}, \tag{48}$$

$$\kappa_1\kappa_2 = \frac{p^2+(p+\tau_0 p^2)}{1+a_2 p+R_H}. \tag{49}$$

The Taylor series expansion for the matrix exponential in Equation (43) is given by

$$\exp\left(-\sqrt{A(p)}(r-R)\right)=\sum_{n=0}^{\infty}\frac{\left[-\sqrt{A(p)}(r-R)\right]^n}{n!}. \tag{50}$$

Using Cayley–Hamilton theorem, we can express $\sqrt{A}$ and higher order of the matrix A in terms of A and I where I is the unit matrix of second order.

Thus, the infinite series in Equation (50) can be reduced to

$$\exp\left(-\sqrt{A(p)}(r-R)\right)=a_0 I+a_1 A, \tag{51}$$

where, a_0 and a_1 are some coefficients depending on p and r only.

By Cayley–Hamilton theorem, the characteristic roots κ_1 and κ_2 of the matrix A must satisfy Equation (51), thus we have

$$\exp\left(-\sqrt{\kappa_1}(r-R)\right)=a_0+a_1\kappa_1. \tag{52}$$

$$\exp\left(-\sqrt{\kappa_2}(r-R)\right)=a_0+a_1\kappa_2. \tag{53}$$

Solving the above linear system of equations, we get

$$a_o = \frac{\kappa_1 e^{-\sqrt{\kappa_2}(r-R)} - \kappa_2 e^{-\sqrt{\kappa_1}(r-R)}}{\kappa_1-\kappa_2}, \tag{54}$$

$$a_1 = \frac{e^{-\sqrt{\kappa_1}(r-R)} - e^{-\sqrt{\kappa_2}(r-R)}}{\kappa_1-\kappa_2}. \tag{55}$$

From Equations (54) and (55) in (51), we deduce the following matrix

$$\exp\left(-\sqrt{A(p)}(r-R)\right)=[L_{ij}],\quad i,j=1,2 \tag{56}$$

where,

$$L_{11}=\frac{1}{\kappa_1-\kappa_2}\left[\left(\frac{p^2+\varepsilon(p+\tau_0p^2)}{1+a_2p+R_H}-\kappa_2\right)e^{-\sqrt{\kappa_1}(r-R)}-\left(\frac{p^2+\varepsilon(p+\tau_0p^2)}{1+a_2p+R_H}-\kappa_1\right)e^{-\sqrt{\kappa_2}(r-R)}\right]$$

$$L_{12}=\frac{1}{\kappa_1-\kappa_2}\left[\frac{b_1(p+\tau_0p^2)}{1+a_2p+R_H}e^{-\sqrt{\kappa_1}(r-R)}-\frac{b_1(p+\tau_0p^2)}{1+a_2p+R_H}e^{-\sqrt{\kappa_2}(r-R)}\right]$$

$$L_{21}=\frac{1}{\kappa_1-\kappa_2}\left[g(p+\tau_0p^2)e^{-\sqrt{\kappa_1}(r-R)}-g(p+\tau_0p^2)e^{-\sqrt{\kappa_2}(r-R)}\right]$$

$$L_{22}=\frac{1}{\kappa_1-\kappa_2}\left[(p+\tau_0p^2-\kappa_2)e^{-\sqrt{\kappa_1}(r-R)}-(p+\tau_0p^2-\kappa_1)e^{-\sqrt{\kappa_2}(r-R)}\right]. \tag{57}$$

We can write the solution in the form

$$\bar{V}(r,p)=\frac{R}{r}[L_{ij}]\bar{V}(R,p). \tag{58}$$

Finally, we have

$$\bar{e}(r,p)=\frac{R}{(\kappa_1-\kappa_2)r}\left[L_1 e^{-\sqrt{\kappa_1}(r-R)}-L_2 e^{-\sqrt{\kappa_2}(r-R)}\right], \tag{59}$$

$$\bar{\theta}(r,p)=\frac{R}{(\kappa_1-\kappa_2)r}\left[M_1 e^{-\sqrt{\kappa_1}(r-R)}-M_2 e^{-\sqrt{\kappa_2}(r-R)}\right], \tag{60}$$

$$\bar{u}=\frac{-R}{(\kappa_1-\kappa_2)r}\left[L_1\left(\frac{1+\sqrt{\kappa_1}\,r}{\kappa_1\,r}\right)e^{-\sqrt{\kappa_1}(r-R)}-L_2\left(\frac{1+\sqrt{\kappa_2}\,r}{\kappa_2\,r}\right)e^{-\sqrt{\kappa_2}(r-R)}\right] \tag{61}$$

$$\bar{\tau}_{rr}=\frac{R}{(\kappa_1-\kappa_2)r}\left\{\left(\left[\beta_e^2+p\beta_v^2+(4+2aP)\left(\frac{1+\sqrt{\kappa_1}\,r}{\kappa_1 r^2}\right)\right]L_1-bM_1\right)e^{-\sqrt{\kappa_1}(r-R)}-\left(\left[\beta_e^2+p\beta_v^2+(4+2ap)\left(\frac{1+\sqrt{\kappa_2}\,r}{\kappa_2 r^2}\right)\right]L_2-bM_2\right)e^{-\sqrt{\kappa_2}(r-R)}\right\} \tag{62}$$

$$\bar{\tau}_{\varphi\varphi}=\bar{\tau}_{\psi\psi}=\frac{R}{(\kappa_1-\kappa_2)r}\left\{\left(\left[\beta_e^2-2a_1p+(2+ap)\left(\frac{1+\sqrt{\kappa_1}r}{\kappa_1 r^2}\right)\right]L_1-bM_1\right)e^{-\sqrt{\kappa_1}(r-R)}-\left(\left[\beta_e^2-2a_1p+(2+ap)\left(\frac{1+\sqrt{\kappa_2}r}{\kappa_2 r^2}\right)\right]L_2-bM_2\right)e^{-\sqrt{\kappa_2}(r-R)}\right\} \tag{63}$$

where,

$$L_1=\left(\frac{p^2+\varepsilon(p+\tau_0p^2)}{1+a_2p+R_H}-\kappa_2\right)\bar{e}(R,p)+\left(\frac{b_1(p+\tau_0p^2)}{1+a_2p+R_H}\right)\bar{\theta}(R,p)$$

$$L_2=\left(\frac{p^2+\varepsilon(p+\tau_0p^2)}{1+a_2p+R_H}-\kappa_1\right)\bar{e}(R,p)+\left(\frac{b_1(p+\tau_0p^2)}{1+a_2p+R_H}\right)\bar{\theta}(R,p)$$

$$M_1=g(p+\tau_0p^2)\bar{e}(R,p)+(p+\tau_0p^2-\kappa_2)\bar{\theta}(R,p)$$

$$M_2=g(p+\tau_0p^2)\bar{e}(R,p)+(p+\tau_0p^2-\kappa_1)\bar{\theta}(R,p)$$

Application

In order to evaluate the unknown parameters L_1, L_2, M_1 and M_2, we will use the boundary conditions on the internal surface of the shell, r = R which is given by:

Thermal boundary condition

The internal surface with $r = R$ is subjected to ramp – type heating in the form

$$\theta(R,t)=\begin{cases}0 & t\le 0\\ \dfrac{\theta_1}{t_o}t & 0<t\le t_o\\ \theta_1 & t>t_o\end{cases}, \tag{64}$$

where, θ_1 is constant and to is called the ramping parameter. Applying the Laplace transform, we get

$$\bar{\theta}(R,P)=\frac{\theta_1(1-e^{-pt_0})}{t_0p^2} \tag{65}$$

Mechanical boundary condition

The internal surface $r = R$ has a rigid foundation, which is rigid enough to prevent any strain, yield
e (R, t) = 0.

Applying the Laplace transform, we get

$$\bar{e}(R,p)=0 \tag{66}$$

Using the conditions (65) and (66) in to Equations (59) – (63), we get

$$\bar{e}(r,p)=\frac{RL}{(\kappa_1-\kappa_2)r}\left[e^{-\sqrt{\kappa_1}(r-R)}-e^{-\sqrt{\kappa_2}(r-R)}\right], \tag{67}$$

$$\bar{\theta}(r,p)=\frac{R}{(\kappa_1-\kappa_2)r}\left[M_1'e^{-\sqrt{\kappa_1}(r-R)}-M_2'e^{-\sqrt{\kappa_2}(r-R)}\right], \tag{68}$$

$$\bar{u}=\frac{-LR}{(\kappa_1-\kappa_2)r}\left[\left(\frac{1+\sqrt{\kappa_1}\,r}{\kappa_1\,r}\right)e^{-\sqrt{\kappa_1}(r-R)}-\left(\frac{1+\sqrt{\kappa_2}\,r}{\kappa_2\,r}\right)e^{-\sqrt{\kappa_2}(r-R)}\right] \tag{69}$$

$$\bar{\tau}_{rr}=\frac{R}{(\kappa_1-\kappa_2)r}\left\{\left(\left[\beta_e^2+p\beta_v^2+(4+2ap)\left(\frac{1+\sqrt{\kappa_1}r}{\kappa_1 r^2}\right)\right]L-bM_1'\right)e^{-\sqrt{\kappa_1}(r-R)}-\left(\left[\beta_e^2+p\beta_v^2+(4+2ap)\left(\frac{1+\sqrt{\kappa_2}r}{\kappa_2 r^2}\right)\right]L-bM_2'\right)e^{-\sqrt{\kappa_2}(r-R)}\right\} \tag{70}$$

$$\bar{\tau}_{\varphi\varphi}=\bar{\tau}_{\psi\psi}=\frac{R}{(\kappa_1-\kappa_2)r}\left\{\left(\left[\beta_e^2-2a_1p+(2+ap)\left(\frac{1+\sqrt{\kappa_1}r}{\kappa_1 r^2}\right)\right]L_1-bM_1'\right)e^{-\sqrt{\kappa_1}(r-R)}-\left(\left[\beta_e^2-2a_1p+(2+ap)\left(\frac{1+\sqrt{\kappa_2}r}{\kappa_2 r^2}\right)\right]L_2-bM_2'\right)e^{-\sqrt{\kappa_2}(r-R)}\right\} \tag{71}$$

where,

$$L = \left(\frac{b_1(p+\tau_0 p^2)}{1+a_2 p + R_H}\right)\bar{\theta}(R,p) \quad (72)$$

$$M_1' = (p+\tau_0 p^2 - K_2)\bar{\theta}(R,p) \quad (73)$$

$$M_2' = (p+\tau_0 p^2 - K_1)\bar{\theta}(R,p) \quad (74)$$

By obtaining θ_1, the temperature increment θ can be obtained by solving Equation (64) to give

$$\theta = \frac{-1+\sqrt{1+2\kappa_1\theta}}{\kappa_1} \quad (75)$$

They complete the solution on the Laplace domain.

Inversion of the Laplace transforms

In order to invert the Laplace transforms, we adopt a numerical inversion method based on a Fourier series expansion [42]. In this method, the inverse g (t) of the Laplace transform $\bar{g}(s)$ is approximated by the relation;

$$g(t) = \frac{e^{ct}}{t_1}\left[\frac{1}{2}\bar{g}(c) + \mathrm{Re}\left(\sum_{k=1}^{\infty} e^{ik\pi t/t_1}\bar{g}(c+ik\pi/t_1)\right)\right], 0 \le t \le 2t_1,$$

where N is a sufficiently large integer representing the number of terms in the truncated infinite Fourier series. N must chosen such that

$$e^{ct}\mathrm{Re}\left[e^{iN\pi t/t_1}\bar{g}(c+iN\pi/t_1)\right] \le \varepsilon_1,$$

where ε_1 is a persecuted small positive number that corresponds to the degree of accuracy to be achieved. The parameter c is a positive free parameter that must be greater than the real parts of all singularities of $\bar{g}(s)$. The optimal choice of c was obtained according to the criteria described by (Honig and Hirdes, 1984).

RESULTS AND DISCUSSION

The copper material was chosen for purpose of numerical evaluations and the constants of the problem were taken as follows (Abd-Alla et al., 2004):

α_t = 1.78 (10^{-5}) K^{-1}, C_E = 383.1 m^2/K, μ_e =3.86(10^{10}) N/m^2, λ_e = 7.76(10^{10}) N/m^2 , ρ = 8954 kg/m^3 , $\tau_0 = 0.02$ sec., T_o= 293 K , $\beta_e^2 = 4$, $\beta_v^2 = 8$.

The computations were carried out for $\theta_1 = 1$. The temperature, stresses, displacement and strain distributions are represented graphically at different value of time *t*. The field quantities , temperature , stresses , displacement and strain depend not only on the state and space variable *t* and *r* , but also depend on t_o , R_H , and Lame's viscoelastic constants (λ_v and μ_v) . It has been observed that, t_o, R_H , and Lame's viscoelastic constants have significant effect on the temperature, stresses, displacement and strain distributions quantities. Here all the variables / parameters are taking in the non-dimensional forms. Taking *r* range from 1.0 to 2.2 has carried out in numerical analysis.

Figure 1 exhibits the space variation of temperature distribution for (MTVE) theory and for different value of t_o, and we observe that:

(1) Some difference in the values of temperature is noticed for different values of t_o. We can see that the ramping parameter t_o has a clear effect on the values of temperature; actually, the value of temperature increase when $t \geq t_o$ and decrease when $t < t_o$ where larger t with respect to t_o means larger heating on the boundary. Figure 2 exhibits the space variation of temperature distribution for the two theories, and we observe that:

(1) Lame's viscoelastic constants (λ_v and μ_v) have a small effect on the values of temperature; actually, the value of temperature increases in the case of (MTE) theory but decreases in the case of (MTVE) theory, and on the boundary the values of temperature have the same values for the two theories .

Figure 3 exhibits the space variation of strain distribution for (MTVE) theory and for different values of R_H and t_o , and we observe the following:

(1) Some differences in the values of strain are noticed for different values of R_H .We can see that, the absolute value of the maximum point of the strain increase in the absence of magnetic field, (that is, the magnetic field causes increasing in the values of the strain).
(2) The ramping parameter t_o has a clear effect on the values of the strain; actually, the absolute value of the strain increase when $t \geq t_o$ and decrease when $t < t_o$.

Figure 4 exhibits the space variation of strain distribution for the two theories and we observe that:

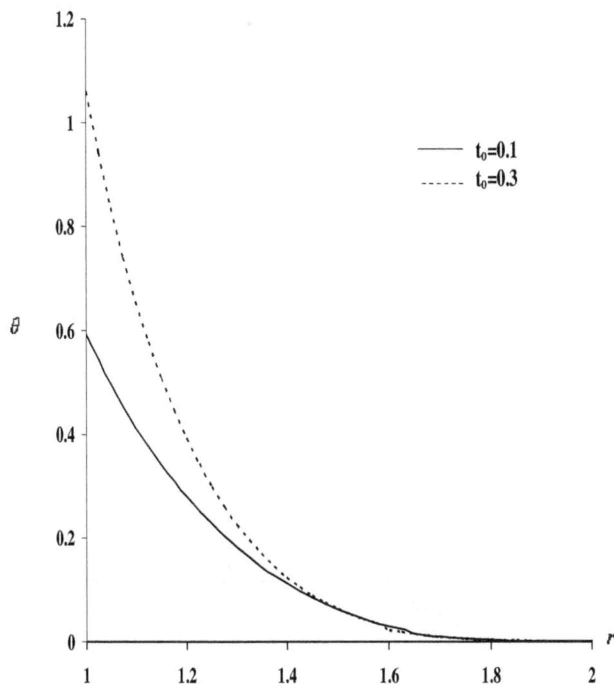

Figure 1. Temperature distribution for MTVE theory at t = 0.2.

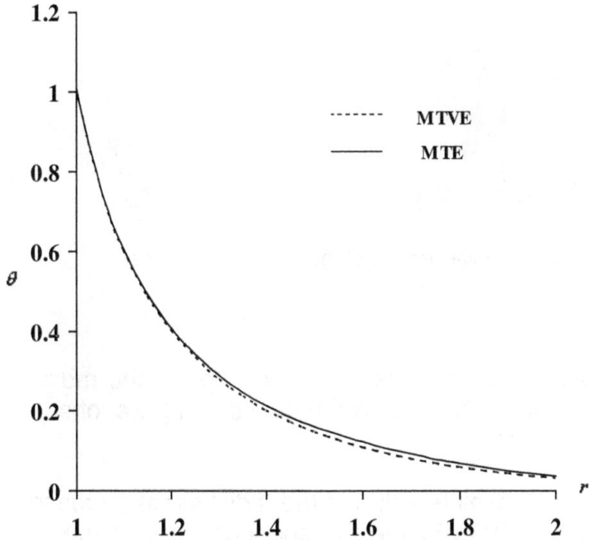

Figure 2. Temperature distribution for two theories at t = 0.3.

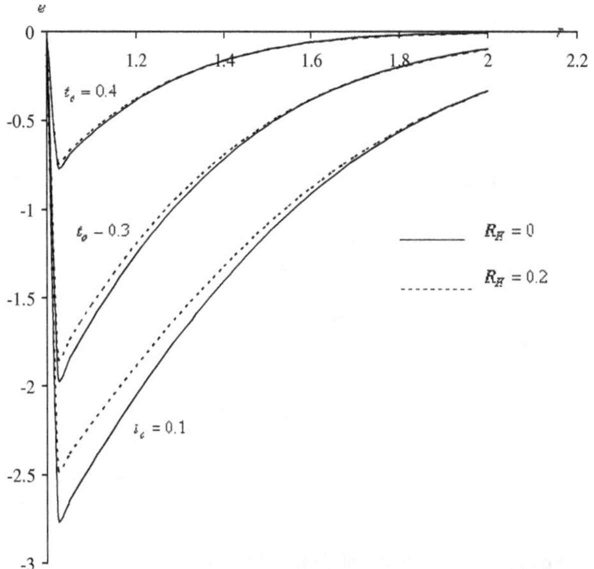

Figure 3. Strain distribution for MTVE theory at t = 0.2.

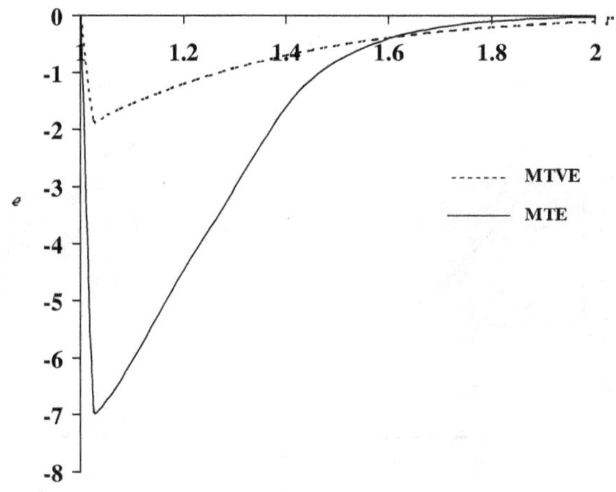

Figure 4. Strain distribution for two theories.

(1) Some difference in the value of strain is noticed for the two theories. We can see that the absolute value of the maximum point of the strain decrease in the case of (MTVE) theory and it increase in the case of (MTE) theory.

Figure 5 exhibits the space variation of the displacement distribution for (MTVE) theory and for different values of R_H, and we observe the following:

(1) Some difference in the values of displacement is noticed for different value of the parameter R_H.

(2) On the boundary, the values of displacement increase in the absence of magnetic field.

Figure 6 exhibits the space variation of displacement distribution for the two theories, and we observe the following:

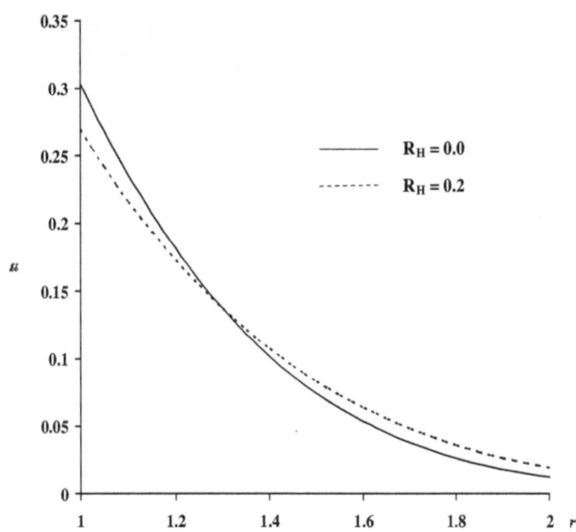

Figure 5. Displacement distribution at t = 0.2.

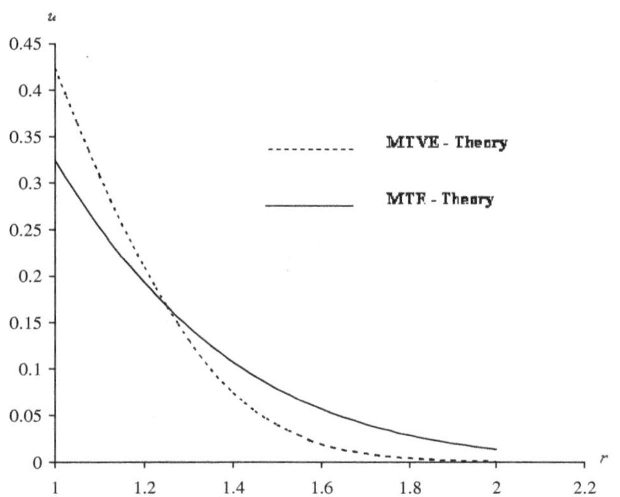

Figure 6. Displacement distribution for two theories.

(1) Some difference in the values of displacement is noticed for the two theories.

(2) On the boundary, the value of displacement increase in the case of (MTVE) theory.

Figure 7 exhibits the space variation of the radial stress distribution for (MTVE) theory and for different values of R_H, and we observe that:

Some differences in the values of the radial stress is noticed for different values of R_H

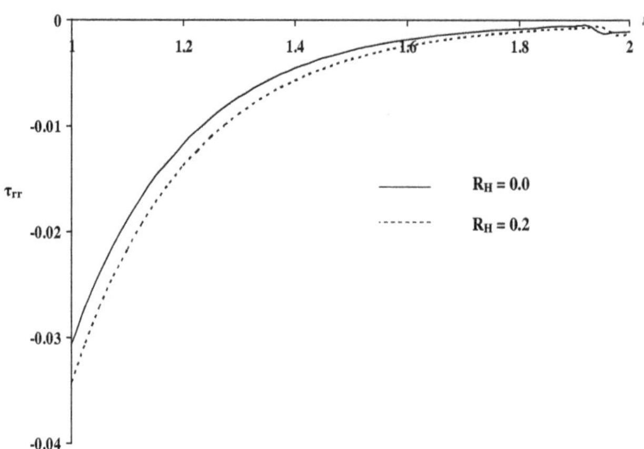

Figure 7. Radial stress distribution for MTVE-theory at t = 0.2.

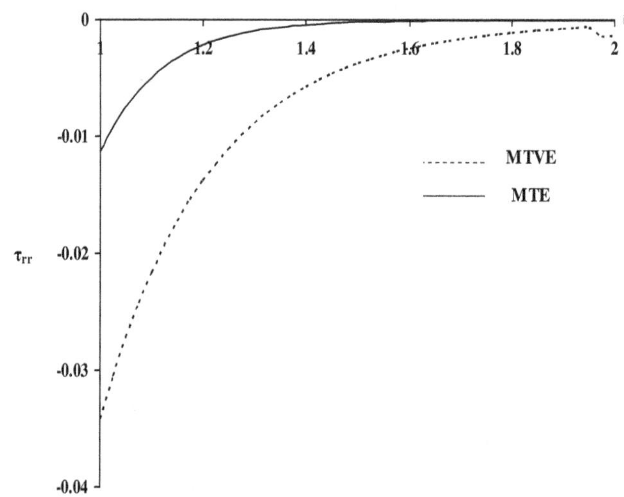

Figure 8. Radial stress distribution for two theories.

Figure 8 exhibits the space variation of the radial stress distribution for the two theories, and we observe the following:

(1) The absolute value of the radial stress increase in the case of (MTVE) theory, and decrease in the case of (MTE) theory, that is, λ_v and μ_v have a clear effect on the values of the radial stress.

(2) On the boundary, the absolute values of the radial stress increase in the case of (MTVE) theory.

One can see that although the curves are smooth and do not have any discontinuous points; they take the same behavior in some range of *r*, where the boundary conditions somewhere are different.

Concluding remarks

The importance of the model used in generalized thermo-viscoelasticity theory is the ability to the separating between the two theories (MTE) and (MTVE). This means that, one able to study the effects of viscosity on the considered functions. Studding the effect of R_H and t_0 on the variables fields, it is found that, all the variables have been affected by the changing of R_H and t_0. Mathematically and physically the ramp-type heating is more realistic than the thermal shock, where the ramping parameter of heating has a clear affect all the fields.

REFERENCES

Abd-Alla AM, Hammad HAH, Abo-Dahab SM (2004). Magneto-thermo-visco-elastic interactions in an unbounded body with a spherical cavity subjected to a periodic loading. Appl. Math. Comput., 155: 235-248.

Alfery T, Gurnee EF (1956). Rheology: Theory and applications, Eirich FR (Ed.), Academic Press, New York, pp. 311–368.

Atkinson C, Craster RV (1995). Theoretical aspects of fracture mechanics. Prog. Aerosp. Sci., 31: 1-83.

Biot M (1956). Thermoelasticity and irreversible thermodynamics. J. Appl. Phys., (27); 240-253.

Biot MA (1954). Theory of stress–strain relations in anisotropic viscoelasticity and relaxation phenomena. J. Appl. Phys., 25: 1385-1391.

Chandrasekhraiah DS (1998). Hyperbolic thermoelasticity, A review of recent literature. Appl. Mech. Rev., 51: 705-729.

El-Karamany AS, Ezzat MA (2002). On the boundary integral formulation of thermo-viscoelasticity theory. Int. J. Eng. Sci., 40: 1943–1956.

El-Karamany AS, Ezzat MA (2004a). Discontinuities in generalized thermo- viscoelasticity under four theories. J. Therm. Stress., 27: 1187-1212.

El-Karamany AS, Ezzat MA (2004b). Thermal shock problem in generalized thermo-viscoelasticity under four theories. Int. J. Eng. Sci., 42: 649-671.

Ezzat MA (1997). Generation of generalized magneto-thermoelastic waves by thermal shock in a perfectly conducting half-space. J. Therm. Stress, 20: 617-633.

Ezzat MA (2008). State space approach to solids and fluids. Can. J. Phys. Rev., 86: 1241-1250.

Ezzat MA, El-Karamany AS (2002).The uniqueness and reciprocity theorems for generalized thermo-visco-elasticity for anisotropic media. J. Therm. Stress, 25: 507-522.

Ezzat MA, Youssef HM (2005). Generalized magneto-thermoelasticity in a perfectly conducting medium. Int. J. Solids Struct., 42: 6319–6334.

Ferry JD (1977). Viscoelastic properties of polymers. J. Wiley and Sons, New York, pp. 1–167.

Green AE, Laws N (1972). On the entropy production inequality. Arch. Rat. Anal., 54: 17-53.

Green AE, Lindsay K (1975). Thermoelasticity. J. Elast., 2: 1-7.

Gross B (1953). Mathematical structure of the theories of visco-elasticity, Hemann, Paris, pp. 198–253.

Gurtin ME (1972). Encyclopedia of Physics, S. Flugge, ed, Springer-Verlag, Berlin, 5a(Part 2): 641.

Hetnarski RB, Ignaczak J (1999). Generalized thermoelasticity. J. Therm. Stress., 22: 451-476.

Honig G, Hirdes U (1984). Methods for the numerical inversion of Laplace transform. J. Comp. Appl. Mech., 10: 113-132.

Ignaczak J (1991). Domain of influence results in generalized thermoelasticity. Surv. Appl. Mech. Rev., 44: 375-382.

Ilioushin A, Pobedria B (1970). Fundamentals of the mathematical theory of thermal Viscoelasticity, Nauka, Moscow, Russian, pp. 45-62.

Joseph D, Preziosi L (1989). Heat waves. Rev. Mod. Phys., 61: 41-73.

Joseph D, Preziosi L (1990). Addendum to heat waves. Rev. Mod. Phys., 62: 75-391.

Lee S, Knauss WG (2000). A note on the determination of relaxation and creep data from ramp tests. Mech. Time-Depend. Mater., 4: 1-7.

Lord H, Shulman Y (1967). A generalized dynamical theory of thermoelasticity. J. Mech. Phys. Solids, 15: 299–309.

Menon S, Tang J (2004). A state-space approach for the dynamic analysis of viscoelastic systems. Comput. Struct., 82: 1123-1130.

Misra S, Samanata S, Chakrabarti A (1992). Transient magneto-thermo-elastic waves in a viscoelastic half-space produced by ramp-heating of its surface. Comput. Struct., 43: 951.

Morland L, Lee E (1960). Stress analysis for linear viscoelastic materials with temperature variation. Trans. Soc. Rheol., 4: 233-263.

Mukhopadhyay B, Bera R (1992). Effect of thermal relaxation on electro-magneto-thermo-visco-elastic plane waves in rotating media. Int. J. Eng. Sci., 30: 459-369.

Muller (1971). The coldness - A universal function in thermoelastic solids. Arch. Rat. Mech. Anal., 41: 319-332.

Öncü TS, Moodie TB (1990). Boundary initiated wave phenomena in thermo-elastic material. Q. Appl. Math., 48: 295–320.

Öncü TS, Moodie TB (1989). Finite speed theromal transients generated by non uniform sources applied to circular boundaries in homogeneous conductor. Int. J. Eng. Sci., 27: 611–621.

Paria G (1967). Magneto – elasticity and magneto-thermo-elasticity. Adv. Appl. Mech., 10: 73–112.

Rakshit M, Mukhopadhyay B (2005). An electro-magneto-thermo-visco-elastic problem in an infinite medium with a cylindrical hole. Int. J. Eng. Sci., 43: 925-936.

Sherief HH, Ezzat MA (1996). A thermal-shock problem in magneto-thermoelasticity with thermal Relaxation. Int. J. Solids Struct., 33: 4449-4459.

Sherief HH, Ezzat MA (1998). A problem in generalized magneto-thermoelasticity for an infinitely long annular cylinder. J. Eng. Math., (34): 387-402.

Staverman AJ, Schwarzl F (1956). In: The physics of high polymers. Stuart HA (ed), Springer-Verlag, New York, 4(1).

Şuhubi ES (1975). Thermoelastic solids. In: Eringen AC (Ed.), Continuum Physics II, Ch. 2, Academic Press, New York, pp. 34-38.

Tschoegl NW (1997). Time dependence in material properties: an overview. Mech. Time-Depend. Mater., (1): 3-31.

Wilson AJ (1963). The propagation of magneto– thermo – elastic plane waves. proc. Camb. Philos. Soc., (59): 483 – 488.

Youssef HM (2005). Dependence of modulus of elasticity and thermal conductivity on reference temperature in generalized thermoelasticity for an infinite material with a spherical cavity. J. Appl. Math. Mech., (26): 470–475.

Youssef HM, Al-Harby AH (2007). State-space approach of two-temperature generalized thermoelasticity of infinite body with a spherical cavity subjected to different types of thermal loading. Arch. Appl. Mech., (77): 675–687.

Influence of hybrid fibres on the post crack performance of high strength concrete: Experimental investigations

A. Sivakumar

Department of Civil Engineering, Vellore Institute of Technology (VIT) University, Vellore – 632007, India.
E-mail: sivakumara@vit.ac.in

This paper discusses the experimental results of tests carried out on the flexural properties of various fibre-reinforced concretes at low volume fractions of fibres upto 0.5%. The flexural properties, namely flexural strength, toughness, and ductility, were measured using four point bending tests on beam specimens. Compared to reference concrete without fibres, fibre addition was seen to enhance the pre-peak as well as post-peak region of the load-deflection curve significantly. The best flexural performance was obtained at the highest volume fraction of 0.5%. At this volume fraction, flexural toughness and ductility of hybrid fibre concretes (incorporating a blend of steel and non-metallic fibres) were comparable to steel fibre concretes. Increased fibre availability in the hybrid fibre systems (due to the lower densities of non-metallic fibres), in addition to the ability of non-metallic fibres of bridging smaller micro cracks, are suggested as the reasons for the enhancement in flexural properties.

Key words: Micro cracking, silica fume, fibre reinforcement, toughness.

INTRODUCTION

Poor toughness, a serious shortcoming of high strength concrete, could be overcome by reinforcing with short discontinuous fibres. Fibres primarily control the propagation of cracks and limit the crack widths (Qian and Stroeven, 2000). High elastic modulus steel fibres also enhance the flexural toughness and ductility of concrete. The contribution of steel fibres can be observed mainly after matrix cracking in concrete, when they help in bridging the propagating cracks (Stroeven and Babut, 1986). The addition of steel fibres at high dosages, however, has potential disadvantages in terms of poor workability and increased cost. In addition, due to the high stiffness of steel fibres, micro-defects such as voids and honeycombs could form during placing as a result of improper consolidation at low workability levels. A compromise to obtain good fresh concrete properties (including workability and reduced early-age cracking) and good ductility of hardened concrete can be achieved by adding two different fibre types, which can function individually at different scales to yield optimum performance (Yao et al., 2003).

The addition of non-metallic fibres such as glass, polyester, polypropylene etc. results in good fresh concrete properties and reduced early age cracking. The beneficial effects of non-metallic fibres could be attributed to their high aspect ratios and increased fibre availability (because of lower density as compared to steel) at a given volume fraction. Because of their lower stiffness, these fibres are particularly effective in controlling the propagation of microcracks in the plastic stage of concrete. However, their contribution to post-cracking behaviour, unlike steel fibres, is not known to be significant.

Use of hybrid combinations of steel and non-metallic fibres can offer potential advantages in improving concrete properties as well as reducing the overall cost of concrete production (Bentur and Mindess, 1990). When fibre fractions are increased, it results in a denser and more uniform distribution of fibres throughout the concrete, which reduces shrinkage cracks and improves post-crack strength of concrete. It is important to have a combination of low and high modulus fibres to arrest the micro and macro cracks respectively. Another beneficial combination of fibres is that of long and short fibres. Once again, different lengths of fibres would control different scales of cracking. A number of studies indicate the overall benefits of using combinations of fibres (Pierre et al., 1999; Soroushianp et al., 1992; Bayasi and Zeng,

Table 1. Concrete mixture proportions used in the study.

Mix Id	Cement (kg/m^3)	Silica-fume (kg/m^3)	Fine aggregate (kg/m^3)	Coarse aggregate (kg/m^3)		Water (kg/m^3)	Superplasticizer dosage (kg/m^3)
				10 mm	20 mm		
Controlled concrete (C1)	372	28	750	570	570	160	8
All Fibre concrete mixtures	372	28	750	570	570	160	8

Table 2. Properties of the different fibres used.

Property	Hooked steel	Polypropylene	Glass	Polyester
Length (mm)	30	20	12	12
Diameter (mm)	0.5	0.12	0.01	0.03
Aspect ratio (l/d)	60	166	1200	400
Density (kg/m^3)	7800	900	2720	1350
Tensile strength (GPa)	1.7	0.45	2.5	0.92
Elastic modulus (GPa)	200	5	80	15
Failure strain (%)	3.5	18	3.6	12

1993; Banthia and Nandakumar, 2003).

The objective of this study was to evaluate the flexural properties, namely, flexural strength, toughness, and ductility, of different fibre reinforced concrete systems, containing individual steel fibres and hybrid combinations of steel and non-metallic fibres such as glass, polyester and polypropylene. The fibres were added at low dosages, primarily from the point of view of providing good workability, and the overall volume fraction varied between 0.3 and 0.5%. A factorial experimental design was carried out and the flexural properties of various concretes were evaluated.

MATERIALS AND METHODS

Materials used

Ordinary Portland Cement conforming to IS 12269 (Indian Standard Designation, IS 12269-1987) was used for the concrete mixtures. Silica fume, obtained from Elkem Materials, India, was also used for the high strength concrete mixtures. River sand with a specific gravity of 2.65 and fineness modulus of 2.64 was used as the fine aggregate, while crushed granite of specific gravity 2.82 was used as coarse aggregate. A naphthalene sulphonate based superplasticizer was used to obtain the desired workability. The fibres used in the study were hooked steel, polypropylene, polyester, and glass, from local manufacturers.

Mixture proportioning

Trial mixtures were prepared to obtain target strength of 60 MPa at 28 days, along with a workability of 75 to 120 mm. In order to obtain the desired workability, only the superplasticizer dosage was varied. The detailed mixture proportions for the study are presented in Table 1, while the properties and volume fractions of various fibres used in the mixtures are given in Tables 2 and 3.

Mixing and casting details

The coarse aggregate, fine aggregate, cement, and silica fume were first mixed dry in a pan mixer of capacity 100 kg for a period of 2 min. The superplasticizer was then mixed thoroughly with the mixing water and added to the mixer. Fibres were dispersed by hand in the mixture to achieve a uniform distribution throughout the concrete, which was mixed for a total of 4 min. Fresh concrete was cast in steel moulds and compacted on a vibrating table. The following specimens were prepared:

i) 100 mm cubes (for compressive strength as per IS 516 - 1999 (Indian Standard Designation, IS 516-1999)
ii) 100 x 200 mm cylinders (for split tensile strength as per IS 5816 - 1999 (Indian Standard Designation, IS5816-1999)
iii) 100 x 100 x 500 mm beam specimens for flexural tests based on ASTM C1018 (ASTM Standard Designation C 1018-97).

Testing methodology

A universal testing machine of capacity 100 tonnes was used for testing the compressive strengths of cube specimens at 3, 7 and 28 days from casting at a loading rate of 140 kg/cm^2/min, as well as split tensile strengths of cylindrical specimens at 28 days at a loading rate of 1.8 N/mm^2/min. Beams were tested as per ASTM C-1018, on a servo-controlled universal testing machine at a displacement-controlled rate of 0.05 mm/min. The support and mid span deflections were recorded on to a computer connected through an electronic digital controller system. A snapshot of the experimental setup is shown in Figure 1. The load versus displacement curve for each specimen was obtained and the toughness parameters, namely, absolute toughness, toughness indices (I_5, I_{10}, and I_{20}), and residual strength factors ($R_{5,\ 10}$ and $R_{10,\ 20}$) were calculated based on ASTM C1018. The load-deflection plots obtained for different fibre volume fractions are given in Figures 2, 3 and 4. The toughness indices and residual strength factors are calculated using the following equations:

I_5 = Area up to 3.0 times the first crack deflection / area up to first crack (1)

Table 3. Volume fractions of different fibre combinations used in the study.

Mixture ID	Hooked steel (%)	Percentage replacement of steel fibre by non-metallic fibre	Polypropylene (%)	Glass (%)	Polyester (%)	Total fibre dosage (V_f in %)	Steel to non-metallic fibre ratio
C1	0	0	0	0	0	0	-
HSPP1	0.21	30	0.09	-	-	0.3	2.33
HSG2	0.24	20	-	0.06	-	0.3	4
HSPO3	0.27	10	-	-	0.03	0.3	9
HSPP4	0.32	20	0.08	-	-	0.4	4
HSG5	0.36	10	-	0.04	-	0.4	9
HSPO6	0.28	30	-	-	0.12	0.4	2.33
HSPP7	0.45	10	0.05	-	-	0.5	9
HSG8	0.35	30	-	0.15		0.5	2.33
HSPO9	0.40	20	-	-	0.1	0.5	4
HSPP10	0.24	20	0.06		-	0.3	4
HSG11	0.27	10	-	0.03	-	0.3	9
HSPO12	0.21	30	-	-	0.09	0.3	2.33
HSPP13	0.36	10	0.04		-	0.4	9
HSG14	0.28	30	-	0.12	-	0.4	2.33
HSPO15	0.32	20	-	-	0.08	0.4	4
HSPP16	0.35	30	0.15	-	-	0.5	2.33
HSG17	0.40	20	-	0.1		0.5	4
HSPO18	0.45	10	-	-	0.05	0.5	9
HSPP19	0.27	10	0.03	-	-	0.3	9
HSG20	0.21	30	-	0.09	-	0.3	2.33
HSPO21	0.24	20	-	-	0.06	0.3	4
HSPP22	0.28	30	0.12	-		0.4	2.33
HSG23	0.32	20		0.08	-	0.4	4
HSPO24	0.36	10	-	-	0.04	0.4	9
HSPP25	0.40	20	0.1	-	-	0.5	4
HSG26	0.45	10	-	0.05	-	0.5	9
HSPO27	0.35	30	-	-	0.15	0.5	2.33
HST1	0.5	0	-	-	-	0.5	-
HST2	0.4	0	-	-	-	0.4	-
HST3	0.3	0	-	-	-	0.3	-

I_{10} = Area up to 5.5 times the first crack deflection / area up to first crack (2)

I_{20} = Area up to 10.5 times the first crack deflection / area up to first crack (3)

$$R_{5,\,10} = 20\,(I_{10} - I_5) \quad (4)$$

$$R_{10,\,20} = 10\,(I_{20} - I_{10}) \quad (5)$$

Experimental design

A factorial experimental design was adopted in this study. The governing factors chosen in this study were (1) total fibre dosage (TFD), (2) steel to non-metallic fibre ratio (SNMFR), and (3) type of fibre combinations (that is steel-glass, steel-polypropylene etc.). These factors were set at three levels each. The various factors and their levels are given in Table 4. The full factorial experimental design consisted of 3^3 (= 27) experimental points with two replicates. In addition to the 27 main experiments, an additional four concrete mixtures were cast for reference, which included one controlled concrete without fibres and three steel fibre concretes at three dosage levels (0.3, 0.4, and 0.5%). The flexural parameters measured were toughness, ductility and flexural strength, and nine experimental design points were evaluated at each level.

TEST RESULTS AND DISCUSSION

Compressive, split tensile and flexural strength

Results for compressive, split tensile and flexural strength for all mixtures are presented in Table 5. It can be seen that for all the fibre concrete mixtures, the compressive

Figure 1. Experimental setup of flexural test.

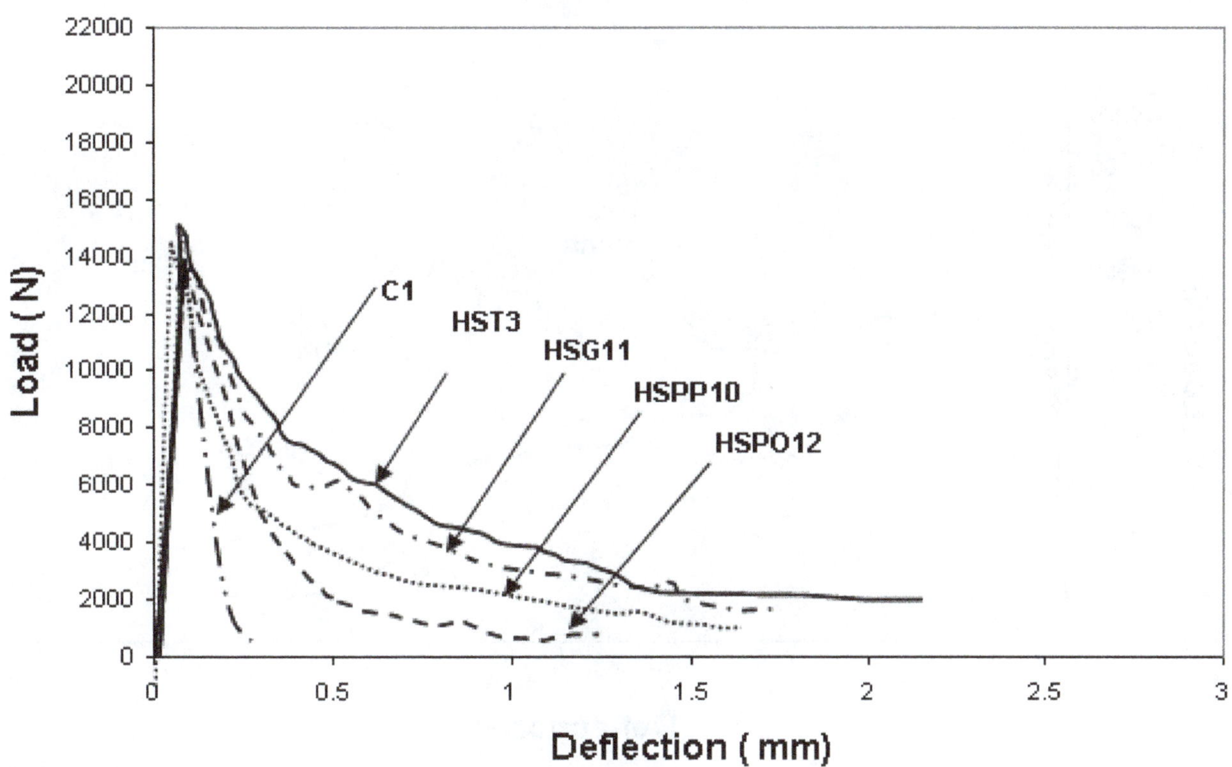

Figure 2. Plot of load versus deflection for various hybrid fibre reinforced concretes at total fibre volume fraction of 0.3 %.

strength at 28 days lies in the range of 58 to 66 MPa, and does not show an appreciable increase compared to controlled concrete. Generally, fibre addition does not affect the compressive strength of concrete significantly, since the mode of crack opening could be other than fracture, in which case the crack bridging effect of the

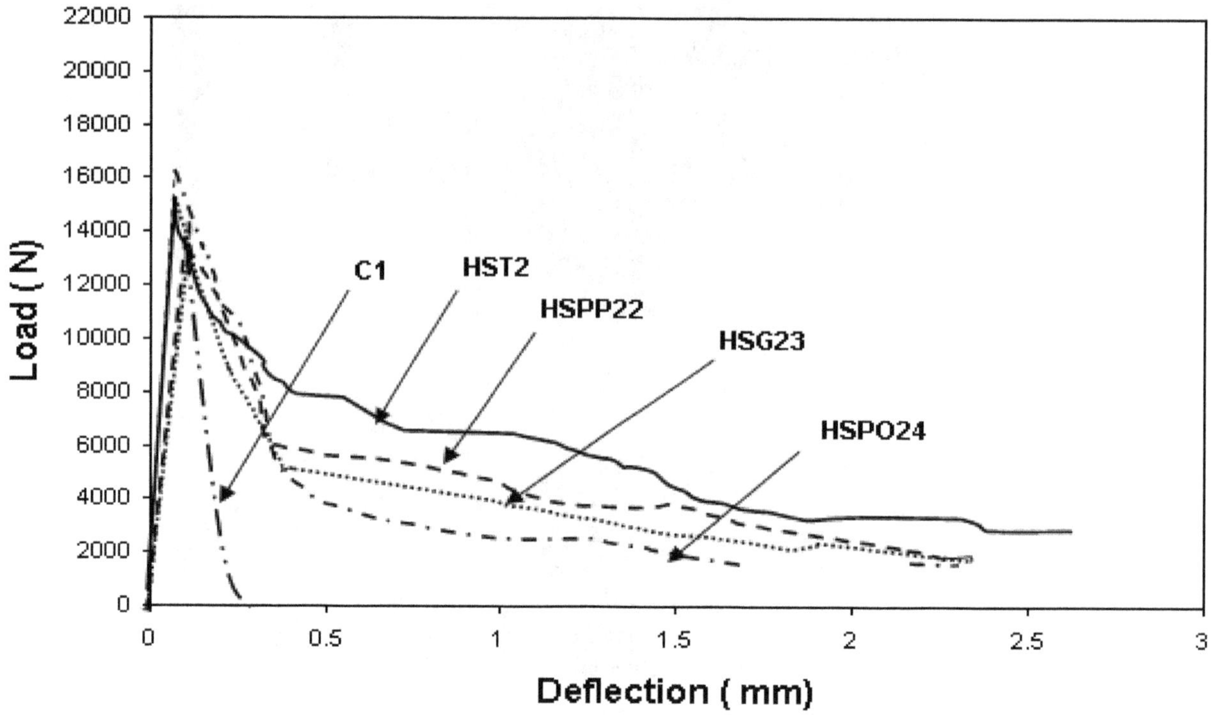

Figure 3. Plot of load versus deflection for various hybrid fibre reinforced concretes at total fibre volume fraction of 0.4%.

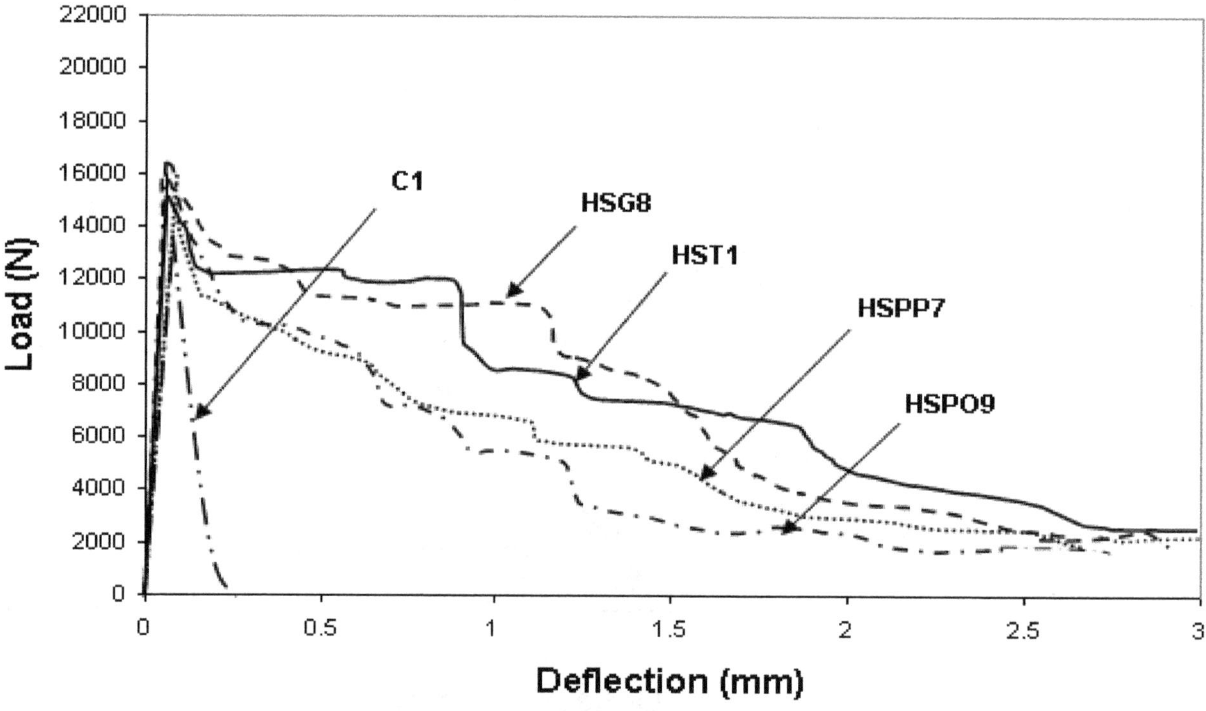

Figure 4. Plot of load versus deflection for various hybrid fibre reinforced concretes at total fibre volume fraction of 0.5%.

fibres is not efficient.

The split tensile strength of hybrid fibre concrete was found to be higher compared to reference and mono steel fibre concrete. From Table 5, it can be observed that the hybrid fibre concretes containing steel and glass at all volume fractions show the best split tensile strength among all concretes at all dosages. Among other combinations, only the steel-polypropylene combination

Table 4. Details of factorial experimental design.

Factors	Level 1	Level 2	Level 3
Total Fibre dosage TFD (% by volume of concrete)	0.3	0.4	0.5
Steel to Non-metallic fibre ratio SNMFR	2.33	4	9
Hybrid Fibre Combinations	Steel + polyester	Steel + glass	Steel + polypropylene
No. of experiments	9	9	9

Table 5. Strength resultsof various concrete mixtures at 28 days.

Mix ID	Fibre volume fraction (%)	Compressive strength (MPa)	Splitting tensile strength (MPa)	Flexural strength (MPa)
C1	0	62.9	5.87	5.72
HSPP1	0.3	63.4	6.10	6.40
HSG2	0.3	64.2	6.31	6.19
HSPO3	0.3	62.0	5.89	6.05
HSPP4	0.4	66.7	6.46	6.52
HSG5	0.4	58.7	6.28	6.40
HSPO6	0.4	60.4	5.91	6.42
HSPP7	0.5	61.2	7.30	6.58
HSG8	0.5	64.4	7.67	7.75
HSPO9	0.5	59.3	6.68	6.74
HSPP10	0.3	62.0	6.19	6.35
HSG11	0.3	65.4	6.60	6.11
HSPO12	0.3	61.4	6.21	6.18
HSPP13	0.4	66.7	6.89	6.46
HSG14	0.4	64.2	6.33	6.71
HSPO15	0.4	60.2	6.17	6.31
HSPP16	0.5	64.6	7.71	7.78
HSG17	0.5	64.4	7.56	7.52
HSPO18	0.5	59.3	6.76	6.68
HSPP19	0.3	62.1	6.18	6.12
HSG20	0.3	65.7	6.69	6.32
HSPO21	0.3	62.9	6.15	6.11
HSPP22	0.4	64.4	6.51	6.55
HSG23	0.4	60.2	6.55	6.48
HSPO24	0.4	59.2	6.24	6.29
HSPP25	0.5	61.2	7.45	7.53
HSG26	0.5	64.4	7.91	7.40
HSPO27	0.5	59.3	6.81	6.92
HST1	0.5	60.3	7.46	7.14
HST2	0.4	63.6	6.68	6.50
HST3	0.3	65.3	6.11	6.40

at a dosage of 0.5% (with 30% polypropylene fibres) gave strength higher than the mono-steel fibre concrete. Enhancement in split tensile strength is expected with fibres since the plane of failure is well defined (diametric). The higher the number of fibres bridging the diametric 'splitting' crack, the higher would be the split tensile strength. However, fibre availability is not the only parameter governing the strength; the stiffness of the fibre is also a major parameter affecting the strength. Thus, although in terms of availability, the glass and polypropylene fibres in hybrid combinations with steel result in higher fibre availability, only the glass fibres are able to enhance the strength at all dosages owing to their high stiffness. Polyester fibres, however, resulted in lowering of strengths; this might be because of difficulty in dispersing these fibres uniformly into the concrete mixture.

Table 6. Ductility, toughness indices and residual strength of concrete mixtures at a total fibre volume fraction of 0.5%.

Mix ID	Fibre combination	First crack deflection δ (mm)	Ductility $D=\delta_R - \delta$ (mm)	Toughness Indices			Absolute toughness (N m)	Residual strength factors		Post-crack strength (MPa)
				I_5	I_{10}	I_{20}		$(R_{5,10})$	$(R_{10,20})$	
C1	-	0.21	0	-	-	-	4.45	-	-	-
HSPP7	S – PP(90 – 10)	0.24	2.62	4.05	7.23	18.65	19.28	63.6	114.2	4.10
HSPP25	S – PP(80 – 20)	0.21	2.69	4.26	7.42	18.32	20.61	63.2	109.0	4.25
HSPP16	S – PP(70 – 30)	0.24	2.75	4.48	7.75	18.06	21.20	65.4	103.1	4.47
HSG26	S – G(90 – 10)	0.20	2.74	4.67	7.61	18.82	19.63	58.8	112.1	4.16
HSG17	S – G(80 – 20)	0.25	2.78	4.88	7.86	18.57	20.78	59.6	107.1	4.33
HSG8	S – G(70 – 30)	0.24	2.87	5.04	8.35	18.44	21.92	66.2	100.9	4.67
HSPO18	S – PO(90 – 10)	0.21	2.52	4.20	6.70	14.78	18.20	50.0	80.8	3.56
HSPO9	S – PO(80 – 20)	0.23	2.44	3.53	5.35	12.92	17.69	36.4	75.7	3.11
HSPO27	S – PO(70 – 30)	0.24	2.29	3.19	4.78	11.68	17.02	31.8	69.0	2.82
HST1	S(100)	0.26	2.92	4.80	7.78	19.28	21.36	59.6	115.0	4.55

Note: S – Steel fibre, PP – Polypropylene fibre, PO – Polyester fibre, and G – Glass fibre.

Compared to the control concrete without fibres, all fibre-reinforced concretes showed higher flexural strengths. Among all fibre concretes, the hybrid combination of steel and glass was once again the best. In certain combinations, the steel-polypropylene combination also performed better than the mono-steel mixtures. Additionally, the steel-polyester combination resulted in lower strengths than other fibre concretes. These trends can be explained using the same concepts of fibre availability and stiffness, as in the case of the splitting tensile strength.

Toughness indices and absolute toughness

The absolute toughness of fibre reinforced concrete is a measure of the strain energy stored in the material. It is characterized by the area under the entire load-deflection plot and realized appropriately in terms of toughness indices. The various toughness indices (I_5, I_{10}, and I_{20}), calculated as per ASTM C-1018 and absolute toughness values of hybrid fibre concrete mixtures and mono fibre concrete mixtures at a volume fraction of 0.5% are given in Table 6.

The trends observed in Figure 5 indicate that all the fibre concretes yield a higher absolute toughness compared to the controlled concrete without fibres. Compared to the mono-steel fibre concrete, both the steel-glass combination and the steel-polypropylene combination gave similar toughness values. On the other hand, the toughness of steel-polyester concretes is lower compared to the mono-steel fibre concrete. In the case of the toughness indices (I_5 and I_{10}) also, the same trends are observed (Figures 6 and 7), in that the steel-glass and steel-polypropylene combinations give a comparable performance to the mono-steel fibre concrete, while the toughness indices for the steel-polyester combination are lower at higher fractions of polyester fibres. Once again, the poor performance of the steel-polyester combination could be attributed to the insufficient dispersion of polyester fibres.

Another notable trend that emerges from the results is that, for the steel-glass and steel-polypropylene combination, the higher the replacement of steel with the non-metallic fibre, the higher the toughness (and the indices). However, this trend is just the opposite for the steel-polyester combinations. It is difficult to explain this trend with the available data. However, it can be safely concluded that some of the steel fibres could be effectively replaced using non-metallic fibres without compromising on the toughness. The contribution to the I_5 is at a low deflection level, and the cracks at this stage, being small in width, are effectively bridged by the non-metallic fibres like glass and polypropylene. On the other hand, in the calculation of I_{20}, a high level of deflection is used, and the wide cracks at

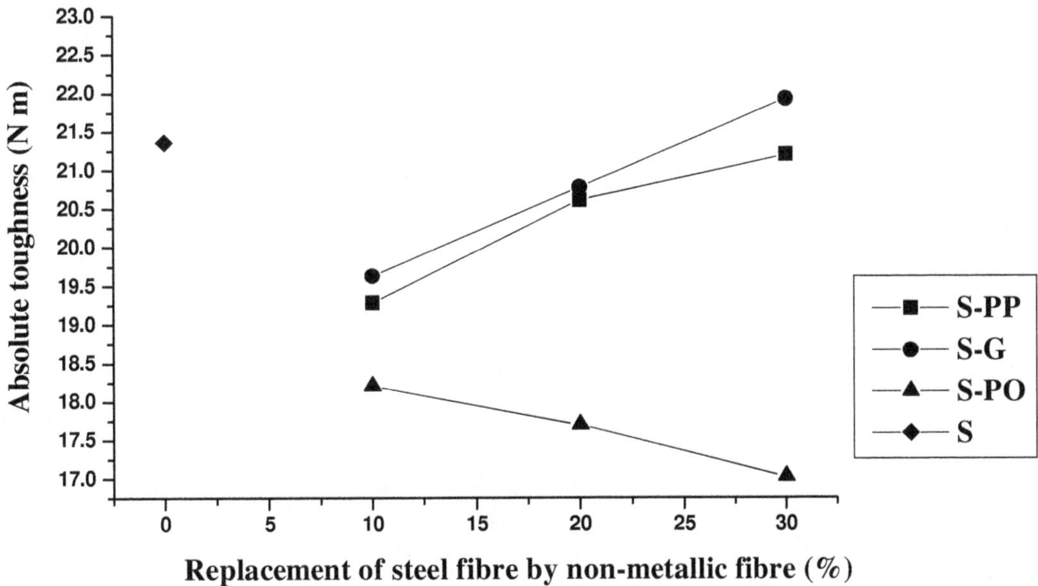

Figure 5. Plot of absolute toughness versus replacement of steel fibre by non-metallic fibre for various hybrid fibre concretes at total fibre volume fraction of 0.5%.

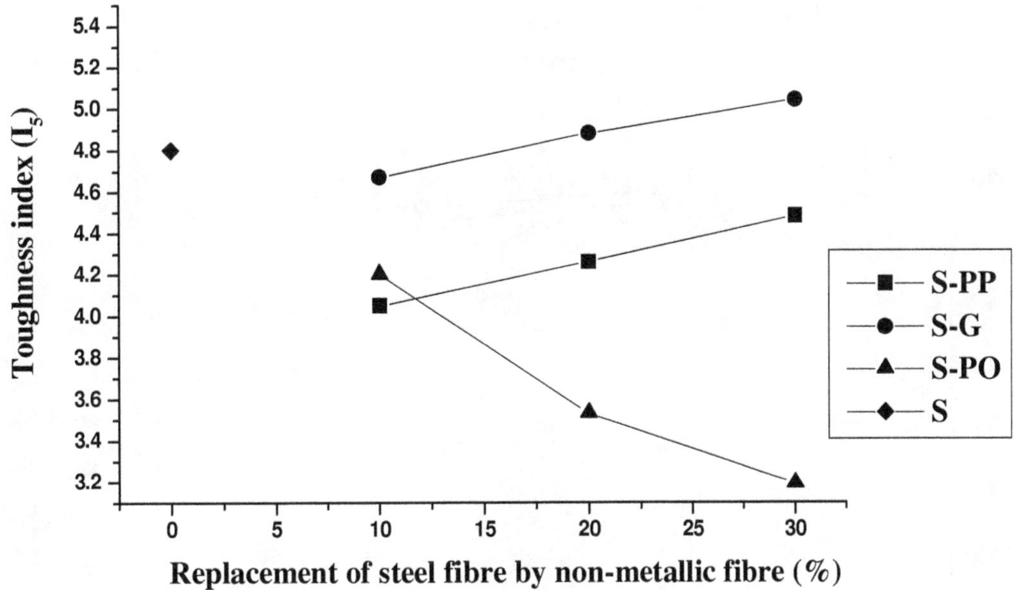

Figure 6. Plot of toughness index (I_5) versus replacement of steel fibre by non-metallic fibre for various hybrid fibre concretes at total fibre volume fraction of 0.5%.

this stage possibly need the action of steel fibres; this can be seen from Figure 8.

Residual strength factors (RSF)

RSFis obtained directly from toughness indices and represent the level of strength retained after first crack (ASTM Standard Designation C 1018-97). The residual strength factors calculated based on ASTM C-1018 for the various concretes are given in Table 6. The observed trends for the residual strength factors are shown in Figures 9 and 10. It is observed from Figure 9 that the RSF ($R_{5,10}$) for steel–glass and steel-polypropylene fibre concretes were higher than mono-steel fibre concrete at higher fractions of glass and polypropylene fibres, which dominated the post peak region of the load –deflection plot up to 5.5δ (where δ is the first crack deflection). However, it can be seen from Figure 10 that the RSF ($R_{10,\ 20}$) of mono steel fibre concrete was higher than

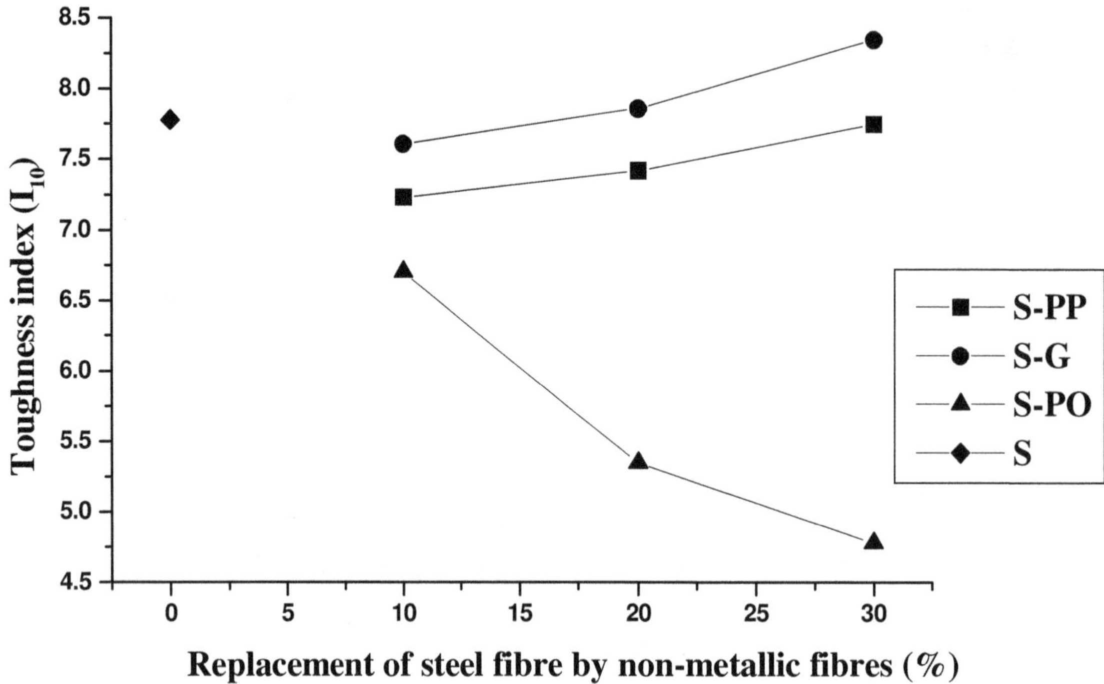

Figure 7. Plot of toughness index (I_{10}) versus replacement of steel fibre by non-metallic fibre for various hybrid fibre concretes at total fibre volume fraction of 0.5%.

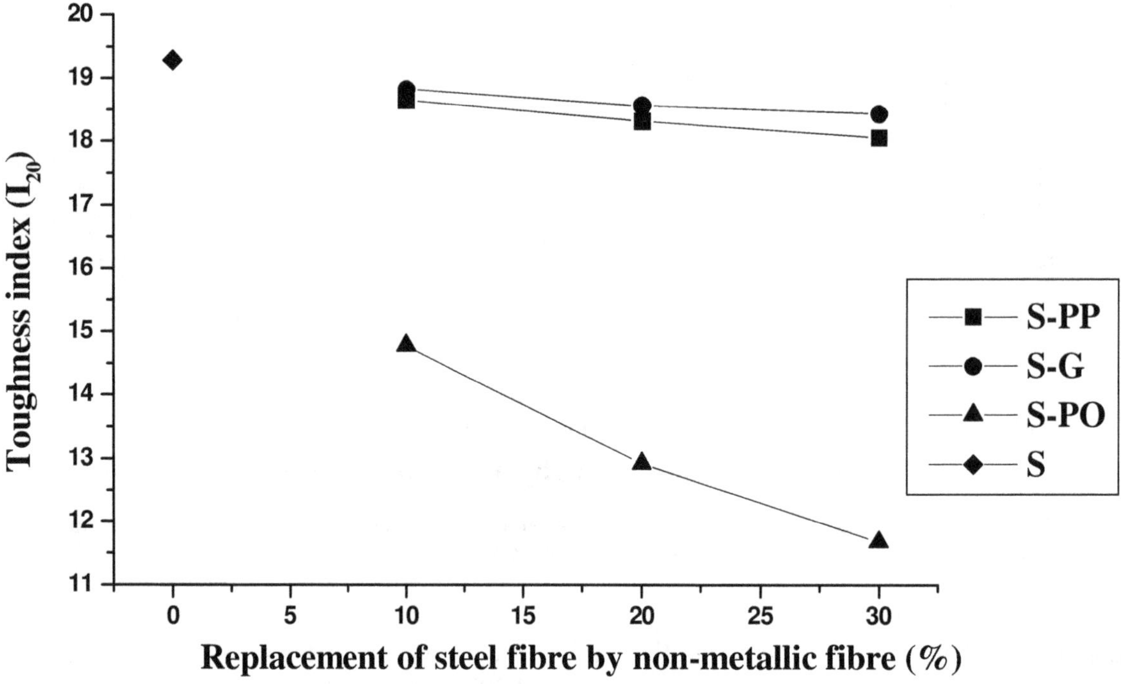

Figure 8. Plot of toughness index (I_{20}) versus replacement of steel fibre by non-metallic fibre for various hybrid fibre concretes at total fibre volume fraction of 0.5%.

that of other hybrid fibre concretes, which reveals the fact that at high strain levels; only the steel fibres were effective in bridging the cracks, while the non-metallic fibres probably ruptured due to the high crack widths. In the case of steel-polyester fibre concretes, both the RSFs were found to be abruptly low compared to mono-steel fibre concrete and other hybrid fibre concretes due to the defects arising from poor dispersion of polyester fibres.

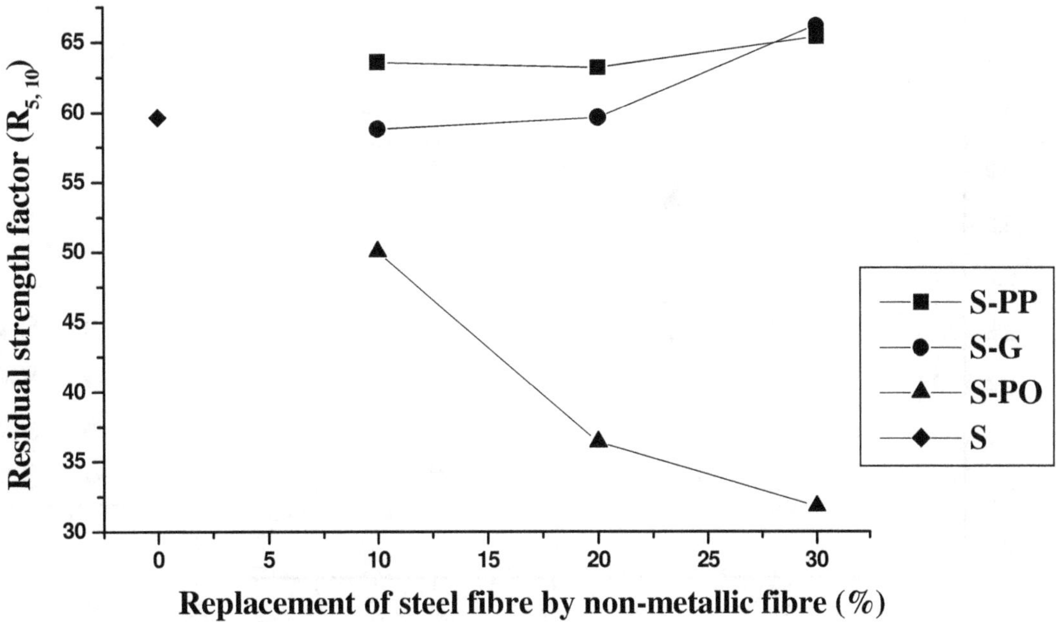

Figure 9. Plot of residual strength ($R_{5, 10}$) versus replacement of steel fibre by non-metallic fibre for various hybrid fibre concretes at total fibre volume fraction of 0.5%.

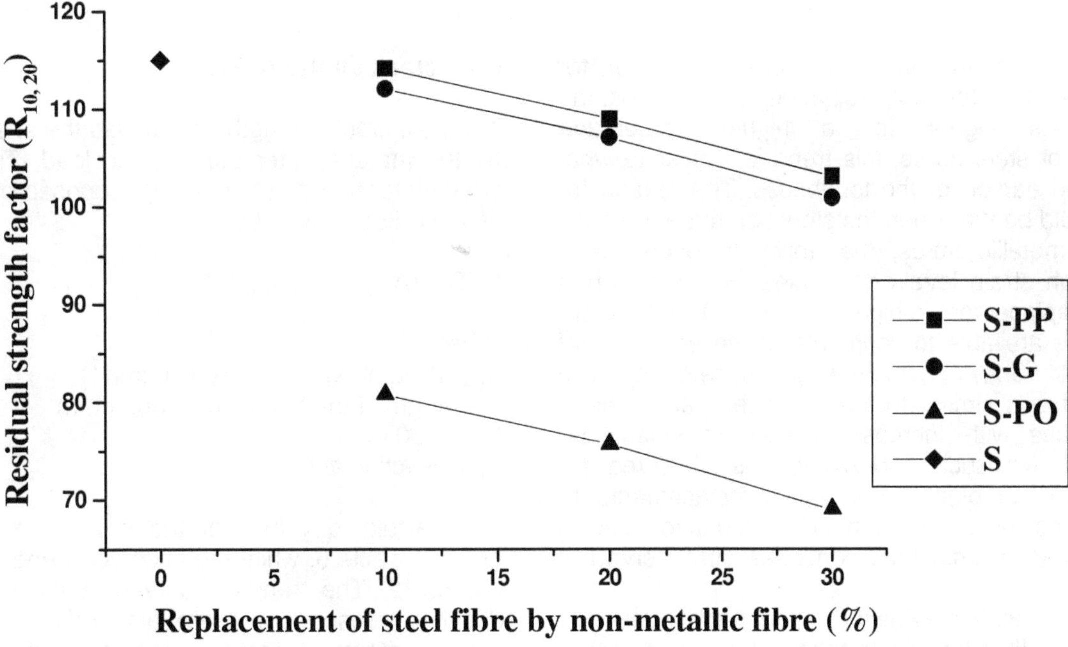

Figure 10. Plot of residual strength ($R_{10, 20}$) versus replacement of steel fibre by non-metallic fibre for various hybrid fibre concretes at total fibre volume fraction of 0.5%.

Ductility

Ductility is a significant property, which characterizes the post cracking behavior of high strength concrete in terms of the deformation sustained after reaching the ultimate load. It is measured in the load deflection plot as the elongation from the point of first crack deflection (δ) to the point where there is no resistance (δ_R) offered by the concrete beam upon further loading. The calculated ductility values are given in Table 6, while the observed trends are plotted in Figure 11. The trends in Figure 11 indicate that the maximum deformation is obtained for the mono-steel fibre concrete. This is followed by the steel-glass combination, steel-polypropylene combination, and

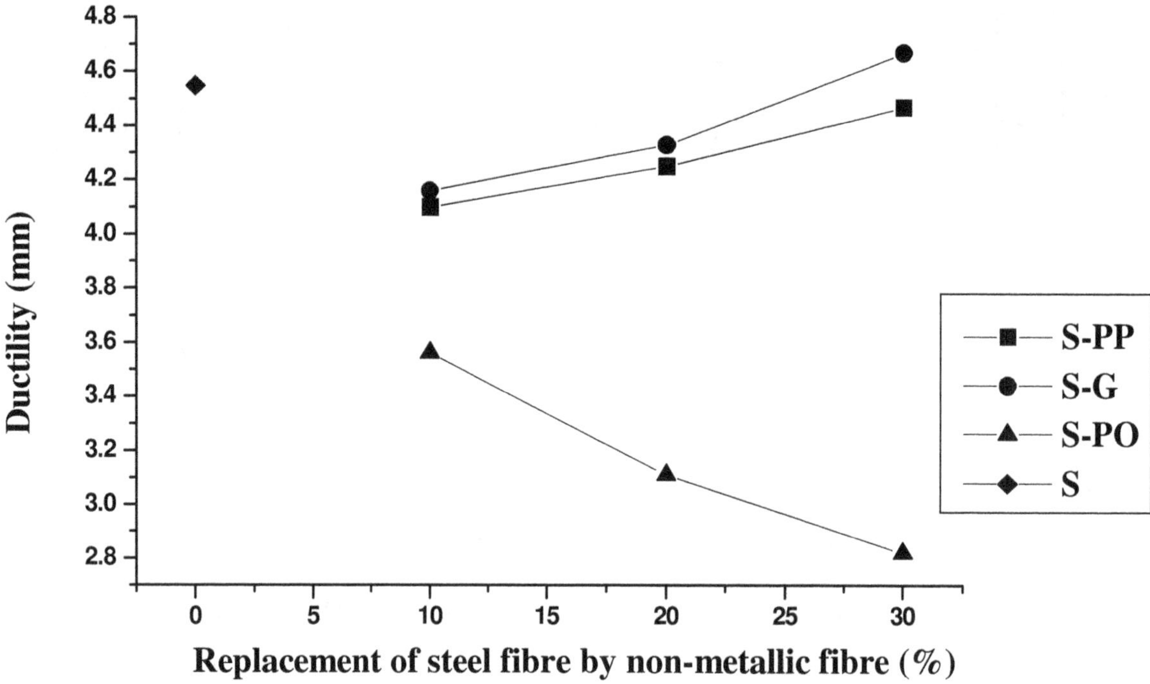

Figure 11. Plot of ductility versus replacement of steel fibre by non-metallic fibre for various hybrid fibre concretes at total fibre volume fraction of 0.5%.

steel-polyester combination, in that order. Moreover, for the steel-glass and steel-polypropylene combinations, the deformation is higher for a higher percentage replacement of steel fibres; this trend is similar to what was observed earlier for the toughness. The reason for this trend could be that when the steel fibres are replaced by the non-metallic fibres, the ability to bridge wide cracks at high strain levels decreases. However, when the level of replacement is high enough (~30%), the non-metallic fibres are able to contribute to the early part of the post-peak behaviour, increasing the deformation in that range (this is similar to the increase seen in the I_5 and I_{10} indices with increasing levels of steel fibre replacement). No such improvement is observed for steel-polyester combination at higher replacements of polyester fibres; on the other hand, a negative trend is seen compared to other fibre concretes as observed in Figure 11.

From the results, it is evident that the ductility depends primarily on the fibre's ability to take high levels of strain, in other words, on the stiffness of the fibre. Although the glass fibres have a reasonably high stiffness and tensile strength, they rupture at wide crack openings due to their high aspect ratio. Because of this reason, they are not able to contribute much to the ductility at the higher end of deflections. However, at high enough replacement of the steel fibres, the glass fibres are able to contribute to the ductility by controlling the thinner cracks. The same argument could be used for the polypropylene fibres; however, these fail at much lower loads, and are not able to take up too much strain because of their low stiffness.

Post crack strength (PCS)

The post crack strength determines the strength retained by the material after the ultimate load. The post crack strength for four points bending proposed by Banthia and Jean (1985) is given by:

$$PCS = (A_{post}\, L) / (D - D_p\, b_w^2) \quad (6)$$

Where,
A_{post}-Area of post crack region (mm^2)
L - Length of the beam specimen (mm)
D - L/150 (mm)
D_p-Deflection at first crack (mm)

The calculated values of the post crack- strength are given in Table 6, while the observed trends are shown in Figure 12. The trends observed in the toughness and ductility measurements are also reflected in the PCS measurements. It can be observed from Figure 12 that the higher the replacement of steel fibres with glass or polypropylene fibres, the higher the post crack strength. At high replacement levels, the post-crack strengths for these two combinations match or better the PCS of mono-steel fibre concrete. These results can once again be explained by the suggestion that at high replacement levels of steel fibres, there is a significant contribution to the early part of the post-peak load-deflection curve from the non-metallic fibres. In the case of the steel-polyester combination, however, there is a drop in the PCS with an increase in the level of polyester fibres. This indicates

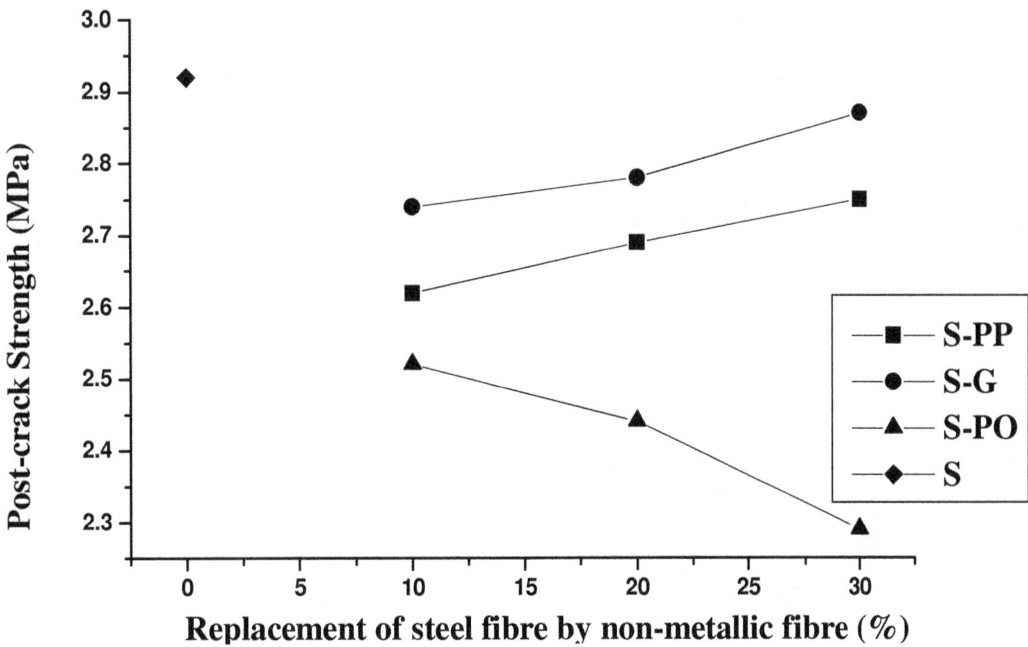

Figure 12. Plot of post crack strength versus replacement of steel fibre for various hybrid fibre concretes at total fibre volume fraction of 0.5%.

that the polyester fibres are not contributing much to the post crack behaviour, or, as said earlier, their dispersion into the concrete mixture is not proper.

Conclusions

The primary objective of this study was to evaluate the effectiveness of hybrid fibre combinations at low volume fractions in improving the post-peak behaviour of high strength concrete. Results from the study indicate the following:

i. It is possible to produce hybrid fibre concretes using glass, polyester, and polypropylene fibres in combination with steel fibres, with an enhanced ductility compared to controlled concrete without fibres.
ii. The steel-glass and steel-polypropylene hybrid combinations in concrete result in comparable levels of flexural strength, toughness, and ductility to mono-steel fibre concrete; steel-polyester combinations, however, yield a poorer composite than the mono-steel fibre concrete.
iii. The experimental observations for toughness and ductility reveal that the best performance of steel-glass and steel-polypropylene hybrid combinations is obtained at a high level of non-metallic fibre; the reason could be that at high levels of non-metallic fibres, there is significant enhancement in the early part of the post-peak behaviour.
iv. Increased fibre availability in the hybrid fibre systems (due to the lower densities of non-metallic fibres), in addition to the ability of non-metallic fibres of bridging smaller microcracks, could be the reasons for the enhancement in flexural properties.

A major significance of these findings is that steel fibres in concrete could be partially replaced with non-metallic fibres without compromising the ductility. This, in combination with the improved early age crack resistance that is made possible by the non-metallic fibres, make hybrid fibre combinations highly competitive as far as applications in high strength or high performance concrete are concerned.

REFERENCES

ASTM Standard Designation C 1018-97, Standard test method for flexural toughness and first crack strength of fiber reinforced concrete (using beam with third-point loading), Annual book of ASTM standards, Pennsylvania, United states, 4(02): 533-540.

Banthia N, Jean FT (1985). Test methods for flexural toughness of fibre reinforced concrete: some concerns and proposition. ACI. Mater. J., pp. 48-57.

Banthia N, Nandakumar N (2003). Crack growth resistance of hybrid fibre reinforced cement composites. Cement Concr. Compos., 25: 3-9.

Bayasi Z, Zeng J (1993). Properties of polypropylene fiber reinforced concrete. ACI Mater. J., 90: 605-610.

Bentur A, Mindess S (1990). Fiber Reinforced Cementitious Composites. Elsevier, London, pp. 221-240.

Indian Standard Designation, IS 12269-1987, Specification for 53 grade ordinary Portland cement, Bureau of Indian Standards, New Delhi, India, pp. 1-3.

Indian Standard Designation, IS 516-1999, Methods of tests for strength of concrete, Bureau of Indian Standards, New Delhi, India, pp. 15-19.

Indian Standard Designation, IS5816-1999, Methods of test for splitting

tensile strength on concrete cylinders, Bureau of Indian Standards, New Delhi, India, pp. 23-25.
Pierre P, Pleau R, Pigeon M (1999). Mechanical properties of steel micro fiber reinforced cement pastes and mortars. J. Mater. Civ. Eng., 11: 317-324.
Qian CX, Stroeven P (2000). Development of hybrid polypropylene-steel fibre-reinforced concrete. Cement Concrete Res., 30: 63-68.
Soroushian P, Khan A, Hsu JW (1992). Mechanical properties of concrete materials reinforced with polypropylene or polyethylene fibers, ACI. Mater. J., 89: 535-540.
Stroeven P, Babut R (1986). Fracture Mechanics and Structural Aspects of Concrete. Heron, 31(2): 15-44.
Yao W, Lib J, Wu K (2003). Mechanical properties of hybrid fiber-reinforced concrete at low fiber volume fraction. Cement Concr., Res., 33: 27-30.

A multilayer perceptron for predicting the ultimate shear strength of reinforced concrete beams

Ayman Ahmed Seleemah

Structural Engineering Department, Faculty of Engineering, Tanta University, Egypt.
E-mail: seleemah55@yahoo.com.

A number of codes of practice exist that predict the maximum shear capacity of reinforced concrete beams. Since the behavior of reinforced concrete (RC) beams with non-homogeneous, non-isotropic, and nonlinear material under a combined shear and bending state of stress is very difficult to establish, these codes seem to under or over-estimate this shear capacity for many cases. This is attributed to the fact that the factors that affect the shear strength of RC beams are too many, making modeling of its actual behavior a hard task. In this paper, several multilayer perceptrons were constructed as an analytical alternative to existing expressions for predicting the shear capacity of RC beams. A large database of experimental tests of beams (574 samples) was utilized to train and test the networks. Both multilayer perceptrons' predictions and four different codes of practice for the shear capacity of RC beams were examined. It was found that, the predictions of multilayer perceptrons are superior to those of any of the current available code relationships.

Key words: Multilayer perceptron, neural networks, shear capacity, reinforced concrete (RC) beams, shear reinforcement.

INTRODUCTION

Structural engineers always attempt to improve the analysis, design, and control of the behavior of structural systems. Such behavior, however, is complex and often governed by both known and unknown multiple variables with their interrelationship generally unknown, nonlinear and sometimes very complicated. The traditional approach used in most research in modeling generally depends on performing a multi variable nonlinear regression analysis so that the major parameters are calibrated to fit the experimental results and to derive the relationships among the involved parameters.

By contrast, the use of multilayer preceptron technique, or alternatively called neural network (NN), provides an alternative method that may be more accurate in predicting the actual response. A neural network is a computational tool that attempts to simulate the architecture and internal operational features of the human brain and neurons systems. In a strict mathematical sense, neural networks do not provide closed form solutions for modeling problems but offer a complex and accurate solution based on a representative set of examples of the relationship.

Neural network modeling techniques have been widely applied in structural engineering fields in recent years. For example, it has been applied to the area of structural analysis, design, and modeling (Senouci, 2000; Zhao et al., 2001; Sirca and Adeli, 2001; Oreta and Kawashima, 2003; Yun et al., 2008); structural dynamic problems (Chang and Zhou, 2002; Taysi, 2010); structural performance evaluation (Pannirselvam et al., 2008, 2010; Noorzaei et al., 2008); damage detection of structures (Zang and Imregun, 2001; Tsai and Hsu, 2002; Seleemah et al., 2012); and in geotechnical engineering applications (Jan et al., 2002; Juang and Jiang, 2003; Juang et al., 2003). The common features of many of the successful applications of neural networks in prediction and modeling are that the quantities being modeled are governed by multivariate interrelationships and the data available are "noisy" or incomplete. Moreover, when neural network models are developed, there is no need to assume any functional relationship among the various variables unlike in regression analysis. Neural networks

Table 1. Ranges of parameters in database.

Parameter	Minimum	Maximum	Parameter	Minimum	Maximum
b_w (mm)	76.2	1000.0	ρ_l (%)	0.14	6.64
d (mm)	110.0	2000.0	f_{yl} (MPa)	275.86	1779.31
a/d	1.56	8.03	ρ_v %	0.00	1.47
$f_c^{'}$ (MPa)	11.97	105.36	f_{yv} (MPa)	250	1431

automatically construct the relationships and adapt based on the data used for training.

A number of codes of practice exist that gives equations to predict the shear behavior of RC beams both with and without shear reinforcement. While these major codes are very important, there prediction seems to under or over-estimate this shear capacity in many cases. This is due to the fact that the behavior of RC beams with nonlinear, non isotropic and nonhomogeneous material under combined shear forces and bending moments is very complicated. Moreover, the shear strength of RC beams is affected by too many factors. This adds to the complexity of modeling its shear behavior.

While many efforts have been conducted to understand the shear behavior of reinforced concrete beams and/or to drive equations for estimating such shear capacity, some researchers explored the application of neural networks for such predictions. For example, Sanad and Saka (2001) applied the neural networks to predict the ultimate shear capacity of reinforced concrete deep beams; Mansour et al. (2004) successfully used the neural networks for prediction of the shear capacity of reinforced concrete beams with shear reinforcement; Oreta (2004) applied neural networks on a set of 155 experimental tests to simulate the size effect on the shear strength of reinforced concrete beams without shear reinforcement. Seleemah (2005) compared the predictions of neural networks with eight different equations for predicting the shear capacity of beams without shear reinforcement.

Rao and Babu (2007) constructed a hybrid neural network model which combines the features of feed forward neural networks and genetic algorithms for the design of beam subjected to moment and shear. Yang et al. (2007) built optimum multi-layered feed-forward neural network models using a resilient back-propagation algorithm to predict the shear capacity of reinforced concrete deep and slender beams. Dopico et al. (2008) applied the neural network technique to predict the shear strength of high and normal strength reinforced concrete beams with or without shear reinforcement. Kumar and Yadav (2008) applied a three-layer feed forward neural network with back propagation algorithm on 194 test results of reinforced concrete rectangular beams with web reinforcement failing in shear.

The intended aims of this study are: (i) to explore the feasibility of using neural networks in predicting the ultimate shear capacity of RC beams having a wide range of different variables including the existence or absence of shear reinforcement; (ii) to compare the results of neural network predictions with both the experimental values and those obtained using four major codes of practice, namely the ACI 318M-08, EC2, NZS 3101, and CSA codes; and (iii) to highlight the specific reasons that leads to over or under-estimation of the beams' shear capacity by these codes.

For this, a large database of 574 specimens that include normal and high-strength concrete beams with different percentages of longitudinal and shear reinforcement was retrieved from existing literature. The shear capacity of these beams was calculated utilizing both the afro mentioned codes and several neural networks having different architectures. Finally a comparison of the predictions obtained by the most successful network and those of the four codes of practice is presented.

METHODOLOGY

A databank of beams that satisfy agreed upon criteria was established and called the evaluation shear databank, Reineck et al. (2003). This databank contained 439 shear tests collected from 64 references. All beams in this database have a rectangular cross section, do not contain shear reinforcement, and were subjected to point loads. Extensive discussions and review on this databank was then conducted by the ACI subcommittee 445-F "Beam Shear" which led to extraction of a revised version (398 tests) that was intended to serve as a basis for any code changes. This database was combined with the database reported by Mansour et al. (2004) that contains a total of 176 test samples of beams with shear reinforcement collected from 15 references to have a total of 574 samples containing beams with or without shear reinforcement. This database was utilized in this study to evaluate and demonstrate the capability of the multilayer perceptron technique for predicting the shear capacity of reinforced concrete beams. Moreover, it was also used for evaluating four different codes of practice that exist for predicting such shear capacity.

It should be pointed out that the aforementioned database covers a very wide range of beam depths, breadths, shear span to depth ratios, maximum aggregate size used in concrete and its tensile and compressive strengths, main reinforcement percentage, shear reinforcement percentage, and yield stresses of longitudinal and shear reinforcement. Table 1 summarizes the ranges of the parameters covered by this database.

Shear strength of RC beams using building codes

Four major codes for estimation of the shear capacity of reinforced concrete beams were examined. These are ACI 318M-08, EC2, NZS 3101, and CSA. The expressions utilized by each of these codes are presented as follows:

ACI 318M-08(2008)

The shear provisions of the ACI building code states that the nominal shear strength V_n of a reinforced concrete beam can be computed as:

$$V_n = V_c + V_s \tag{1}$$

where V_c is nominal shear strength provided by concrete, and V_s is nominal shear strength provided by shear reinforcement. The concrete contribution to the shear strength can be calculated for members subject to shear and flexure only, from the following relationship:

$$V_c = 0.17\lambda\sqrt{f_c^{'}}.b_w d \tag{2}$$

Where λ equals 1.0 for normal weight concrete, b_w and d are the beam width and depth, respectively; and $f_c^{'}$ is the concrete cylinder compressive strength. The steel contribution can be calculated as:

$$V_s = \frac{A_v f_{yv} d}{s} \tag{3}$$

where A_v is the area of shear reinforcement within spacing s; and f_{yv} is the tensile yield stress of the shear reinforcement perpendicular to axis of the member and shall not exceed 420 MPa.

EC2 (2004)

This code gives the following relationship for beams without web reinforcement

$$V_n = V_c = \left[0.18k(100\rho_l f_c^{'})^{\frac{1}{3}}\right] b_w d \geq 0.035k^{1.5}\sqrt{f_c^{'}} b_w d \tag{4}$$

Where $k = 1 + \sqrt{\frac{200}{d}} \leq 2.0$ (d in mm) and $\rho_l \leq 0.02$

For the case of beams with web reinforcement

$$V_n = Max\left[V_c, (\rho_v f_{yv} \cot\theta) b_w d\right] \leq v_e \frac{f_c^{'}}{1.5} / (\cot\theta + \tan\theta) \tag{5}$$

Where $v_e = 0.6(1 - f_c^{'}/250)$ and $1 \leq \cot\theta \leq 2.5$

Where ρ_l and ρ_v are the longitudinal and shear steel reinforcement ratios, respectively; and θ is the angle of web reinforcement to longitudinal axis of beam.

NZS 3101(1995)

This code is applicable for members with concrete strength up to 100 MPa. The concrete contribution to the shear capacity is expressed as:

$$V_n = [(0.07 + 10\rho_l)\sqrt{f_c^{'}}].b_w d \tag{6}$$

The concrete stress term should be within the minimum and maximum limits of $0.08\sqrt{f_c^{'}}$ and $0.2\sqrt{f_c^{'}}$, respectively. The shear reinforcement contribution is similar to that of the ACI code given by Equation (3).

CSA (Canadian Standard Association) building code(1994)

The simplified method of this code considers the shear reinforcement contribution as that of the ACI code given by Equation (3) and the concrete shear contribution, V_c, is given as dependent on the shear reinforcement and the effective depth of the beam, d, as following

$$V_c = 0.2\sqrt{f_c^{'}}.b_w d \tag{7}$$

$A_v \geq 0.06\sqrt{f_c^{'}} b_w s / f_{yv}$ or $d \leq 300mm$

$$V_c = \left(\frac{260}{1000 + d}\right)\sqrt{f_c^{'}}.b_w d \geq 0.1\sqrt{f_c^{'}}.b_w d \tag{8}$$

$A_v < 0.06\sqrt{f_c^{'}} b_w s / f_{yv}$ and $d > 300mm$

Multilayer perceptron architecture

A multilayer perceptron is a nonlinear dynamic system consisting of a large number of highly interconnected processing units, called artificial neurons. The main computational characteristics of multi-layer perceptrons are their ability to learn functional relationships from examples and to discover patterns and regularities in data through self-organization. So, they are very suitable for modeling the nonlinear mapping type of problems.

A multilayer perceptron consists of a collection of simple processing units or nodes connected through links called connections. The topology or architecture of a multilayer perceptron is presented schematically in Figure 1, In which, a four-layered feed-forward multilayer perceptron is represented in the form of

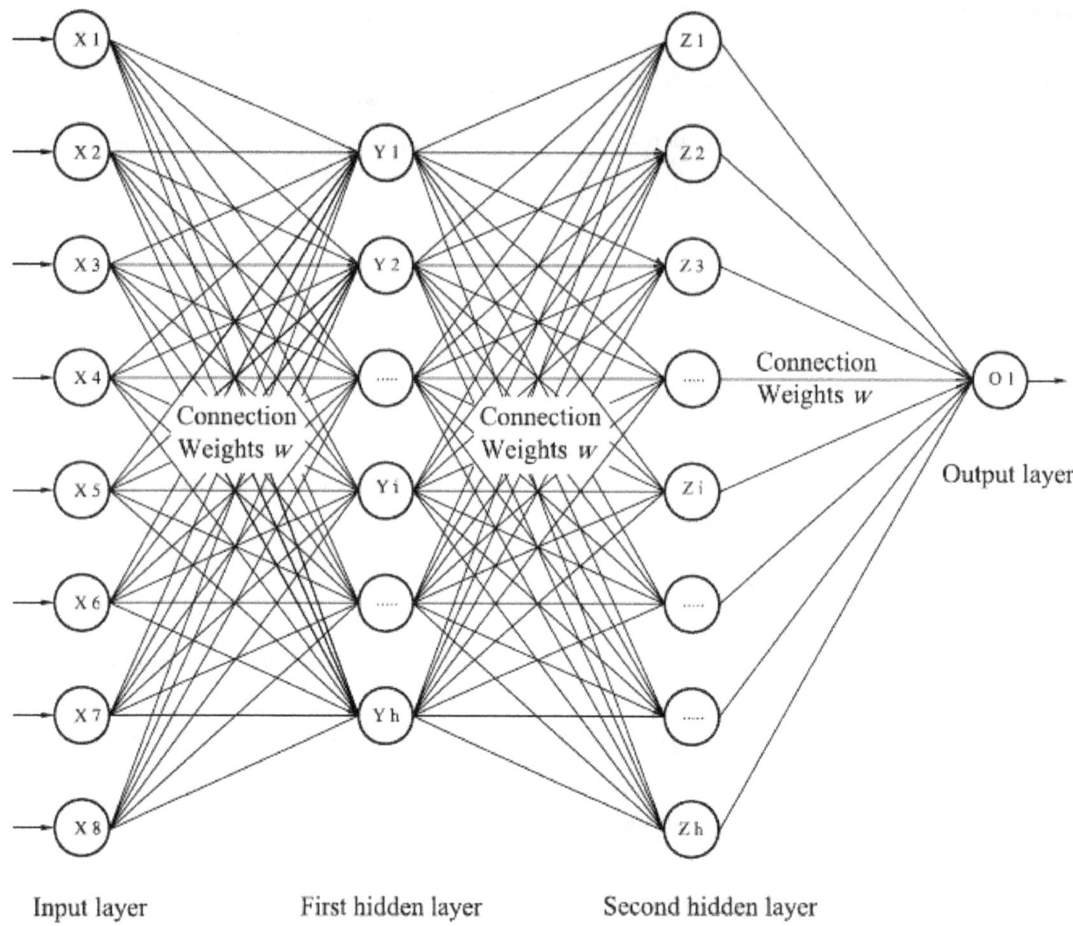

Figure 1. Typical architecture of a multilayer perceptron.

a directed graph, where the nodes represent the processing unit, the arrows represent the connections, and the arrowheads indicate the normal direction of signal flow.

The processing units may be grouped into layers of input, hidden, and output processing units. The main tasks of processing units are to receive input from its neighboring units which provide incoming activations, compute an output, and send that output to its neighbors receiving its output. The strength of the connections among the processing units is provided by a set of weights (w_i) which affect the magnitude of the input that will be received by the neighboring units. The output produced by the output processing units is compared to the target output data, and the weights are appropriately modified or adjusted based on training or learning rule. Eventually, if the problem can be learned, a stable set of weights adaptively evolves that will produce good results.

The connections between the neurons are individually weighed so that the total of i inputs (X_i) to the single neuron is:

$$\text{Input}=\sum_i w_i X_i \tag{9}$$

The weights may be positive or negative such that some inputs will be excitatory and others will be inhibitory. This input passes through an activation function to produce the values of Y_i or Z_i of the hidden layer(s) or O_i of the output layer. The activation function may have many forms. The most used and effective form in our case is the sigmoid function defined as:

$$\text{Output}=1/\left[1+e^{-\alpha(Input)}\right] \tag{10}$$

where α is a constant that typically varies between 0.01 and 1.0.

Signals are received at the input layer, pass through the hidden layers, and reach the output layer, producing the output of the network. The learning process primarily involves the determination of connection weight matrix and the pattern of connections. It is through the presentation of examples, or training cases, and application of the learning rule that the network obtains the relationship embedded in the data.

Input and output layers of neural networks

In this study, the neural networks were designed to have an input layer that consists of eight input nodes representing the most important parameters that affect the shear capacity of reinforced concrete beams. Based on careful study of recent approaches for the shear phenomena in concrete members (refer to ASCE-ACI shear and torsion committee 445, 1998; Reineck et al., 2003) , it was decided to design the input layer to consist of 1- beam depth (d); 2- beam breadth (b_w); 3-shear span to depth ratio (a/d); 4- concrete cylinder compressive strength f_c'; 5- percentage of Main longitudinal reinforcement (ρ_l); 6- longitudinal steel yield stress(f_{yl}); 7- percentage of shear reinforcement (ρ_v); and 8- shear reinforcement yield stress (f_{yv}). The output layer consisted of one node representing the ultimate shear capacity of the beam.

Training of the network

In a multilayer feed forward neural networks, *training* refers to the iterative process involving the presentation of training data to the network, the invocation of learning rules to modify the connection weights, and, usually, the evolution of the network architecture, such that the knowledge embedded in the training data is appropriately captured by the weight structure of the network. During the training phase, the training data consists of input and associated output pairs representing the problem that the network should learn.

Training data and test data

An important factor that can significantly influence a network's ability to learn and generalize is the number of patterns in the training set. Increasing the number of training patterns, though increases the time required to train a network, provides more information about the shape of the solution surface, and thus increases the potential level of accuracy that can be achieved by the network. Since a total of 574 data patterns were available in this study, it was decided to use 50% of the data for the training process (287 beams) and save the other 50% for testing or validation. Training data were checked to make sure that they satisfy a good distribution over the problem domain.

Back propagation network learning algorithm

The training phase of the network is implemented by using a learning algorithm such as the popular and effective back propagation algorithm. The training phase of the algorithm consists of two passes. The forward pass computes the network output for a given set of connection weights and input data. The backward pass computes the error of the network with respect to the target outputs and this error is passed backward to the network and is used to modify the connection weights. Usually, an error criterion for the network output is chosen and the maximum number of cycles is set to provide a condition for terminating the simulations. The performance of the neural networks can be monitored by observing the convergence behavior of the error with respect to the number of cycles. If the network "learns," the error will approach a minimum value. After the training phase, the neural networks can be tested for other input data where the final values of the weights obtained in the training phase are used. No weight modification is involved in the testing phase.

Network data preparation

Neural networks are very sensitive to absolute magnitudes. To minimize the influence of absolute scale and numerical overflow, all inputs and outputs to neural networks were scaled so that they correspond roughly to the same range of values. Because of the property of the sigmoid function which is asymptotic to values 0 and 1, the derivative at or close to values 0 and 1 will approach zero producing very small signal error which results in slow learning. To avoid the slow rate of learning near the end points specifically of the output range, the data were scaled between the interval 0.1 and 0.9.

Network validation and error analysis

A neural network model, after it has been trained by presenting it with a set of training patterns, has to be empirically validated. The usual practice for neural network model validation is to evaluate the network performance measure using a selected error metric based on data (referred to as test data) that was not used in the training phase. Aside from validating the trained network, this performance measure is often used in research to show the superiority of certain network architecture. The evaluation and validation of neural network prediction model can be done by using common error metrics such as the root mean squared error (RMS). The definitions for RMS error is given by,

$$RMS = \sqrt{\sum_{i=1}^{n}\sum_{j=1}^{m}(O_{ij} - Y_{ij})^2 / nm} \tag{11}$$

Where n is the number of patterns in the validation set; m is the number of components in the output vector; O is the output of a single neuron; and Y is the target output for the single neuron.

Network topology

Since there is no direct and precise way of determining the most appropriate number of hidden layers and optimum number of neurons to include in each hidden layer for a specific problem, a trial and error procedure is typically used to approach the best network topology for a particular problem. In this paper, several network topologies were examined. These included networks with one hidden layer containing from one to seventeen neurons in the single hidden layer (a total of seventeen networks); and networks with two hidden layers containing from two to eight neurons in the first hidden layer; and from two to ten neurons in the second hidden layer (a total of twenty networks). The target network would be the one that produces minimum error for both the training and testing patterns based on statistical evaluation.

RESULTS AND DISCUSSION

Achieving best network

The performance of different networks in terms of RMS error of both training and testing patterns is shown in Figure 2. Clearly, increasing the number of neurons causes the RMS training error to decrease.

Six of the best networks that perform well during training and testing phases were selected and highered for an extra evaluation. These networks are NN 8-2-1, NN 8-5-1, and NN 8-14-1 which have one hidden layer containing 2, 5, and 14 neurons, respectively; and NN 8-4-2-1, NN 8-6-10-1, and NN 8-8-10-1 which have two hidden layers containing 4 and 2, 6 and 10, and 8 and 10 neurons in the first and second hidden layers, respectively. It is worth mentioning that networks NN 8-2-1, NN 8-5-1, and NN 8-4-2-1 were selected to represent small-sized networks that perform relatively well during both training and testing phases.

To judge which network performs better, the ratios of experimental to model predicted shear capacity

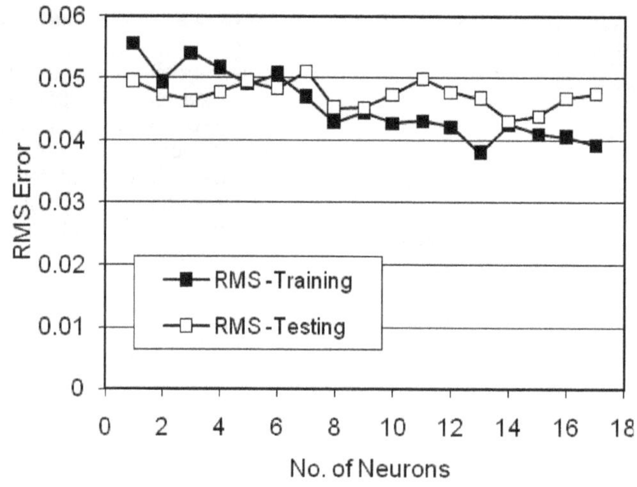

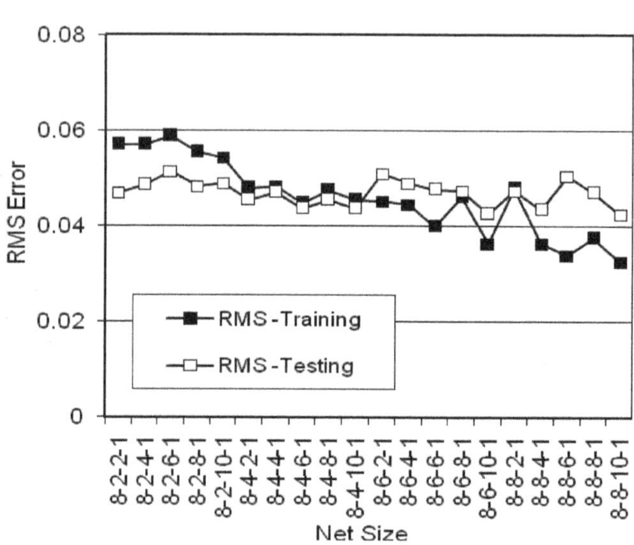

Figure 2. RMS error for different networks.

$\gamma_{mod}=V_{exp}/V_{pred}$ were calculated for all patterns, that is 574 pattern sets. The results are shown in Figure 3, which shows the maximum, minimum, average, standard deviation (SD), and the coefficient of variance (COV), for each of the tested networks. Moreover, a comparison of the experimental versus predicted shear capacity by the aforementioned six networks for all 574 data patterns is shown in Figure 4. The line of equality and the lines of plus or minus 10% error are also plotted on the figures to facilitate the visualization and judgment on the results. The network which gives results closer to the equality line is of course better. Using judgment on both figures, it was decided to select NN 8-8-10-1 as the most successful network. The histogram of ratio between experimental and predicted shear capacity for this network is shown in Figure 5. The histogram looks very close to the bell shape and it is nearly symmetrical with its maximum occurring at γ_{mod} = 1.0. However, very few odd results appeared at γ_{mod} = 1.80 (3 cases) and γ_{mod} = 2.20 (2 cases).

Comparison with different models

The prediction of the shear capacity of all 574 beam specimens were calculated using NN 8-8-10-1, together

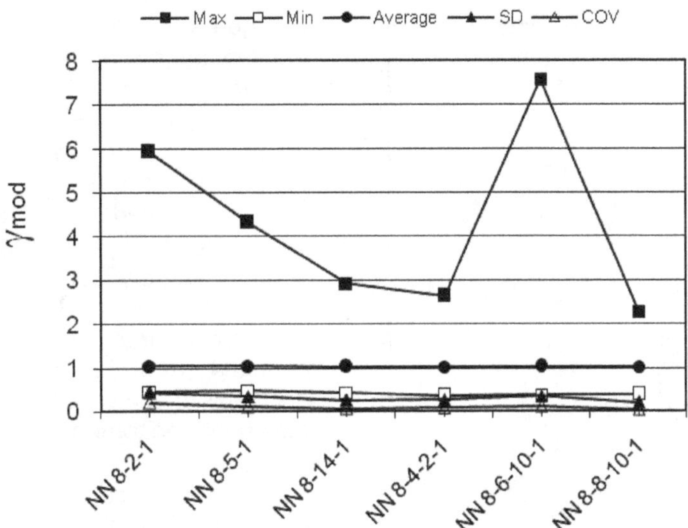

Figure 3. Statistical data for the ratio between experimental and predicted shear capacity for different networks

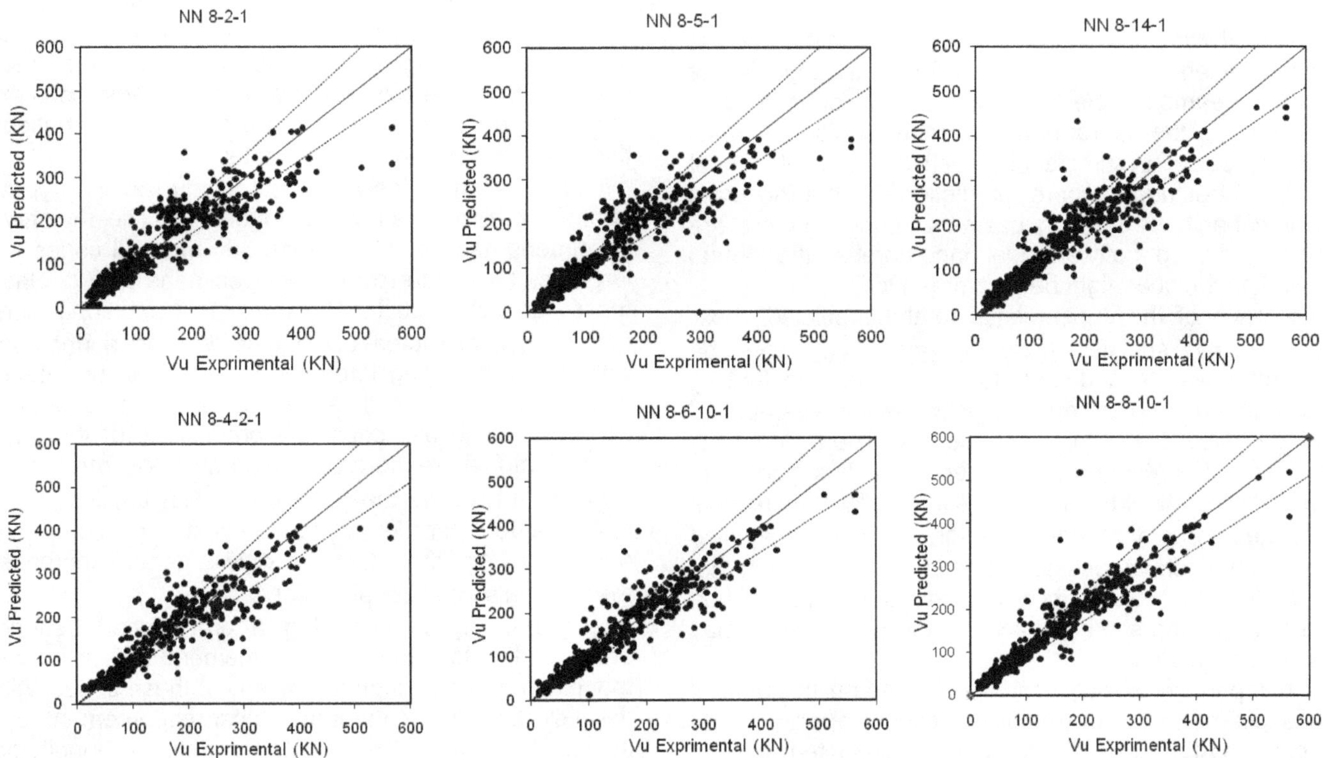

Figure 4. Comparison of experimental and predicted shear capacity for different networks.

with the four different codes mentioned earlier. The obtained results are shown in Figure 6, which shows, on separate plots, comparisons between experimental versus predicted shear capacity. It is clear from Figure 6 that the different codes give wide variations on either sides of the equality line. This means that these codes underestimate the shear capacity for some specimens and overestimate the shear capacity for other specimens. On contrast, the predictions of the neural network are much better with most results laying on or very close to

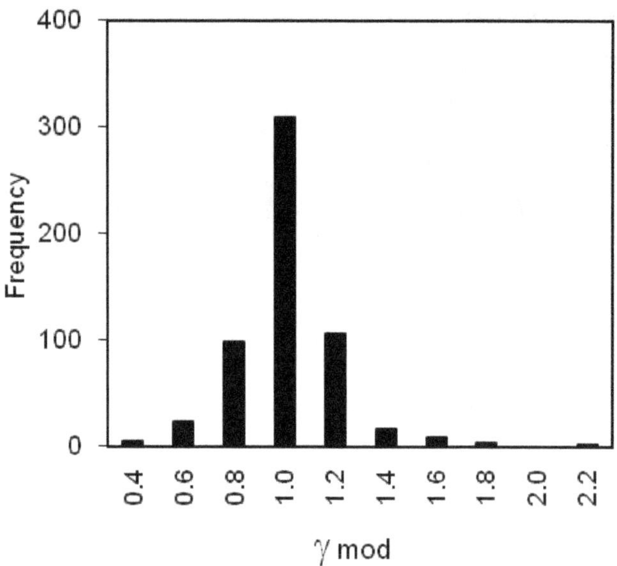

Figure 5. Histogram of ratio between experimental and predicted shear capacity for NN 8-8-10-1.

the equality line. This accuracy suggests that the most critical variables that control the shear capacity of concrete beams are the variables that have been used as input data to the neural network model. It also suggests that, in special circumstances such as rehabilitation of existing structures, where prediction of the accurate shear capacity of building beams or bridge girders are crucial, using a software that incorporates the neural network technique might be recommended.

The ratio of the experimental to the predicted shear capacity by specific code $\gamma_{mod}=V_{exp}/V_{pred}$ can be interpreted as the additional factor of safety implied by this code since no strength reduction factor was applied during calculations of any code's shear capacity prediction. For economy considerations, this additional safety factor should not be very large since there is a concrete strength reduction factor, γ_m, that is applied during any design process. The ratio γ_{mod} also should not be too less than unity since this means that the method overestimate the shear capacity of the beam which may lead to unsafe design.

A comparison of statistical calculations on the ratio of the experimental to the predicted shear capacity for all models and for all data patterns was conducted and the results are shown in Figure 7. Clearly, the neural network method gives best results with maximum ratio of 2.27, minimum of 0.38, and an average of 1.01. All codes have maximum ratio laying between 5.2 and 6.4, minimum between 0.15 and 0.19, with an average between 1.25 and 1.52. Both the SD and SV are minimum for the neural network method.

To describe the influence of dominant parameters on prediction of each model, the ratio of actual to predicted shear capacity is plotted versus the primary parameters in Figures 8 to 12. A common feature of all the plots is the very large scatter in the predictions by different codes in which γ_{mod} ranged between 0.15 to 6.4.

Figure 8 shows the plot of γ_{mod} with the shear span to depth ratio *(a/d)*. Most of the experimental tests were conducted for *a/d* ratio less or equal to 4.0. A large scatter is noticed in the predictions made by all codes for *a/d* ratios laying between 2.0 and 4.0. For all codes, the additional safety factor, γ_{mod}, increase with the decreases in a/d ratio indicating a beneficial influence due to direct load transfer that is not captured by these codes. Results obtained from the neural network indicate consistent accuracy in all ranges of *a/d* indicating a well capture of the shear phenomena.

The variation of the additional safety factor, γ_{mod}, with the concrete compressive strength for the models considered in this study is shown in Figure 9. Most of the testes in the database were carried out for normal strength concrete (NSC) having compressive strength less than 45.0 MPa. The scatter of the predictions by different codes for NSC specimens is very apparent. For concrete strength more than 45.0 MPA, the ACI, NZS, and CSA codes are generally conservative. The EC2 seems to overestimate the shear capacity for many specimens. Among all the shown models, the results obtained from the neural network is the most consistent one, having values close to unity for a wide variation of the concrete compressive strengths.

The variation of the additional safety factor, γ_{mod}, with beam depth, *d,* is shown in Figure 10. The majority of specimens had depth less than 500 mm. All codes yield large scatter of the results for specimens with depths in the range (d = 200 to 400 mm). The additional safety factor, γ_{mod}, predicted using different codes decreases with depth indicating improper capturing of the effect of depth on the overall shear capacity of beams. In contradiction to all codes, the predictions of the neural network model are more consistent and closer to unity.

Figure 11 shows the plot of γ_{mod} with the longitudinal reinforcement tensile strength indicator ($f_{yl} .\rho_l$). Results obtained using different codes show a general increase in the model's additional safety factor, γ_{mod}, with the increase in ($f_{yl} .\rho_l$) indicating a pronounced beneficial effect of the longitudinal reinforcement tensile strength that is not properly captured by any of these codes. While the yield strength of the longitudinal reinforcement is not included in any of the selected codes, the longitudinal reinforcement ratio, (ρ_l), is included only in the EC2 and the NZS codes but it seems that the beneficial effect of reinforcement exceeds what is included in their relationships. The performance of the neural network model is superior to all other models. It is not only unaffected by the change in the longitudinal reinforcement tensile strength, but also gives values of γ_{mod} very close to unity.

Lastly, Figure 12 shows the plot of γ_{mod} with the shear reinforcement tensile strength indicator ($f_{yv} .\rho_v$). Results

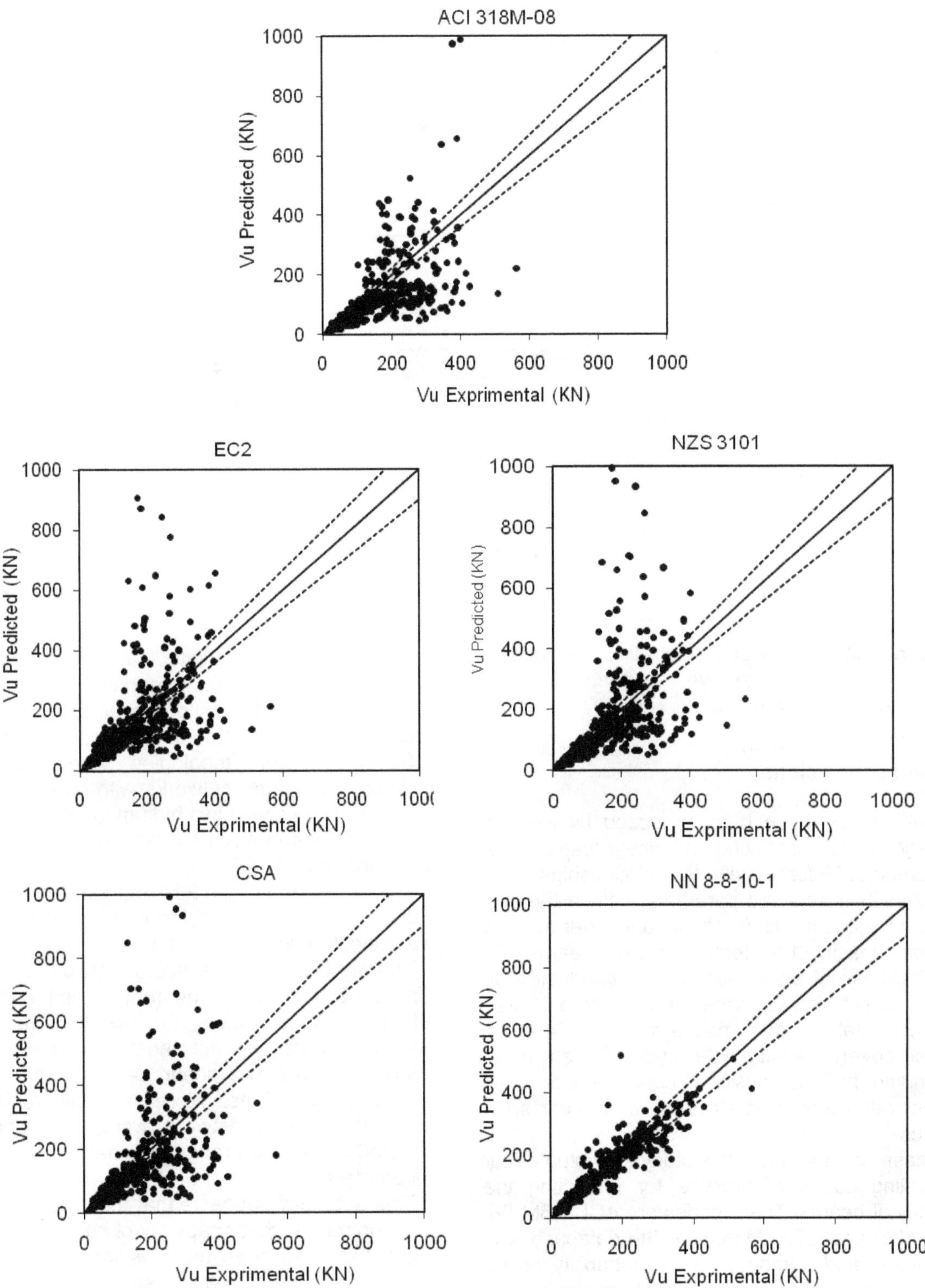

Figure 6. Comparison of experimental and predicted shear capacity for 573 specimens.

obtained using different codes show a serious decrease in the model's additional safety factor, γ_{mod}, with the increase in ($f_{yv} .\rho_v$) indicating an overestimation of the effect of the shear reinforcement effect. This overestimation is less pronounced in the CSA. This is due to the fact that this specific code puts limitation on the concrete shear contribution based on the shear reinforcement ratio. Once again, the performance of the

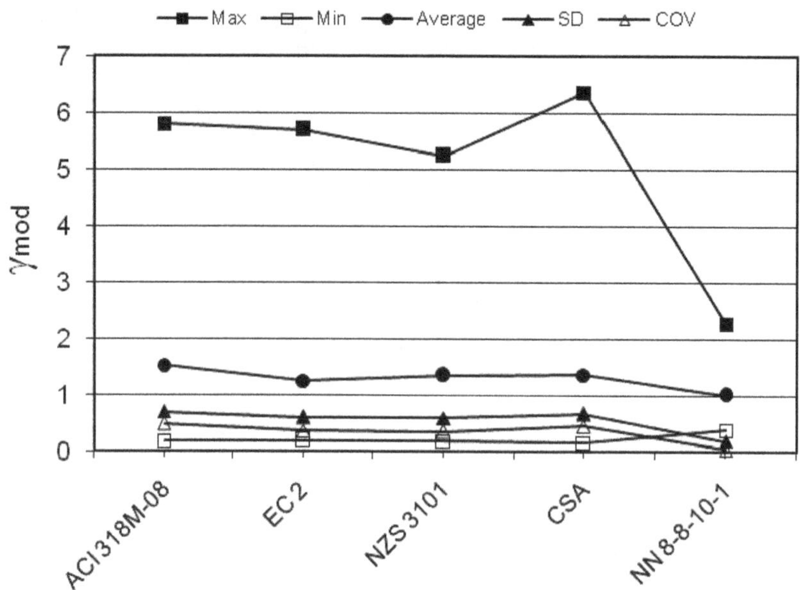

Figure 7. Statistical data for the ratio between experimental and predicted shear capacity for different shear proposals.

neural network model is superior to all other models indicating a good capability to give reasonable solution for such complicated phenomena.

Summary and conclusions

Different relationships have been proposed by existing codes of practice for predicting the shear capacity of concrete members. Unfortunately, the relationships differ considerably in their selected parameters since there is no generally accepted model for the load transfer and the ultimate shear capacity of reinforced concrete beams.

A Large database of experimental work conducted on beams with and without shear reinforcement was collected. This database contained a total of 574 tests. The database covered a very wide range of beam parameters including their dimensions, concrete strengths, reinforcement ratios and yield stresses, and shear span to depth ratios.

This database was utilized in this study to evaluate four different existing codes of practice for predicting the shear capacity of beams. These codes are ACI 318M-08, EC2, NZS 3101, and CSA. Moreover, the database was used to evaluate and demonstrate the capability of the feed-forward back propagation neural networks for predicting such shear capacity.

Based on careful study of recent approaches for the shear phenomena in concrete members, it was decided to design the neural network to have eight nodes in the input layer containing data regarding the beam depth (d), breadth (b_w), shear span to depth ratio (a/d), concrete cylinder compressive strength $f_c^{'}$, percentage of main reinforcement (ρ_l), longitudinal steel yield stress (f_{yl}), percentage of shear reinforcement (ρ_v), and shear steel yield stress (f_{yv}). The output layer consisted of one node representing the ultimate shear capacity of the beam.

Several network topologies were examined. These included seventeen networks with one hidden layer and twenty networks with two hidden layers containing from 2 to 10 neurons in each hidden layer. All these networks were trained on 287 shear tests and performance of these networks on both 287 training and 287 testing data sets was compared and the one with the best prediction was selected as the successful one.

The predictions of the neural network and those of the different codes were compared. It was found that, among all the existing models, the results obtained from the neural network are the most accurate results, giving values of maximum shear capacity very close to the experimental values. Moreover, the results obtained using the neural network were very consistent and covered a very wide range of variation of any of the input parameters.

This accuracy suggests that the most critical variables that control the shear capacity of concrete beams are the eight that have been used as input data to the neural network model. It also suggests that, in special structures, where prediction of the accurate shear capacity of building beams or bridge girders are crucial, using a software that incorporates the neural network technique might be recommended.

It was observed that for small *a/d* ratios there is a beneficial influence on the shear capacity due to direct load transfer. Moreover, there is a pronounced beneficial effect of the longitudinal reinforcement tensile strength on

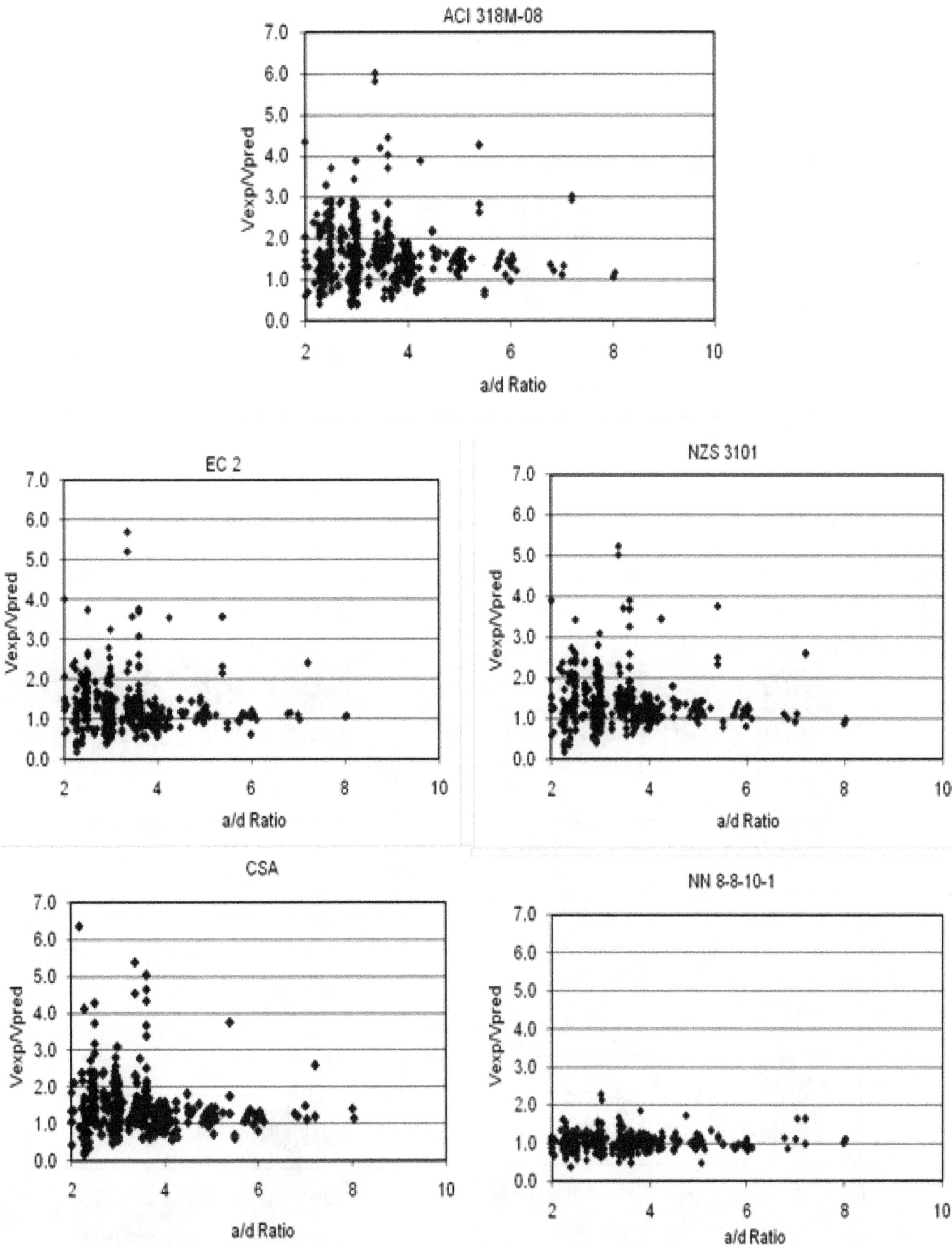

Figure 8. Experimental to predicted shear capacity versus a/d ratio for different shear proposals.

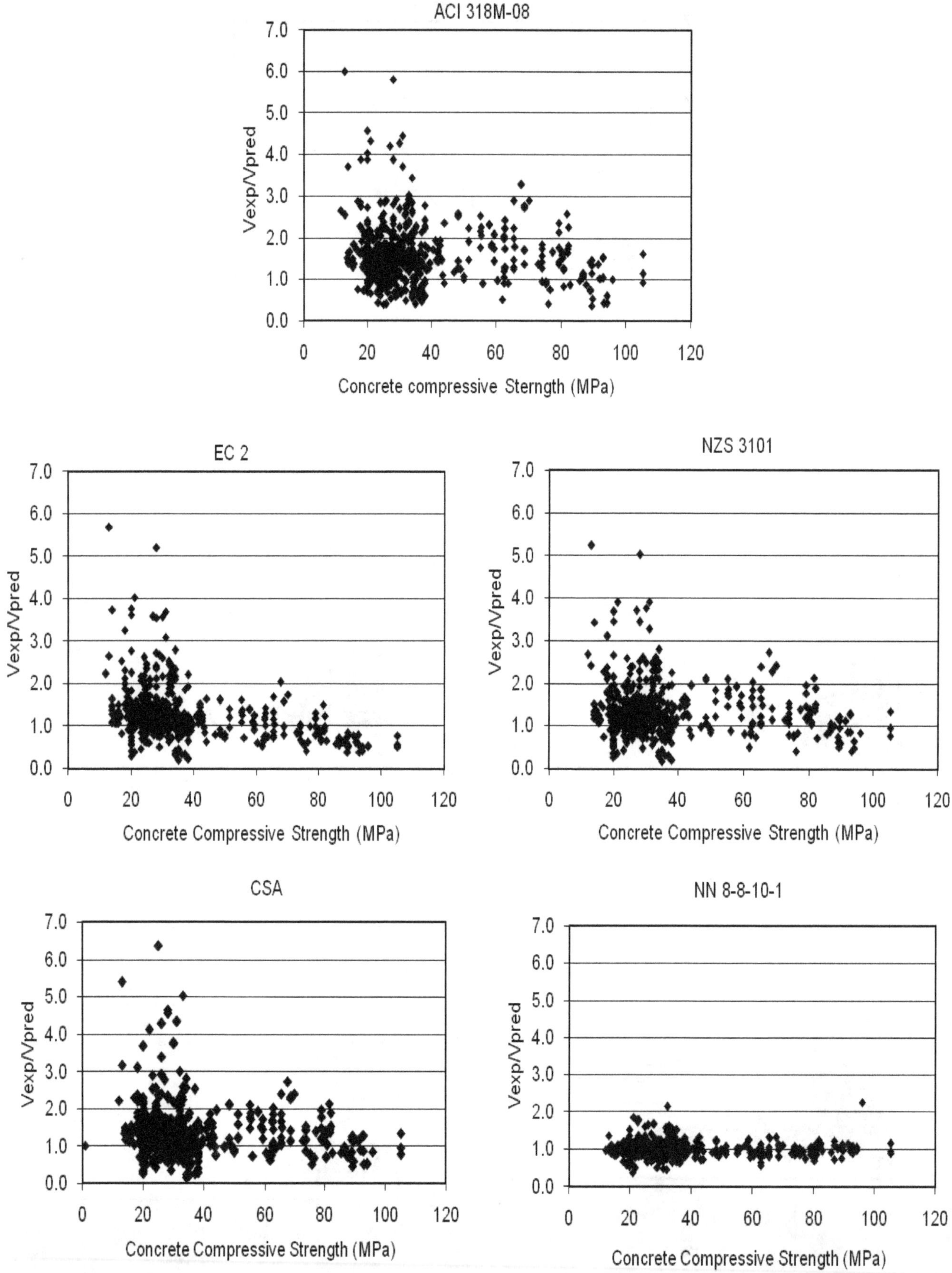

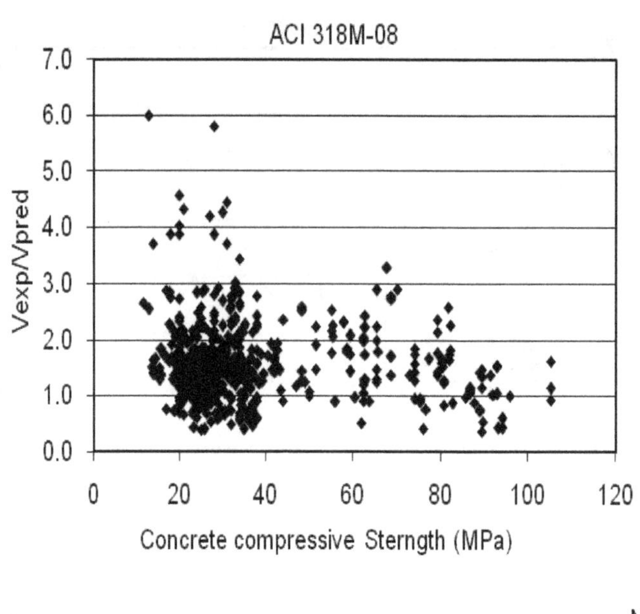

Figure 9. Experimental to predicted shear capacity versus concrete compressive strength for different shear proposals.

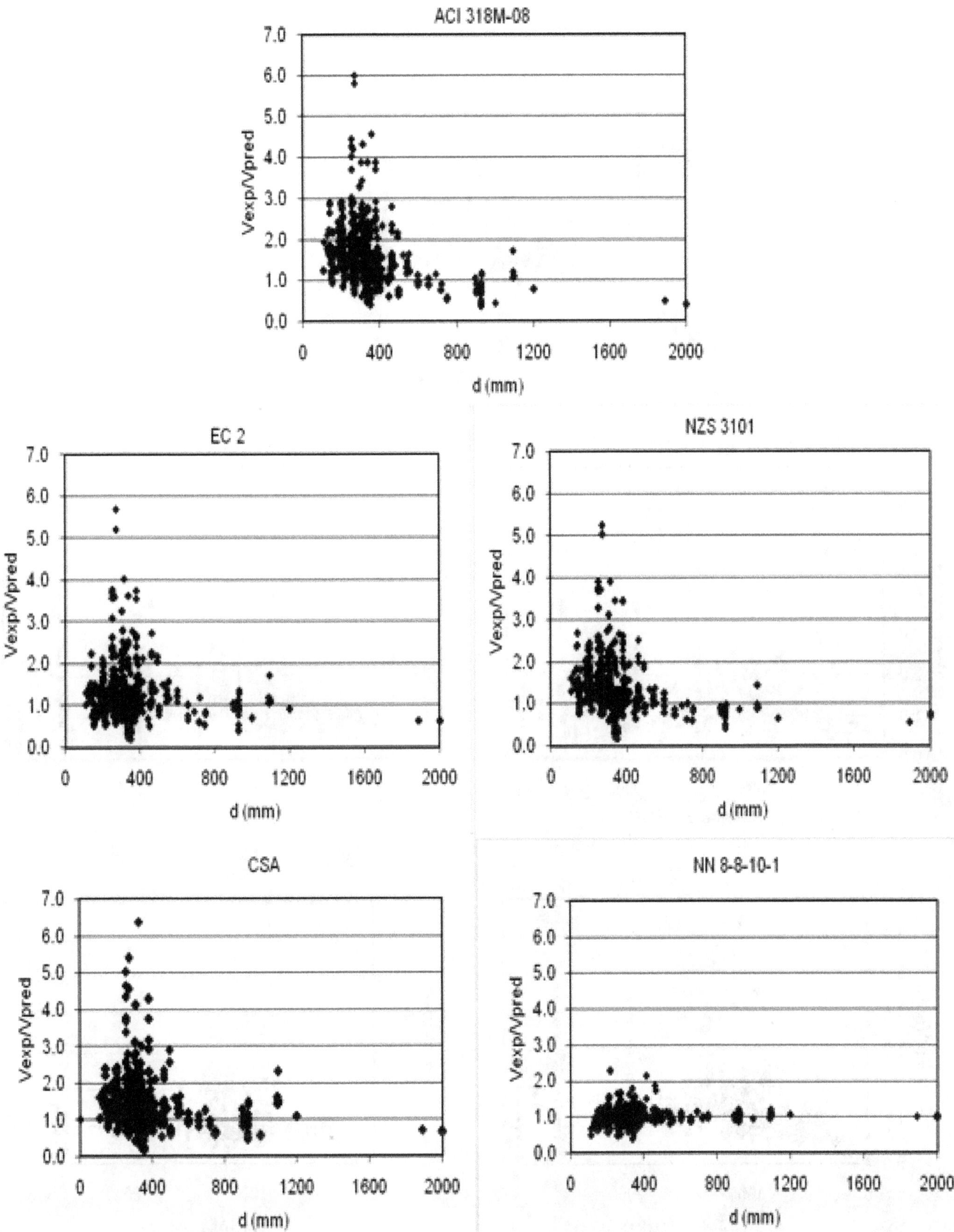

Figure 10. Experimental to predicted shear capacity versus beam depth for different shear proposals.

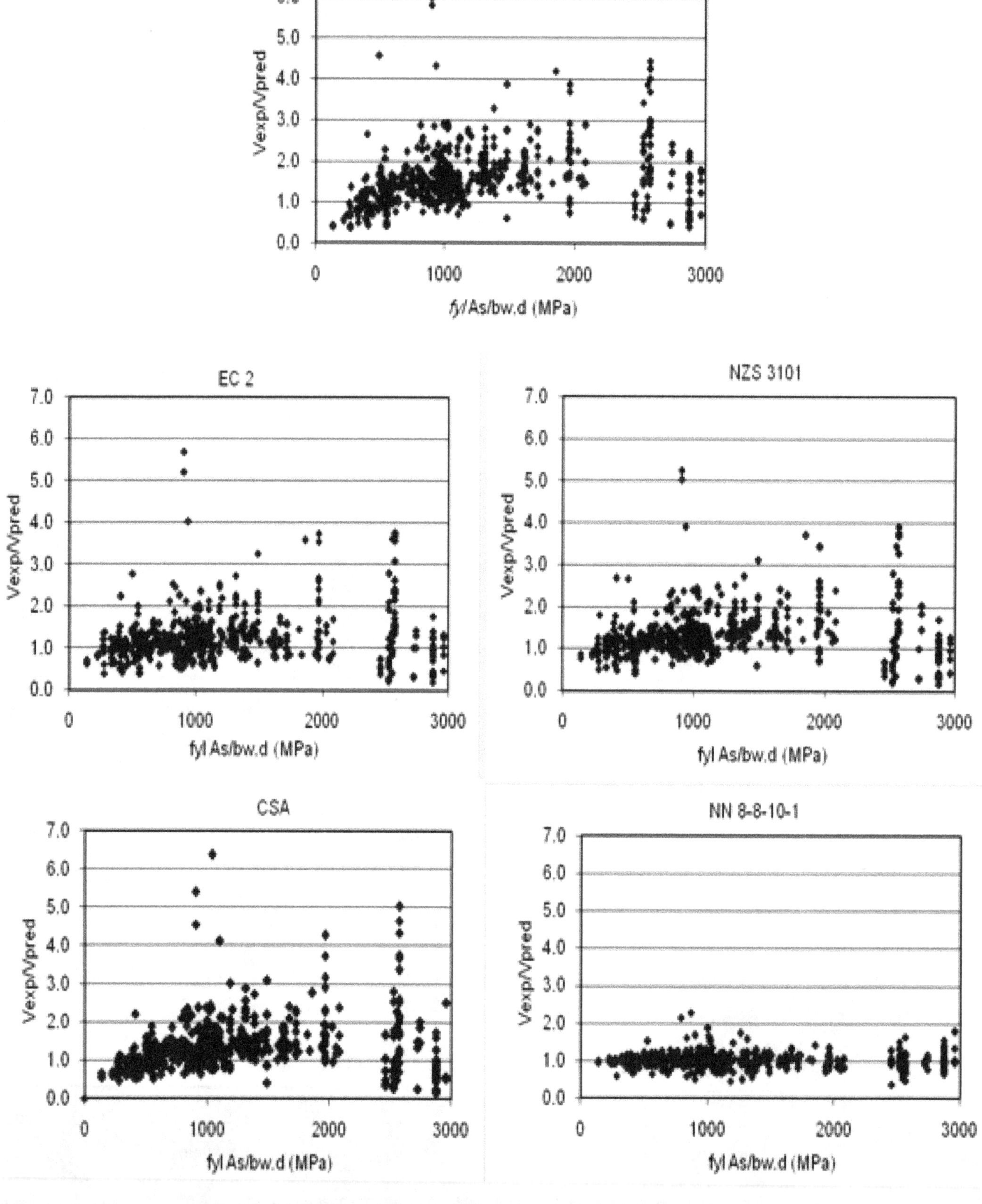

Figure 11. Experimental to predicted shear capacity versus f_{yl}. ρ_l for different shear proposals.

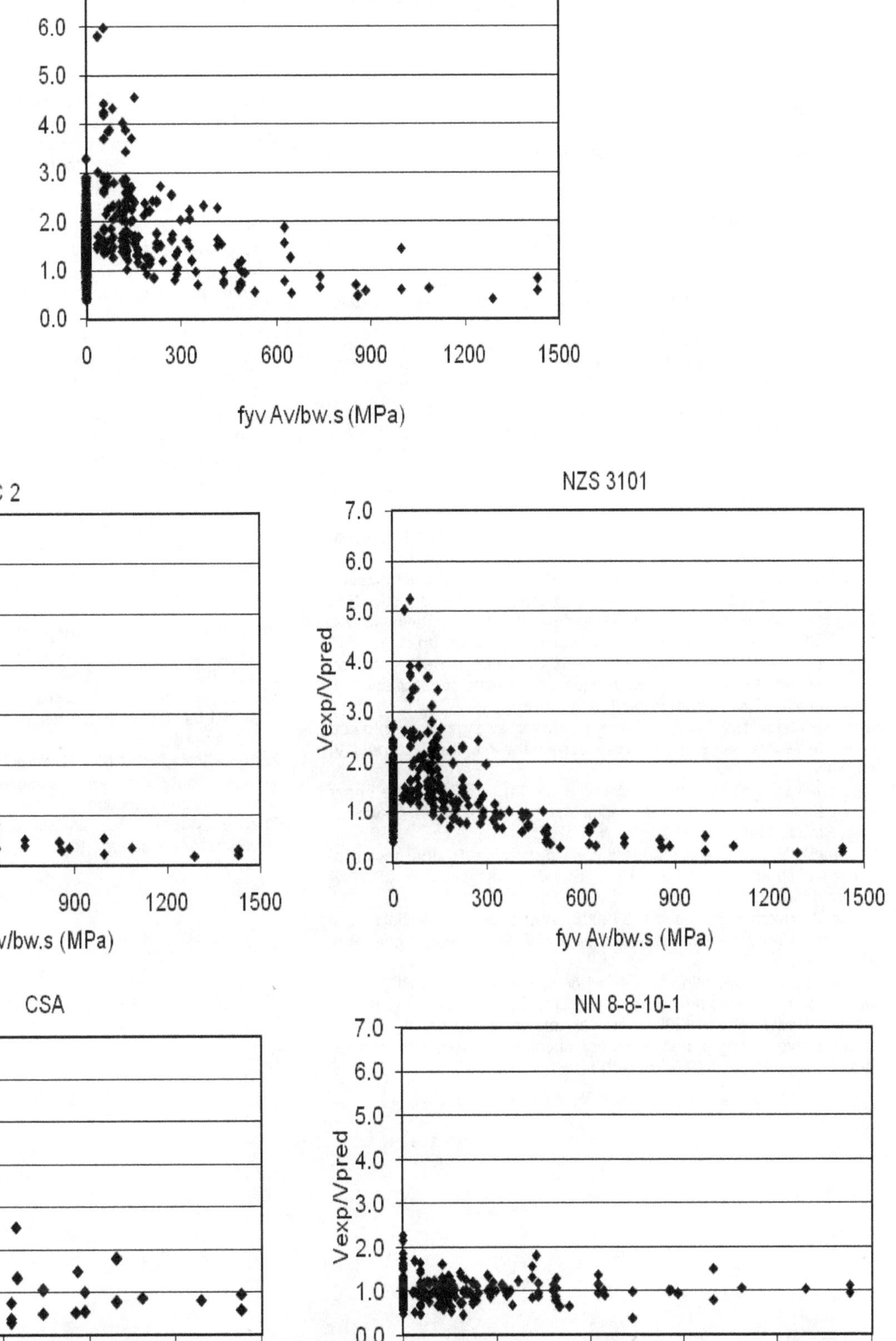

Figure 12. Experimental to predicted shear capacity versus fyv. ρv for different shear proposals.

the shear capacity. These effects are not properly captured by the examined codes. Also, the examined codes were observed to overestimate the effect of the shear reinforcement.

REFERENCES

ACI Committee 318M (2008). Building code requirements for structural concrete (ACI 318M-99) and commentary, American Concrete Institute, Farmington Hills. Mich.

ASCE-ACI Committee 445 on Shear and Torsion (1998). Recent approaches to shear design of structural concrete, J. Struct. Eng., ASCE, 124 (12), 1375-1416.

Canadian Standards Association (CSA) (1994). Design of concrete structures, A23.3-94, Canadian Standards Association, Rexdale, Ontario.

Chang CC, Zhou Li (2002). Neural network emulation of inverse dynamics for a magnetorheological damper, J. Struct. Eng., *ASCE,* 128 (2): 231-239.

Dopico JR, Ordóñez JL, Bohigas AC, Fonteboa BG, Abella FM (2008). Shear and bond analysis on structural concrete using artificial neural networks, 5th International Engineering and Construction Conference, ASCE, August, pp. 27-29.

EC2 (2004). Eurocode 2: Design of concrete structures-Part 1: General rules and rules for building, BS EN 1992-1-1.

Jan JC, Hung SL, Chi SY, Chen JC (2002). Neural network forecast model in deep excavation, J. Comput. Civ. Eng., 16 (1):59-65.

Juang CH, Jiang T (2003). A New approach to site characterization using generalized regression neural networks, Probabilistic Site Catigorization at the National Geotechnical Experimentation Sites, ASCE/GEO Institute, Geotechnical Special Publication, pp. 121.

Juang CH, Yuan H, Lee DH, Lin PS (2003). Simplified cone penetration test-based method for evaluating liquefaction resistance of soils, J. Geotechnical Geoenviron. Eng., 129 (1):66-80.

Kumar S, Yadav NK (2008). Predicting shear strength of reinforced concrete beams with stirrups using neural networks, IE(I) Journal-CV, 89, May.

Mansour MY, Dicleli M, Lee JY, Zhang J (2004). Predicting the shear strength of reinforced concrete beams using artificial neural networks, Eng. Struct., 26(6):781-799.

NZS (1995). New Zealand standard code of practice for the design of concrete structures (NZS 3101). Standard Association of New Zealand (NZS), Wellington, New Zealand.

Noorzaei J, Hakim SJS, Jaafar MS (2008). Application of artificial neural network to predict compressive strength of high strength concrete, ICCBT-A (04), 57-68.

Oreta AC (2004). Simulating size effect on shear strength of RC beams without stirrups using neural networks, Eng. Struct., 26 (5):681-691.

Oreta AC, Kawashima K (2003). Neural network modeling of confined compressive strength and strain of circular concrete columns, J. Struct. Eng., ASCE, 129 (4):554-561.

Pannirselvam N, Nagaradjane V, Chandramouli K, Ravindrakrishna M (2010). Artificial neural network model for performance evaluation of RC rectangular beams with externally bonded glass fiber reinforced polymer reinforcement, ARPN, J. Eng. Appl. Sci., 5 (3):77-85.

Pannirselvam N, Raghunath PN, Suguna K (2008). Neural network for performance of glass fiber reinforced polymer plated RC beam, Am. J. Eng. Appl. Sci., 1(1):82-88.

Rao HS, Babu BR (2007). Hybrid neural network model for the design of beam subjected to bending and shear, Sadhana, 32 (5):577-586.

Reineck KH, Kuchma D, Kim KS, Marx S (2003). Shear database for reinforced concrete members without shear reinforcement, ACI Struct. J., 100 (2):240-249.

Sanad A, Saka MP (2001). Prediction of ultimate shear strength of reinforced concrete deep beams using neural networks, J. Struct. Eng., ASCE, 127 (7):818-827.

Seleemah AA (2005). A neural network model for predicting maximum shear capacity of concrete beams without shear reinforcement, Can. J. Civ. Eng., 32:644–657.

Seleemah AA, Aburayan A, Samy M (2012). A Neural Network Model for Damage Detection of El-Ferdan Bridge. Fourth International Conference on Structural Stability and Dynamics, ICSSD-2012, India. pp. 775-783.

Senouci AB (2000). Preliminary design of reinforced concrete beams using neural networks, Eng. J. Univ. Qatar, 13:107-122.

Sirca GFJ, Adeli H (2001). Neural network model for uplift load capacity of metal roof panels, J. Struct. Eng., ASCE, 127 (11):1276-1285.

Taysi N (2010). Application of neural network models on analysis of prismatic structures, Sci. Res. Essays, 5(9):978-989.

Tsai CH, Hsu DS (2002). Diagnosis of reinforced concrete structural damage based on displacement time history using the back-propagation neural network technique, J. Comput. Civ. Eng., 16 (1): 49-58.

Yang KH, Ashour AF, Song JK (2007). Shear capacity of reinforced concrete beams using neural network, Int. J. Concrete Struct. Mater., 1 (1):63-73.

Yun GJ, Ghaboussi J, Elnashai AS (2008). A new neural network-based model for hysteretic behavior of materials, Int. J. Numer. Meth. Eng., 73:447–469.

Zang C, Imregun M (2001). Structural damage detection using artificial neural networks and measured FRF data reduced via principal component projection, J. Sound Vib., 242 (5): 813-827.

Zhao Z, Wenwei H, Fan SC (2001). Preliminary design system for concrete box girder bridges, J. Comput. Civ. Eng., 15 (3):184-192.

4

Flexural behavior of cantilever concrete beams reinforced with glass fiber reinforced polymers (GFRP) bars

Mohamed S. Issa* **and S. M. Elzeiny**

Housing and Building National Research Center, Giza, Egypt.

The objective of the current study is to investigate and evaluate the flexural behavior of concrete cantilever beams when using locally produced GFRP bars as a longitudinal main reinforcement. The experimental program includes six concrete cantilever beams. The main parameters were the type of rebars (steel or GFRP), strength of concrete and ratios of GFRP rebars. The results of experiments were evaluated and discussed. The ultimate flexural capacities were calculated theoretically. Then a comparison between both experimental and theoretical results was done. This comparison indicated that the theoretical analysis gives results which are about 30% lower than the experimental ultimate flexural capacity for GFRP-reinforced cantilever beams. Also, deflections were calculated and it was found that the model of Brown and Bartholomew for I_e (effective moment of inertia) gives the best predictions for deflections.

Key words: Flexure, cantilever, deflection, glass fiber reinforced polymers (GFRP) bars.

INTRODUCTION

The use of fiber reinforced polymer (FRP) reinforcements in concrete structures has increased rapidly in the last 10 years due to their excellent corrosion resistance, high tensile strength, and good non-magnetization properties. However, the low modulus of elasticity of the FRP materials and their non-yielding characteristics results in large deflection and wide cracks in FRP reinforced concrete members. Consequently, in many cases, serviceability requirements may govern the design of such members. In particular, FRP rebar offers great potential for use in reinforced concrete construction under conditions in which conventional steel-reinforced concrete has yielded unacceptable service. If correctly applied in the infrastructure area, composites can result in significant benefits related to both overall cost and durability. Other advantages include high strength and stiffness to weight ratios, resistance to corrosion and chemical attack, controllable thermal expansion and damping characteristics, and electromagnetic neutrality. These advantages could lead to increased safety and life cycle as well as providing savings in fabrication, equipment, and maintenance costs. FRPs generally consist of synthetic or organic high strength fibers in a resin matrix.

FRP composite made with resin-impregnated continuous fibers is considered a promising alternative to the traditional steel reinforcements because of its inherent corrosive resistance, though the long-term performance of some types of fiber in certain environments is still questionable. Other appealing characteristics of FRP include high tensile strength, good fatigue and damping response, high strength-to-weight ratio, and electromagnetic transparency. The wide-ranging application of FRP reinforcements, especially as a main reinforcement, however, has been rather limited. This may be partially attributed to the high initial cost of materials and the lack of design guidelines, but more essentially can be attributed to two major engineering drawbacks of FRP materials: Low modulus of elasticity and the lack of ductility of most commercially available FRPs.

FRP generally exhibits a linear elastic tensile stress-strain relationship up to failure which, in comparison to steel-reinforced members, may result in poor structural

*Corresponding author. E-mail: drmsisssa@yahoo.com.

ductility even in properly designed (according to standard reinforced concrete design guidelines) FRP-reinforced members. Essentially, ductility is the ability of inelastic energy dissipation. For conventional steel-reinforced concrete members, ductility is primarily achieved by the yielding of steel reinforcement, thus consuming a substantial amount of energy while allowing the full compressive strain capacity of the concrete to develop. For an FRP-reinforced concrete member, however, such an inelastic energy consumption mechanism does not exist (Benmokrane et al., 1996; Mota et al., 2006; Toutanji and Saafi 2000; ACI Committee 440, 2006; ACI Committee 318, 2008).

The most commonly used FRPs for civil engineering applications are carbon (CFRP), aramid (AFRP), and glass (GFRP). However, their extensive use in reinforced concrete structural engineering has been very limited, due to lack of research data and design specifications. This has been the main impetus of carrying out this study which focuses on the flexural behavior of concrete beams reinforced with GFRP rebars.

The paper also attempts to present the properties of GFRP and to give an oversight of relevant research activities involving GFRP rebars as reinforcement (Benmokrane et al., 1996; Mota et al., 2006; Toutanji and Saafi, 2000; ACI Committee 440, 2006; ACI Committee 318, 2008; Alsayed et al., 2000; Wang and Belarbi, 2005; Grace et al., 1998; Lee, 2002; Solano-Carrillo, 2009; Chen, 2010; ECP 203, 2006; ECP 208, 2005; Metwally, 2009; Vijay and GangaRao, 2001; Ashour, 2006).

Some research was done on FRP-reinforced concrete beams of more than one span. Grace et al. (1998) presented research about the behavior of simply and continuously supported beams reinforced with fiber reinforced polymer (FRP) bars. Seven continuous T-section beams were tested. Reinforcing bars and stirrups were made of steel, carbon, or glass fiber reinforced polymer. They concluded that the use of GFRP stirrups increased the shear deformations and as a result deflection increased. The use of GFRP stirrups resulted in shear or flexural-shear mode of failure. Also, the use of FRP reinforcement in continuous beams increased deformation.

Different failure modes and ductilities were noted for FRP-reinforced beams when compared to steel-reinforced beams. Habeeb et al. (2008) tested two simply and three continuously supported concrete beams reinforced with GFRP bars. Three different GFRP reinforcement combinations of over and under reinforcement ratios were used for the top and bottom layers of the continuous concrete beams. They found that over-reinforcing the bottom layer is a key factor in controlling the width and propagation of cracks, enhancing the load capacity, and reducing the deflection. Their research revealed that ACI 440.1R-06 equations can reasonably predict the load capacity and deflection of The simply and continuously supported GFRP reinforced concrete beams.

EXPERIMENTAL WORK

The current research program was carried out to investigate the flexural behavior of cantilever concrete beams with main reinforcement of GFRP bars.

Test program

The experimental program consists of testing six cantilever concrete beams. The tested beams were classified into three groups as shown in Tables 1 and 2. They had cross section dimensions of 150 x 250 mm and 2000 mm total length as shown in Figure 1. The cantilever length was 600 mm. Figure 1 and Table 2 give the reinforcement details of all the tested beams. The beams of group (A) were reinforced with two high grade longitudinal steel bars of 10 and 12 mm diameter (top and bottom) for beams SN10-10 and SN 12-12, respectively. While beam SN 8-8 in group (A) was reinforced with two mild steel longitudinal bars of 8 mm diameter (top and bottom). The beams of the group (B) were reinforced with two GFRP longitudinal bars of 8 mm diameter (top and bottom) for beam GN 8-8.

The beam GN 12-10 in group (B) was reinforced with two GFRP longitudinal bars of 12 mm diameter at the top and two 10 mm diameter GFRP bars at the bottom. The beam GM 10-10 in group (C) was reinforced with two GFRP longitudinal bars of 10 mm diameter at the top and at the bottom.

Properties of the used material

Reinforced concrete

The tested reinforced concrete beams were produced using locally manufactured ordinary Portland cement, natural sand and crushed dolomite with a maximum size of 10 mm. The beams were demolded after 24 h from casting, covered with wet burlap and stored under the laboratory conditions for 28 days before proceeding to the testing stage. The compressive strength (fcu) was determined by testing six standard cubes (150*150*150mm) from each cast. The equivalent standard cylinder compressive strength (f'c) was calculated as 0.8 times the standard cube compressive strength and is shown in Table 1.

GFRP and steel bars

The stirrups used in all the tested beams were of mild steel of 8 mm diameter. The tested beams were reinforced with two types of bars. The first type was steel bars of 8 mm diameter (mild steel) and 10 and 12 mm diameter (high grade). The second type was GFRP bars of 8, 10 and 12 mm diameter as shown in Figure 1 and Table 2. Tables 3 and 4 show the mechanical properties of the used GFRP and steel bars, respectively. The GFRP bars are locally manufactured. They are coated by sand to improve their bond characteristics.

Test setup, procedure and measurements

The beams were tested in the reinforced concrete lab of the Housing and Building National Research Center. They were supported over two rigid supports as shown in Figure 2. One point load was applied to all the beams and was monotonically increased

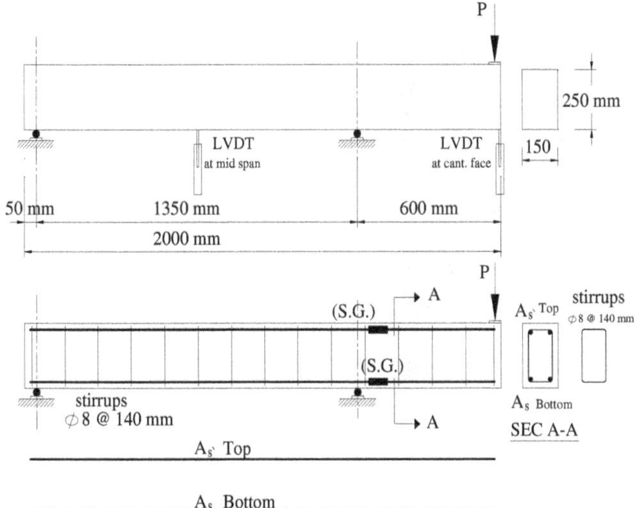

Figure 1. Concrete dimensions of tested beams.

Table 1. Equivalent standard cylinder compressive strength (f'c).

Group No.	Beam notation	f'_c (MPa)
A	SN8-8	33.96
A	SN10-10	33.96
A	SN12-12	33.96
B	GN8-8	33.96
B	GN12-10	33.96
C	GM10-10	47.41

until failure. The load was applied vertically over the top of the cantilever end. The deflections were measured at the mid span and under the loading point (cantilever tip) by using ± 200 mm linear variable differential transducers (LVDTs). Strain gages (S.G.) of 10mm length were installed at the top and bottom longitudinal reinforcement bars at the section over the support on the cantilever side. The load cell used for measuring the driving force was of 500 kN capacity.

The measured data were recorded by a data logger connected to computer system programmed using "lab view software".

RESULTS AND DISCUSSION

Failure modes and cracking patterns

The observed mode of failure of the GFRP-reinforced cantilever beams (GN8-8, GM10-10, and GN12-10) was flexural failure (tension failure). For the steel reinforced beams a flexural-shear mode of failure was noticed for beams SN10-10 and SN12-12 while a flexural failure was noticed for beam SN8-8. Figure 3 shows the crack pattern for all the tested beams. In general cracking consisted predominantly of major vertical flexural crack on top of the cantilever side support and a number of small vertical flexural cracks in the nearby zone of the beam. For beams SN10-10 and SN12-12, as the load was increased shear stresses increased and resulted in inclined cracks. It was noticed that the initiation and propagation of cracks depends on the type of reinforcement. The cracking loads for all the beams are shown in Table 5.

Generally, the cracking loads for the GFRP-reinforced beams were smaller than that of the corresponding steel reinforced beams which could be attributed to the lower modulus of elasticity of GFRP bars than that of steel bars. For example, the cracking load for specimen GM10-10 was 26.2% smaller than that of beam SN10-10 while the cracking load for specimen GN12-10 was 67.5% lower than that of beam SN12-12. In average, the measured cracking loads were about 31.9% of the experimental ultimate loads shown in Table 6.

Reinforcement strains

Table 5 presents the top reinforcement strains at ultimate load. It is clear from the data in the table that the average strain in the case of GFRP-reinforced cantilever beams is 0.0083 which is much less than the rupture strains given in Table 3. This indicates that local debonding between top GFRP bars and concrete has occurred. The maximum strain was recorded for beam GM10-10 which had the highest concrete compressive strength. The strains for all the steel reinforcement at ultimate load exceeded the yield.

From Figure 4, the strains of the GFRP reinforcement increased significantly after concrete cracking and they were higher than those in the steel before yielding. Both load-strain curves consist of two mainly linear stages.

Load-deflection relation

Figure 5 shows the applied load against the vertical deflection at the cantilever tip for the GFRP-reinforced cantilever beams. It is very clear that the three cantilever beams GN8-8, GM10-10, and GN12-10 had similar load-deflection behavior up to close to the ultimate loads. After the ultimate load, differences are noted between the three cantilever beams. The ultimate load for GN8-8 was 28.8 kN as shown in Table 6. Cantilever beam GM10-10 recorded an increase of 8.7% in the ultimate load while cantilever beam GN12-12 recorded an increase of 18.8% in the ultimate load. From Figure 6 similar trend is noticed for the steel reinforced beams. However, the increase in the load capacity with the increase in reinforcement was higher. For example, cantilever beam SN10-10 recorded an increase of 163.2% in the ultimate load over cantilever beam SN8-8 and cantilever beam SN12-12 recorded an increase of 297.6% in the ultimate load over cantilever beam SN8-8.

Table 2. The tested beams.

Group No.	Beam No.	Cross Sec. Dim. bxh (mm)	Cantilever length (mm)	Long. bars type	Stir. type	Top long. bars	Bott. long. bars
	SN 8-8	150x250	600	Steel	Steel	2ø8	2ø8
A	SN 10-10	150x250	600	Steel	Steel	2ø10	2ø10
	SN 12-12	150x250	600	Steel	Steel	2ø12	2ø12
B	GN 8-8	150x250	600	GFRP	Steel	2ø8	2ø8
	GN 12-10	150x250	600	GFRP	Steel	2ø12	2ø10
C	GM 10-10	150x250	600	GFRP	Steel	2ø10	2ø10

Table 3. Mechanical properties of GFRP bars.

Diameter (mm)	F_u (N/mm^2)	Modulus of elasticity E_f (GPa)	Rupture strain ε_{fu}
8	416.30	34.30	0.025
10	407.40	33.81	0.029
12	347.50	32.67	0.05

Table 4. Mechanical properties of steel reinforcement.

Diameter (mm)	F_y (N/mm^2)	F_u (N/mm^2)	Elongation %
	300	460	19.4
8	300	460	27.9
	310	460	28.4
	450	680	16.8
10	440	680	18.2
	450	680	18.2
	430	690	13.6
12	450	690	14.2
	450	690	19.3

As noted from Table 7, the three GFRP-reinforced beams had close deflections at ultimate loads. Cantilever beam GN8-8 recorded 19.0 mm deflection at ultimate load while cantilever beams GM10-10 and GN12-10 recorded an increase of 23.2 and 24.2% in the deflection at ultimate load, respectively. The deflections at ultimate loads for the steel reinforced cantilever beams were higher than those of the GFRP reinforced cantilever beams. However, it has to be noted that the steel reinforced cantilever beams generally recorded higher experimental ultimate loads.

Flexural capacity predictions

The failure loads for the GFRP-reinforced cantilever beams were calculated using the formulas of ACI440.1 R-06 (ACI Committee 440, 2006). For the steel-reinforced cantilever beams, the formulas of ACI318-08 (ACI Committee 318, 2008) were used. In the calculations, the self-weight of the tested beams was neglected. The predictions of the equations of both codes for the failure loads are listed in Table 6. Both codes gave values which are on the conservative side with an average error of 33% for the GFRP-reinforced concrete beams. This is because of ignoring the reinforcement on the compression side.

Deflection predictions

The design of GFRP reinforced concrete beams is typically governed by serviceability limit state requirements. This is because the modulus of elasticity of

Figure 2. General view of the test setup.

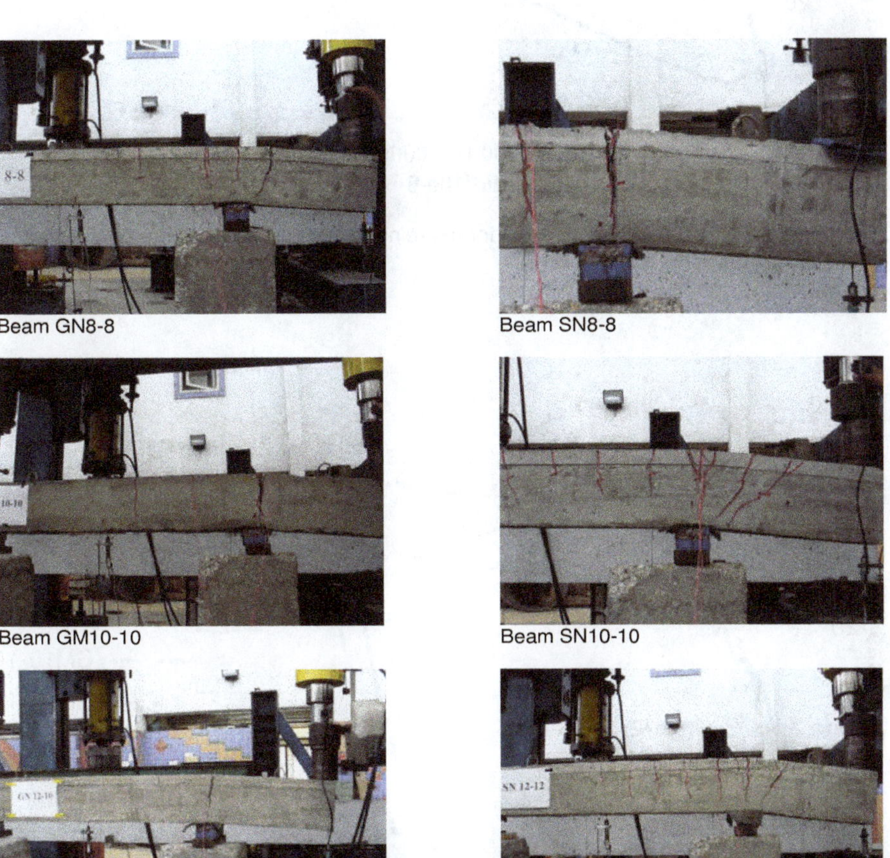

Figure 3. Patterns of crack at failure.

Table 5. Cracking loads and strains of GFRP bars.

Beam notation	First visible cracking load (kN)	Experimental top reinforcement strain at ultimate load for section at the support
GN8-8	12.0	0.0073
SN8-8	7.8	0.0015
GM10-10	10.7	0.0091
SN10-10	14.5	>0.011
GN12-10	7.8	0.0086
SN12-12	24.0	>0.010

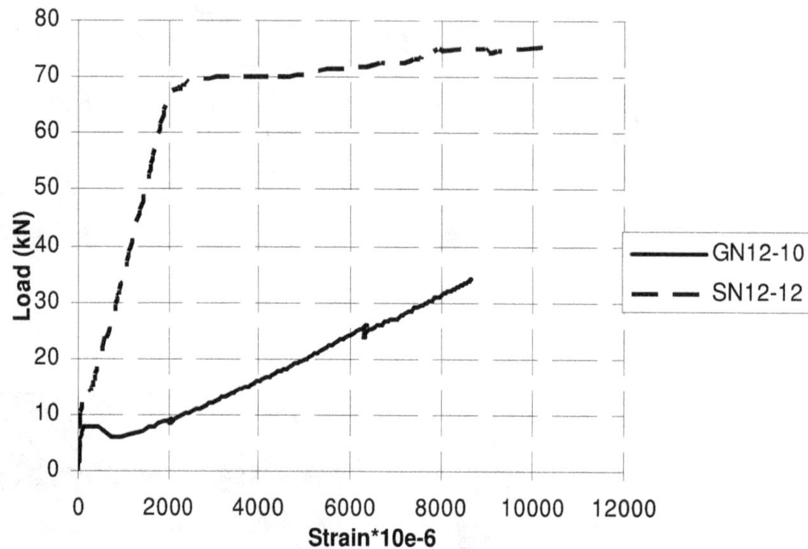

Figure 4. Load-Strain relationship for top reinforcement at sections on the support (Beams GN12-10 and SN12-12).

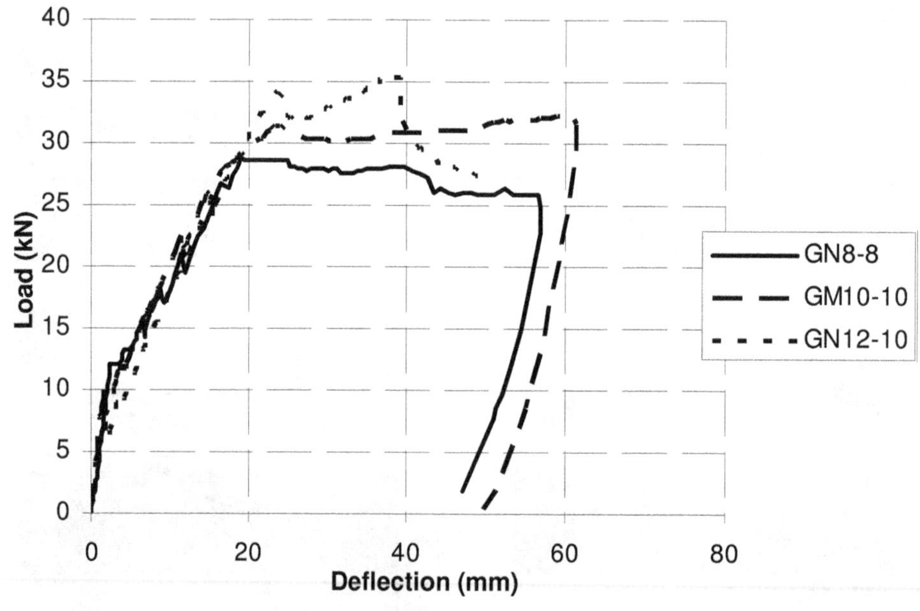

Figure 5. Load-Deflection curves for the GFRP-Reinforced beams.

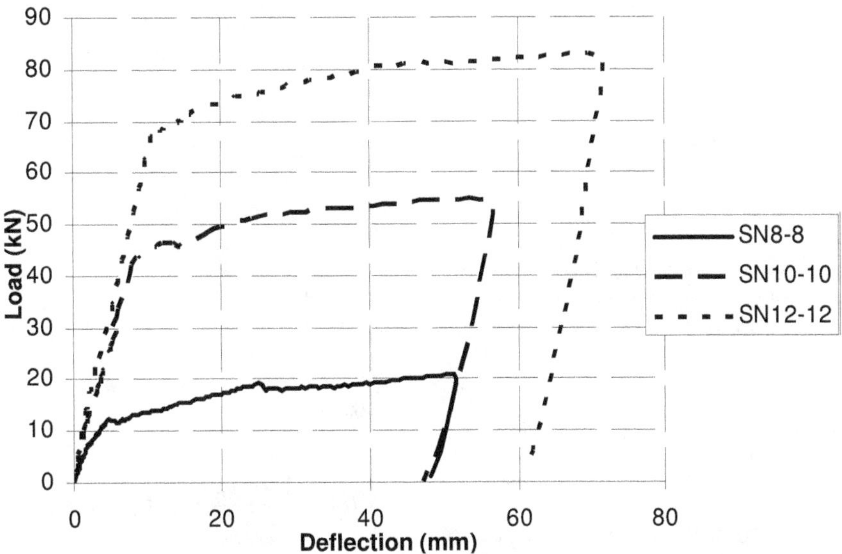

Figure 6. Load-Deflection curves for the Steel-Reinforced beams.

Table 6. Ultimate loads and failure modes.

Beam notation	Experimental ultimate load (kN)	Experimental service load (kN)	ACI 440.1 R-06 [4] and ACI 318M-08 [5] predictions for the failure loads of sections at the cantilever side over supports (kN)	Experimental failure mode
GN8-8	28.8	11.52	14.35	Flexure
SN8-8	20.9	8.36	15.47	Flexure
GM10-10	31.3	12.52	22.17	Flexure
SN10-10	55.0	22.0	23.83	Flexure-shear
GN12-10	34.2	13.68	27.53	Flexure
SN12-12	83.1	33.24	33.73	Flexure-shear

[4] ACI Committee 440.1R-06(2006); [5] ACI Committee 318

Table 7. Experimental deflections at ultimate loads.

Beam notation	Experimental deflection at ultimate load for the cantilever tip (mm)
GN8-8	19.0
SN8-8	50.2
GM10-10	23.4
SN10-10	53.5
GN12-10	23.6
SN12-12	69.6

GFRP bars is much smaller than that of steel. Thus, a method is needed that can accurately calculate the expected deflections of GFRP reinforced members. In the following paragraphs the calculations of deflections will be explained with emphasis on the different methods used to calculate the effective moment of inertia. The supporting system for the tested beams is shown in Figures 1 and 2. The beams were loaded by concentrated loads (P) at the cantilevers' tips. The distance from the load to the center line of the neighboring support is "c" and the span of the rest of the beam is "L".

After calculating an effective uniform moment of inertia (I_e) considering the moment over the support on the cantilever side, the deflection of the cantilever tip can be derived from linear elastic analysis and is:

$$\Delta = \frac{L+c}{3}\frac{Pc^2}{E_c I_e} \quad (1)$$

Where E_c is the modulus of elasticity of concrete and is equal to $4750\sqrt{f_c'}$ MPa.

Based on a comprehensive experimental program, Benmokrane et al. (1996) defined the effective moment of inertia for flexural members reinforced with FRP as:

$$I_e = \left(\frac{M_{cr}}{M_a}\right)^3 \frac{I_g}{7} + 0.84\left[1-\left(\frac{M_{cr}}{M_a}\right)^3\right] I_{cr} \le I_g \quad (2)$$

Where M_{cr} is the cracking moment, M_a is the applied moment, I_g is the moment of inertia of the gross section, and I_{cr} is the moment of inertia of the cracked section transformed to concrete.

Mota et al. (2006) reported that Brown and Bartholomew proposed that a fifth order equation can be used rather than a cubic. The form of the equation is:

$$I_e = \left(\frac{M_{cr}}{M_a}\right)^5 I_g + \left[1-\left(\frac{M_{cr}}{M_a}\right)^5\right] I_{cr} \le I_g \quad (3)$$

Toutanji and Saafi (2000) found that the order of the I_e equation depends on the modulus of elasticity of the FRP as well as the reinforcement ratio. They recommended the following equation:

$$I_e = \left(\frac{M_{cr}}{M_a}\right)^m I_g + \left[1-\left(\frac{M_{cr}}{M_a}\right)^m\right] I_{cr} \le I_g \quad (4)$$

Where $m = 6 - 10\frac{E_{FRP}}{E_s}\rho_{FRP}$ if $\frac{E_{FRP}}{E_s}\rho_{FRP} \succ 0.3$

Otherwise *m=3*

Where E_{FRP} is the modulus of elasticity of the FRP bars, E_s is the modulus of elasticity of the steel bars, and ρ_{FRP} is the FRP reinforcement ratio.

Mota et al. (2006) reported that the ISIS design manual M03-1 suggested the following equation:

$$I_e = \frac{I_T I_{cr}}{I_{cr} + \left[1-0.5\left(\frac{M_{cr}}{M_a}\right)^2\right](I_T - I_{cr})} \le I_g \quad (5)$$

Where I_T is the uncracked moment of inertia of the section transformed to concrete and for simplicity will be taken equal to I_g in the following calculations.

According to ACI 440.1 R06 (ACI Committee 440, 2006) an account for the reduced tension stiffening of FRP-reinforced members needs to be made by using the factor β_d. The form of the equation is:

$$I_e = \left(\frac{M_{cr}}{M_a}\right)^3 \beta_d I_g + \left[1-\left(\frac{M_{cr}}{M_a}\right)^3\right] I_{cr} \le I_g$$

(6)

Where,

$$\beta_d = \frac{1}{5}\frac{\rho_{FRP}}{\rho_{fb}} \le 1.0 \quad (7)$$

ρ_{fb}=balanced FRP reinforcement ratio,

$$\rho_{fb} = 0.85\,\beta_1 \frac{f_c'}{f_{fu}} \frac{E_{FRP}\,\varepsilon_{cu}}{E_{FRP}\,\varepsilon_{cu} + f_{fu}} \quad (8)$$

β_1 = the factor which defines the depth of the equivalent rectangular stress block

f_c' = the standard cylinder (150*300 mm) compressive strength

f_{fu} = the ultimate tensile strength of the FRP bar

ε_{cu} = the ultimate compressive strain in concrete = 0.003

Alsayed et al. (2000) reported that Faza and GangaRao derived an expression for the modified effective moment of inertia, which is referred to as I_m. Despite the fact that this model was derived for beams subjected to two equal concentrated loads, its applicability for cantilever beams is checked here. The model is written as:

$$I_m = \frac{23\, I_{cr}\, I_e}{8\, I_{cr} + 15\, I_e} \quad (9)$$

Where,

$$I_e = \left(\frac{M_{cr}}{M_a}\right)^3 I_g + \left[1-\left(\frac{M_{cr}}{M_a}\right)^3\right] I_{cr} \quad (10)$$

With regard to the cracking moment, it can be calculated from the following equation:

$$M_{cr} = \frac{f_r}{y_t} I_g \quad (11)$$

Where f_r = the modulus of rupture of concrete,

$$fr = 0.62\sqrt{f_c'} \text{ MPa}$$

y_t = the distance from the centroid to the extreme concrete fiber in tension

$y_t = h/2$

The six methods for calculating I_e, explained earlier, were utilized along with Equation 1 to calculate the deflection of the tested GFRP reinforced beams. For the steel reinforced beams, only Equation 1 along with Equation 10 was used to calculate the deflection. The calculated deflection and the experimental deflection for the cantilever tip were compared at two load levels. Tables 8 and 9 give the predicted deflection values at the levels of service loads which are taken equal to 0.40 times the experimental ultimate loads as suggested by Wang and Belarbi (2005) while Tables 10 and 11 present the same values at load levels equal to 1.3 times the service loads. From Tables 8 and 10 it is clear that Equation 3 generally gives the smallest average error and standard deviation for the error for the deflections of GFRP-reinforced beams. The entire relation between the load and deflection for all the specimens were drawn up to the failure load (Figure 7), utilizing Equations 3 and 1 for the GFRP-reinforced beams and Equations 10 and 1 for the steel-reinforced beams.

All the curves showed close correlation between the experimental load-deflection and the predicted curve with limited underestimation of the deflection at early stages of loading and overestimation of the deflection at later stages of loading. Near the ultimate loads larger errors are noticed due to the some non-linearity that takes place at these load levels.

Span-to-Service load deflection

The service load deflection and the span-to-service load deflection ratios are shown in Table 12). As shown in the table, the span-to-service load deflection ratio varies between 87.0 and 250.0. These values are high when compared to the usually accepted ratio of about span/450 which confirms that serviceability limit state generally governs the design of GFRP reinforced beams.

Stiffness

From Table 13, it is clear that generally large reduction in the stiffness occurred after the cracking loads for the tested beams. Both the initial stiffness and the stiffness after cracking for the GFRP-reinforced cantilever beams were either in the same order of magnitude or smaller than these for the corresponding steel-reinforced cantilever beams.

Ductility

Many definitions for ductility are available for steel reinforced beams. The yield point of the reinforcing steel is usually the base in most of these definitions. FRP reinforcement is different because it has linear stress-strain relationship up to the failure point without yield plateau. Spadea et al. (1997) suggested the energy ductility for FRP reinforced beams to be expressed as:

$$\mu = \frac{E_{tot}}{E_{0.75pu}} \qquad (12)$$

Where E_{tot} = the total area under the load-deflection diagram up to the failure load (total energy).

$E_{0.75pu}$ = the area under the load-deflection diagram up to 0.75*the ultimate load.

The aforementioned definition for ductility was employed to calculate the ductilities shown in Table 14. In the calculations, the failure load was taken as 0.95 of the ultimate attained load.

From Table 14, it can be noted that the GFRP-reinforced cantilever beams attained similar level of ductility.

Conclusions

Based on the experimental results and analytical analysis the following conclusions were arrived at:

1. The area of the GFRP reinforcement has a small effect on the load-deflection relation up to close to the ultimate load.
2. The change in the deflection at ultimate load for GFRP-reinforced cantilever beams with the increase of the GFRP reinforcement area and with the increase of the concrete compressive strength was about 24%.
3. Steel reinforced cantilever beams generally recorded higher experimental ultimate loads than the corresponding GFRP reinforced cantilever beams.
4. The average strains in the case of GFRP-reinforced cantilever beams is 0.0083 which is much less than the rupture strains.
5. The strains for all the steel reinforcement at ultimate load exceeded the yield.
6. The measured average cracking loads were about 31.9% of the experimental ultimate loads.
7. The equations of the ACI 440.1 R-06 for calculating the ultimate moment capacity gives about 30% underestimation for the values of the moment capacity of GFRP-reinforced cantilever beams.
8. Utilizing the equation of Brown and Bartholomew for the effective moment of inertia gives the best deflection predictions for GFRP-reinforced cantilever beams.
9. The accuracy of the deflection equations varies at the different load levels with larger errors noticed close to the ultimate loads.

Table 8. Experimental and analytical deflections at service loads (GFRP reinforced beams).

Beam notation	Exp. Def. (mm)	Calculated deflection (mm)						Error%					
		Eq. 2	Eq. 3	Eq. 4	Eq. 5	Eq. 6	Eq. 9	Eq. 2	Eq. 3	Eq. 4	Eq. 5	Eq. 6	Eq. 9
GN8-8	2.40	5.82	1.34	0.918	13.2	12.0	13.0	142.4	44.0	61.7	448.4	399.4	442.9
GM10-10	3.98	4.34	0.83	0.663	8.65	7.96	9.34	9.0	79.1	83.4	117.4	100.0	134.8
GN12-10	6.90	7.91	3.05	1.68	8.84	9.15	8.01	14.7	55.8	75.7	28.1	32.6	16.1
Average error %								55.4	59.6	73.6	198	177.3	197.9
Standard deviation %								61.6	14.6	9.0	180.8	159.4	179.9
Coefficient of variation (ratio)								1.11	0.24	0.12	0.91	0.90	0.91

Table 9. Experimental and analytical deflections at service loads (steel reinforced beams).

Beam notation	Exp. Def. (mm)	Calculated deflection utilizing Eq. 10 (mm)	Error%
SN8-8	2.50	0.371	85.2
SN10-10	4.10	3.85	6.1
SN12-12	5.13	5.40	5.3
Average error %			32.2
Standard deviation %			37.5
Coefficient of variation (ratio)			1.16

Table 10. Experimental and analytical deflections at 1.3*Service loads (GFRP reinforced beams).

Beam notation	Exp. Def. (mm)	Calculated deflection (mm)						Error%					
		Eq. 2	Eq. 3	Eq. 4	Eq. 5	Eq. 6	Eq. 9	Eq. 2	Eq. 3	Eq. 4	Eq. 5	Eq. 6	Eq. 9
GN8-8	6.0	12.8	5.51	2.5		19.8	17.4	113.9	8.2	58.4	241.9	229.5	190.3
GM10-10	7.1	9.45	3.45	1.79	14.1	13.5	12.5	33.1	51.3	74.8	98.3	90.5	75.7
GN12-10	9.78	13.3	8.52	4.08	12.8	13.3	11.1	36.0	12.9	58.3	31.3	36.2	13.2
Average error %								61.0	24.1	63.8	123.8	118.7	93.1
Standard deviation %								37.4	19.3	7.75	87.9	81.4	73.3
Coefficient of variation (ratio)								0.61	0.80	0.12	0.71	0.69	0.79

Table 11. Experimental and analytical deflections at 1.3*Service loads (Steel reinforced beams).

Beam notation	Exp. Def. (mm)	Calculated deflection utilizing Eq. 10 (mm)	Error%
SN8-8	3.85	0.695	82.0
SN10-10	5.04	5.79	15.0
SN12-12	6.72	7.27	8.2

Table 11. Contd.

Average error %	35.1
Standard deviation %	33.3
Coefficient of variation (ratio)	0.95

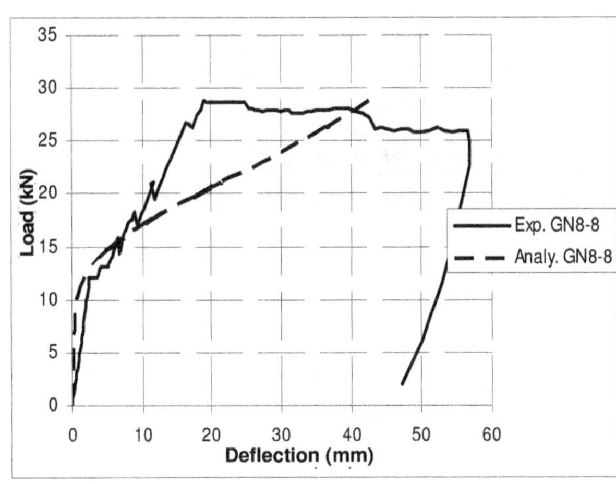

Beam GN8-8

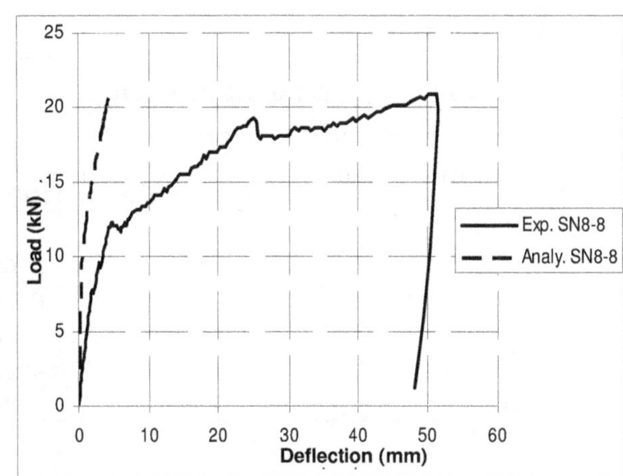

Beam SN8-8

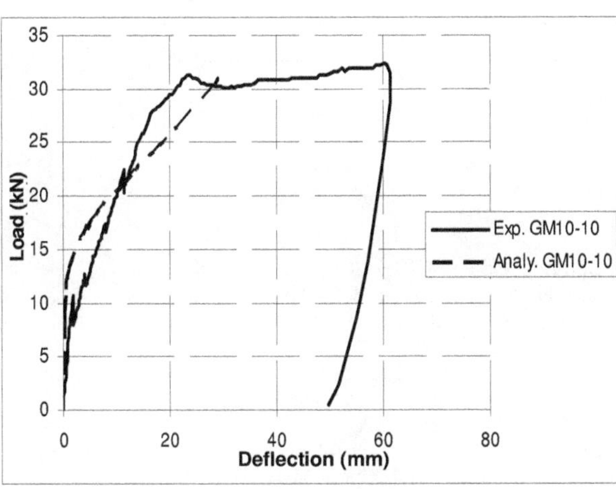

Beam GM10-10

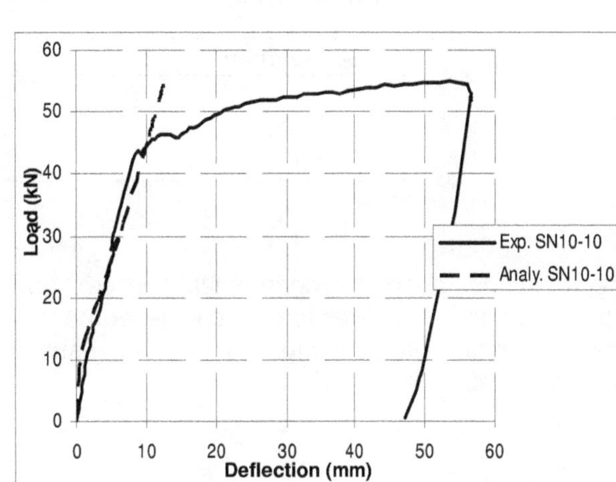

Beam SN10-10

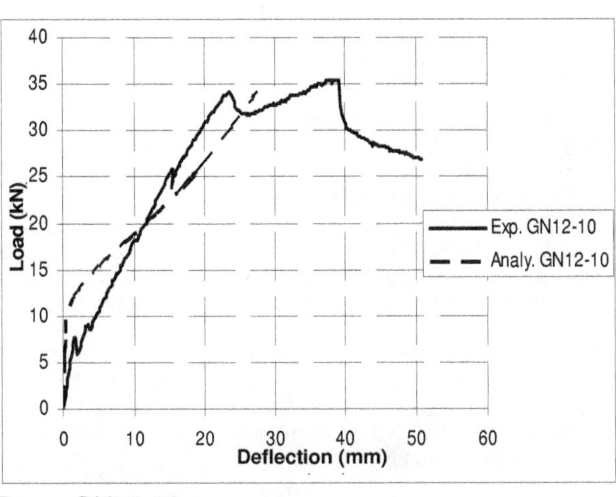

Beam GN12-10

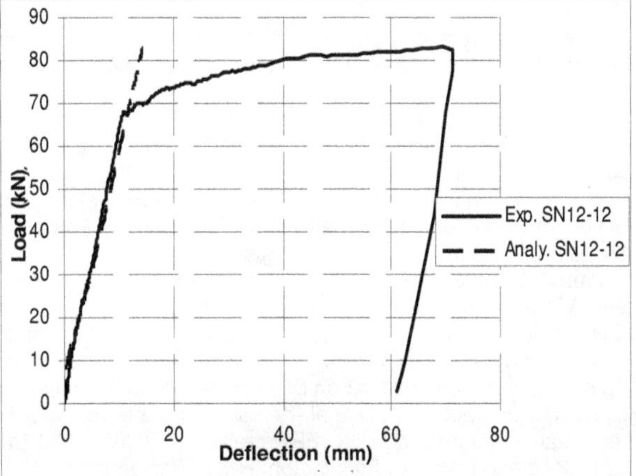

Beam SN12-12

Figure 7. Load-Cantilever tip deflection.

Table 12. Service load deflection at cantilever tip and cantilever span-to-service load deflection.

Beam notation	Service load def. (mm)	Span/service load def.
GN8-8	2.4	250.0
GM10-10	3.98	150.8
GN12-10	6.9	87.0

Table 13. Slopes of the load-deflection curves.

Beam notation	Slope of P-Δ curve (kN/mm)	
	Stiffness (N/mm) $0 \rightarrow P_{cr}$	Stiffness (N/mm) $P_{cr} \rightarrow P_u$
GN8-8	5700	960
SN8-8	5100	550
GM10-10	6000	1200
SN10-10	6700	4000
GN12-10	4900	1400
SN12-12	7800	6100

Table 14. Ductility of GFRP-Reinforced cantilevers.

Beam notation	Ductility
GN8-8	2.4
GM10-10	2.3
GN12-10	2.3

10. The obtained span-to-experimental service load deflection ratios for GFRP-reinforced cantilever beams are relatively high when compared to the usually accepted ratio of about span/450.

REFERENCES

Benmokrane B, Chaallal O, Masmoudi R, (1996). "Flexural response of concrete beams reinforced with FRP reinforcing bars". ACI Struct. J., 91(2): 46-55.

Mota C, Alminar S, Svecova D (2006). "Critical review of deflection formulas for FRP-RC members". J. Compos. Constr. ASCE, 10(3): 183-194.

Toutanji HA, Saafi M (2000). "Flexural behavior of concrete beams reinforced with glass fiber-reinforced polymer (GFRP) bars". ACI Struct. J., 97(5): 712-719.

ACI Committee 440.1R-06, (2006). "Guide for the design and construction of concrete reinforced with FRP bars". American Concrete Institute, Farmington Hills, Michigan, p. 44.

ACI Committee 318 (2008)."Building Code Requirements for Reinforced Concrete (ACI 318-08)", American Concrete Institute.

Alsayed SH, Al-Salloum YA, Almusallam TH (2000). "Performance of glass fiber reinforced plastic bars as a reinforcing material for concrete structures", Composites Part B: Engineering, (31): 555-567.

Wang H, Belarbi A (2005). "Flexural behavior of fiber-reinforced-concrete beams reinforced with FRP rebars", ACI-SP-230-51, American Concrete Institute, Farmington Hills, Michigan, pp. 895-914.

Grace NF, Soliman AK, Abdel-Sayed G, Saleh KR (1998) "Behavior and ductility of simple and continuous FRP reinforced beams", J. Compos. Const. ASCE, 2(4):186-194.

Lee K (2002). "Large deflections of cantilever beams of non-linear elastic material under a combined loading". Int. J. Non Linear Mech., 37(3): 439-443.

Solano-Carrillo E (2009). "Semi-exact solutions for large deflections of cantilever beams of non-linear elastic behavior". Int. J. Non Linear Mech., 44(2): 253-256.

Chen L (2010). An integral approach for large deflection cantilever beams ", Int. J. Non-Linear Mech., 45(3): 301-305.

"Egyptian Code for Design and Construction of Concrete Structures", ECP 203-2006.

"Egyptian Code for Design and Construction of FRP Reinfiorced Concrete Structures", ECP pp. 208-2005.

Metwally IM (2009). "Evaluation of existing model for predicting of flexural behavior of GFRP-reinforced concrete members", HBRC J. Hous. Build. Res. Cent. Cairo, Egypt, 5(1): 46-58.

Vijay P, GangaRao V (2001) "Bending behavior and deformability of glass fiber-reinforced polymer reinforced concrete members". ACI Struct. J., 98(6): 834-842.

Ashour AF (2006) "Flexural and shear capacities of concrete beams reinforced with GFRP bars". Constr. Build. Mater., (20): 1005-1015.

Habeeb MN, Ashour AF (2008) "Flexural behavior of continuous GFRP reinforced concrete beams". J. Compos. Constr. ASCE, 12(2): 115-124.

Spadea G, Bencardino F, Swamy RN (1997) "Strengthening and upgrading structures with bonded CFRP sheets design aspects for structural integrity". Proceedings of the Third International RILEM Symposium (FRPRC-3): Non-Metallic (FRP) for Concrete Structures, Sapporo, Japan, pp. 379-386.

5

Pelletized fly ash lightweight aggregate concrete: A promising material

A. Sivakumar* and P. Gomathi

VIT University, India.

The production of concrete requires aggregate as an inert filler to provide bulk volume as well as stiffness to concrete. Crushed aggregates are commonly used in concrete which can be depleting the natural resources and necessitates an alternative building material. This led to the widespread research on using a viable waste material as aggregates. Fly ash is one promising material which can be used as both supplementary cementitious materials as well as to produce light weight aggregate. Artificial manufactured lightweight aggregates can be produced from industrial by-products such as fly ash, bottom ash, silica fume, blast furnace slag, rice husk, slag or sludge waste or palm oil shell, shale, slate, clay. The use of cost effective construction materials has accelerated in recent times due to the increase in the demand of light weight concrete for mass applications. This necessitates the complete replacement or partial replacement of concrete constituents to bring down the escalating construction costs. In recent times, the addition of artificial aggregates has shown a reasonable cut down in the construction costs and had gained good attention due to quality on par with conventional aggregates.

Key words: Fly ash, aggregate, pelletizer, lightweight concrete.

INTRODUCTION

Presently in India the power sector depends on coal based thermal power stations which produces a huge amount of fly ash and estimated to be around 110 million tonnes annually. The utilization of fly ash is about 30% as various engineering properties requirements that is for low technical applications such as in construction of fills and embankments, backfills, pavement base and sub base course; intermediate technical application such as in producing blended cement, concrete pipes, precast/ prestressed products materials, lightweight concrete bricks/blocks, autoclaved aerated concrete and lightweight aggregate (Baykal and Doven, 2000). Lightweight concrete is produced in different categories based on the no-fines concrete, aerated cellular concrete and lightweight aggregate concrete. With increasing concern over the excessive exploitation of natural aggregates, synthetic lightweight aggregate produced from environmental waste is a viable new source of structural aggregate material. The uses of structural grade lightweight concrete reduce considerably the self-load of a structure and permit larger precast units to be handled. For example the Autoclaved cellular concrete (ACC), is a lightweight concrete material, which can be manufactured using 60 to 75% fly ash by weight (Ahmaruzzaman, 2010).

One of the common techniques while producing the lightweight aggregate is by agglomeration technique. In agglomeration technique the pellet is formed in two ways either by agitation granulation and compaction. The agitation method is not taking any external force rather than rotating force. With increase in the dosage of water in the binder the cohesive force of the particles increase. Sintering, auto claving and cold bonding are three different processes to harden the green pellet (Bijen, 1986). The partial replacement of normal weight aggregate is 20 to 40% by the volume of lightweight aggregate with difference of the compressive strength as 1% only (Behera et al., 2004). This paper is to review on the application of using fly ash lightweight aggregate based on the cost, which method to fallow for manufacturing

*Corresponding author. E-mail: sivakumara@vit.ac.in.

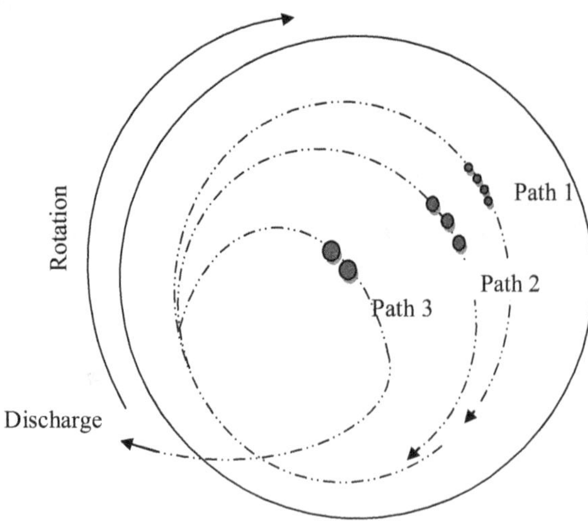

Figure 1. Growing path of pellets (Bijen, 1986).

Figure 2. Disc pelletizer machine.

LWA, mix proportions, strength improving physical and mechanical properties of lightweight aggregate concrete. Lightweight aggregate concrete is used for structurally lightweight structure, it reducing the density of concrete also over all weight of the structure.

PELLETIZING PROCESS

The desired grain size distribution of an artificial lightweight aggregate is either crushed or by means of agglomeration process. The pelletization process is used to manufacture lightweight coarse aggregate; some of the parameters need to be considered for the efficiency of the production of pellet such as speed of revolution of pelletizer disc, moisture content, angle of pelletizer disc and duration of pelletization (Harikrishnan and Ramamurthy, 2006). The different types of pelletizer machine were used to make the pellet such as disc or pan type, drum type, cone type and mixer type. With disc type pelletizer the pellet size distribution is easier to control than drum type pelletizer. With mixer type pelletizer, the small grains are formed initially and are subsequently increased in particle size by disc type pelletization (shown in Figure 1, Bijen, 1986). The disc pelletizer size is 570 mm diameter and side depth of the disc as 250 mm, it is fixed in a flexible frame with adjusting the angle of the disc as 35 to 55° and to control for the rotate disc in vertically manner should varying speed as 35 to 55 rpm shown in Figure 2 (Manikandan and Ramamurthy, 2007). In a cold bonded method is to made the increase the strength of the pellet as to increase the fly ash/cement ratio as 0.2 and above (by weight) (Yang, 1997). Moisture content and angle of the disc parameter influence the size growth of pellets (Harikrishnan and Ramamurthy, 2006). The dosage of binding agent is more important for making flyash balls and the optimum range was found to be around 20% to 25% by the total weight of binders (Bijen, 1986). Initially some percentage of water is added in the binder and then poured in a disc; remaining water is sprayed during the rotating period because while rotating without water in the disc the fly ash powder tends to form lumps and does not increase the distribution of particle size. The pellets are formed approximately in duration of 20 min.

HARDENING PROCESS IN LWA

The two major classes of fly ash are specified in ASTM C 618, namely class-C and class-F classified based on the chemical composition resulting from the different types of coal burning. Class-C fly ash is normally produced from the burning of sub-bituminous coal and lignite and class-C fly ash. The flyash aggregates are porous material and to improve the strength of the pellet the binder material like cement, lime, bentonite, metakaolin, kaolinite, glass powder and ceramic powders are added. The clay binders like metakaolin and kaolinite gives higher fines value (Geetha and Ramamurthy, 2011). The percentage of binder content is taken by the weight of fly ash. Hardening the pellets is done by various process namely cold bonding, sintering and autoclaving. Cold-bonded fly ash aggregates are hardened by different curing process namely normal water curing, steam curing and autoclaving. Autoclave and steam curing method is less effective to improve the properties of aggregate as compare to normal water curing method. Among accelerated cured class c fly ash aggregate, autoclaved aggregates has properties closer to the normal water cured aggregate due to the dense microstructure formation. The curing method is more important to enhance the aggregate strength. Hence, a normal water curing method can be adopted and autoclaving may be adopted for high-early strength (Manikandan and Ramamurthy, 2008). A higher strength of the aggregate can be obtained at 8 to 10 h in autoclave curing (Bekir and Tayfun, 2007).

Sintering process can be defined as burning the cold bonded pellet in a muffle furnace at temperature range of 800 to 1200°C. The mineral particles in the binder fuse together to form the crystalline structure (CSH) and results in higher strength of the aggregate. Therefore sintered fly ash lightweight aggregate production is more convenient while replacing the normal weight aggregate to lightweight aggregate (Verma et al., 1998).

MIX DESIGN OF LWAC

The mix design of lightweight aggregate concrete is not same as the conventional concrete mix design. Since the aggregates are porous and results in compensation of extra water for obtaining more workability. The mix design concepts are usually based on the production of higher strength matrix to low water cement ratio for the weaker aggregate. Therefore in ordinary concrete, the number of batches that are necessary to determine the best composition can be reduced to a minimum. But in a Lightweight aggregate concrete mix design more complicated for adding of water, LWA is a porous aggregate so we need extra water in the concrete (Grubl, 1979). The gradation of aggregate with different aggregate grading size distributions are required to improve the engineering properties in the concrete mix (Sari and Pasamehmetoglu, 2005).

The self-consolidating properties of lightweight aggregate concrete can be obtained by means of densified mixture design algorithm (DMDA) which gives higher strength, flow-ability and excellent durability as compared to the ACI 211.11 method (Chao-Lung and Meng-Feng, 2005). The design of lightweight aggregate are followed in two methods; loose volume calculation and absolute solid volume calculation (Wang et al., 2005). In mix proportion the LWA are mixed in different status while fully saturated condition, partially saturation condition and dry condition. The lightweight aggregate is pre-wetting before addition of concrete mix. The Polyurethane (PUR) foam waste as a lightweight aggregate were prepare before mixing in a concrete mix while LWA were immersed in water of 24 h to improve the workability of concrete (Amor et al., 2010). The selection of sand-aggregate ratio is 28 to 42% in the mix proportion, which can influence the compressive strength and regulate the workability of concreter (Wang et al., 2005).

The strength of concrete is equal to the effective water to binder ratio which is chosen as 0.26. The quantity of the ingredients can be selected the volume of coarse aggregate to total volume of aggregate ratio as 0.6; based on the cold-bonded fly ash aggregate the quantity of cement content as 551 kg/m^3 greater than sintered fly ash aggregate as around 548 kg/m^3. Both type of lightweight aggregate concrete had shown the higher compressive strength (Niyazi and Turan, 2011). Lightweight concrete incorporating the bottom ash and the sintered fly ash in the concrete should increase the permeability; by replacing 30% of OPC with fly ash, to improve the permeability of LWC (Yun Bai et al., 2004). Addition of admixture in the lightweight concrete is to increase the strength and elastic modulus. The addition of silica fume at 5 to 15% in the LWC can improve the strength properties while, replacements of 10% fly ash instead of cement in concrete can decrease strength as compared to without fly ash (Shannag, 2011). A detailed mix proportion of light weight aggregate concrete adopted in different studies are given in Table 1.

PHYSICAL PROPERTIES OF LWAC

The physical characteristics of the lightweight aggregate produced by pelletization are given in Table 2 (Bijen, 1986). The moisture content and amount of binder can affect the size of fly ash aggregates thus formed. The fineness of the fly ash (414 m^2/kg) gives the better pelletization efficiency compared to the coarser fly ash (257 m^2/kg). Therefore finer fly ash needs the addition of the binder material and the addition of clay binder in the coarser fly ash will increase the pelletizing efficiency (Manikandan and Ramamurthy, 2007). The specific gravity of fly ash lightweight aggregate is increase without adding binder and it's a denser structure. The addition of bentonite and glass powder in fly ash is to reducing the specific gravity as compare to lime and cement binder in fly ash (Ramamurthy and Harikrishnan, 2006).

Density of LWAC

The properties of lightweight aggregate can be improved with the addition of different binder at various percentage. Therefore, the percentage of binder increased vice versa density increase. Density of sintered fly ash aggregate with binder is decreased while increased the temperature range between 1150 to 1200°C. The bentonite and glass powder binder is melted and bloating firmly for rising temperature and the glassy particle filled the voids in a crystal form to improve the strength (Niyazi and Turan, 2011). The difference between the density of the pre-wetting and without pre-wetting PUR lightweight aggregate concrete is lower than 12 Kg/m^3 (Amor et al., 2010). The density of shell aggregate is 28% lower than the normal aggregate (Okafor, 1988)

POZZOLANIC REACTIVITY OF LWA MADE WITH FLY ASH

A pozzolanic reaction occurs between dissolved minerals

Table 1. Mix proportion of LWAC ingredients studied from various literatures.

Author	Concrete type	W/b ratio	Cement content (kg/m^3)	Fine aggregate content (kg/m^3)		V_{CA}/V_{TA} ratio	Light weight aggregate content (kg/m^3)	AEA (%)	Admixtures (%)	
				Natural sand	Crushed sand				FA-F	SP
Yannick et al., 2006	LWAC	0.27	475.6	674.4	-	0.6	546.6	-	158.7	-
		0.34	335.3	728.9	-		612.4		110.6	
		0.28	391.2	734.0	-		540		107.0	
Niyazi, 2011 (Niyazi and Turan, 2011)	CLWC	0.26	551	318	318	0.6	592	0.2	-	1.1
	LWBC		548	316	317		567			
	LWGC		549	317	317		580			
Wasserman and Bentur, 1997	SLWAC	0.4	440	49%		0.51	51%	-	-	-

from glass and calcium from portlandite. Hydroxyl ions break down the silica in the glass, which in turn react with the calcium in the portlandite to form CSH paste. This reaction increases the bond strength between the aggregate and the cement matrix. Since an artificial fly ash lightweight aggregate are porous structure and it composed of glass phase, pozzolanic reaction expected on the surrounding of this aggregate. Commonly in fly ash two type of carbonaceous fragment matter (Nambu et al., 2007). The reduction of CH occurs during the sintering process of flyash aggregate at higher temperature (900°C) (Weasserman and Bentur, 1997).

STRENGTH PROPERTIES

The cement, lime and bentonite are used as a binder in 10, 20 and 30% by weight of fly ash for pelletization. It is also observed that the improvement in the 10% fines value and reduction in water absorption of sintered fly ash aggregate. For 10% fineness is used to test strength of lightweight aggregate. The addition of bentonite is to enhance the aggregate strength, cement is to give minimum strength and the lime is for improving the ballability. Therefore, the addition of 20% bentonite gives an optimal strength (Ramamurthy and Harikrishnan, 2006)

The strength of the LWAC with various binder content improves the strength properties of aggregate and given in Table 3. The compressive strength of polypropylene fiber reinforced SLWC is higher than the steel fiber reinforced by 7 Mpa (Kayali et al., 2003). Fiber reinforced concrete increase the tensile strength with low modulus of elasticity as well as reducing the shrinkage cracking in LWAC (Kayali et al., 1999). The lightweight aggregate manufactured using pelletizing process gives a smooth surface after sintering process. The sintered fly ash aggregate (FAA) were crushed that is not invove pelletizing, the structure gives a rough surface and enhancing the compressive strength as 66.76 Mpa (Kayali, 2008).

Expanded clay lightweight aggregate has higher porosity in the transition zone which may show significant effect on the permeability of lightweight concrete. The pre-wetting time of expanded clay lightweight aggregate were critically affected the strength and slump of the concrete (Lo et al., 1999). The pore structure of the sintered pulverized fuel ash lightweight aggregate is approximate range of the pore size from 200 μm down to less than 1μm with all the size had been evenly distributed throughout the pellet and gives the better bond between the pellets and cement matrix (Swamy and Lambert, 1981)

The high resolution optical microscope and image analysis software were used to find out the pore area percentage and pore size distribution in the cement paste and the interfacial zone of concrete cured at 28 days. The transition zone is a weak zone of more porous in nature between the aggregate and cement matrix. The experimental results of lightweight aggregate show large water absorption range from 8.9 to 11% which produce greater pore percentage as 14.4 and 21.7% at the interfacial zone (Lo et al., 2006).

Therefore, lightweight aggregate is more porous from the outer layer and it present dense interfacial zone for the aggregate without any outer layer. So that the aggregate gives better bond appeared due to the mechanical interlocking

Table 2. Physical characteristics of pelletized aggregates from various literatures.

Authors	Type of LWA used	Specific gravity of LWA		BD (Kg/m^3)		Voids (%)		Water absorption (%)		Crushed strength of pellet (Mpa)
		SSD	OD	LBD	RBD	LV	RV	24	48	
	CLWA	1.63	1.3	789	842	39.2	35.1	-	25.5	3.7
Niyazi and Turan, 2011	SFA+1200+10B	1.57	1.56	933	993	40.1	36.2	-	0.7	12
	SFA+1200+10G	1.6	1.59	936	936	41	37	-	0.7	9.6
Ramamurthy, 2006	SFA+20B	-	1.83	850	-	-		15.8	-	-
Amor et al., 2010	Polyurethane foam waste LWA	45		21		13.9		-		-
Chi et al., 2003	CLWA	1.76	1.44	972		-		20.8		8.57

Table 3. Mechanical properties of lightweight aggregate concrete from various literatures.

Author	Concrete type	Comp strength (Mpa)		Split tensile strength (Mpa)		Modulus of elasticity (Gpa)	
		28 d	56 d	28 d	56 d	28 d	56 d
Byung-Wan et al., 2007 (Chi et al., 2003)	AFLAC	26.7	-	-	-	-	-
Kayali et al., 2003 (Behera, 2004)	SFAC	68.0	-	6.6	-	25	-
Santish and Leif, 1983 (Chao-Lung and Meng-Feng, 2005)	LWAC	20.4	-	-	-	-	-
	SFA+1200+10G	55.8	60.4	4.9	5.1	25.7	25.9
Niyazi and Turan, 2011	SFA+1200+10B	53.5	59.5	4.8	5.1	26.0	26.3
	LWCC	42.3	44.6	3.7	3.9	19.6	19.7
Kayali, 2008 (Grubl, 1979)	FAA	66.75	-	3.75	-	25.5	-

AFLAC – Alkali-activated fly ash lightweight aggregate concrete; SFAC – Sintered fly ash aggregate concrete; SFA+1200+10G – Sintered fly ash aggregate with 10% glass powder at 1200°C temperature; SFA+1200+10B - Sintered fly ash aggregate with 10% bentonite at 1200°C temperature; LWCC – Cold-bonded fly ash lightweight aggregate concrete; FAA - Fly ash aggregate manufacture by using sintering without pelletizing aggregate and the procedure is same. That aggregate are crushed in briquette and fired in a kiln.

between aggregate and the cement paste (Min-Hong and Gjorv, 1990). The use of silica fume for adding in LWC is to improve the mechanical properties, but disadvantage of shrinkage performance is less compared to normal weight concrete (Mehmet et al., 2004).

MICROSTRUCTURAL CHARACTERISTICS OF FLYASH AGGREGATE CONCRETE

The mechanical behavior and durability aspects of concrete affected by its aggregate and cement paste as well as the interfacial zone between them. Normal weight concrete the aggregate-cement paste interface is the weakest part of the micro-structural system and the place where cracks begins, strongest component that is normal aggregate (Min-Hong and Gjorv, 1990). But, the lightweight aggregate concrete is different to the

Table 4. Durability properties of different lightweight aggregate concrete.

Authors	Concrete type	Chloride penetration test (coulombs)		Water permeability test (mm)		Accelerated Corrosion test (days)		Freezing and thawing resistance	
								Air entrainment	
		28 d	56 d	28 d	56 d	28 d	56 d	4%	6%
	LWCC	1464	748	36	79	28	49	-	-
(Niyazi and Turan, 2011)	LWBC	586	264	19	39	123	-	-	-
	LWGC			23	41	106	-	-	-
Byung-Wan et al., 2007	ALWA	-		-	-	-	-	78	92

interaction between the cement paste-aggregate is complex and it's vary to the normal aggregate concrete. This type of aggregate are porous in nature, the grains are capable of absorbing water which yielded to the surrounding matrix. The porosity of Lytag aggregate can vary between 25 to 75% depending on the manufacturing process used (Swamy and Lambert, 1981). Many more research work to indentify the internal and external structure of lightweight aggregate, particularly cement matrix- aggregate interface carried out (Shondeep et al., 1992). For applied micromechanical method considered the perfect bonding between the aggregate and mortar (Chung-Chia and Ran, 1998).

Normally sintered fly ash lightweight aggregate were produced by heat and polymer treatment so that to improve their strength, absorption and pozzolanic activity according to their properties of aggregate by change to the microstructure. SEM analysis to observe the higher magnification to see more uniform distribution of small pore size in the sintered fly ash aggregate at the temperature treated aggregate as 1200 to 1300°C (Weasserman and Bentur, 1997). Mechanical interlocking plays an important role for strengthening the interface (Shondeep et al., 1992). The effect of aggregate using is dry and prewetting lightweight aggregates on the ITZ microstructure. The thickness of ITZ around the dry aggregate is 10 μ less than the other prewetted and normal aggregate as 15 μ and beyond 35 μ respectively (Amir Elsharief et al., 2005).

DURABILITY PROPERTIES OF HARDENED CONCRETE

Durability of concrete essentially dictates the permeability resistance of concrete and needs to be assessed for long time sustainability. The durability properties of lightweight aggregate concrete is given in Table 4. Permeable concrete is significantly attack the concrete ingredients and accumulate water inside of concrete it caused deterioration of concrete and reinforcement. Normally permeable of water and chloride will be decrease when increase the age of concrete but in lightweight aggregate concrete will be more permeable than normal concrete. To carry out the chloride penetration test for LWBC gives the best performances compare to other type of lightweight aggregate concrete. Sintering lightweight aggregate concrete showed the low permeablity except cold-bonded lightweight aggregate at 28days. Sintering and cold-bonded aggregate has highest chloride permeability with total charge passed values of 1464 and 586 coulombs at 28 days and 748 and 264 coulombs at 56 days (Niyazi Ugur Kockal and Turan Ozturan, 2011). A sintered lightweight aggregate with bentonite is less water permeable compare to normal aggregate concrete. Almost glass powder, bentonite binder adding in the sintering aggregate which gives the best performance of water permeability test. In a cold-bonded process the water permeability is more than sintered process (Niyazi Ugur Kockal and Turan Ozturan, 2010; Niyazi Ugur Kockal and Turan Ozturan, 2011). The durability factor of the 4% air entrainment specimen gives the marginal freezing and thawing was 78 with compare to the 6% air entrainment specimen gives the good freeze-thaw resistance was 92 (Byung-Wan Jo et al., 2007).

CONCLUSION

The potential applications of light weight aggregate are more phenomenal in terms of the usage as new construction materials. Cost effective construction practices with alternate construction materials are most desired in terms of huge savings in construction cost. Fly ash is not a waste and can be effectively used in concrete either as aggregate fillers, replacement for fine aggregates or as a fly ash brick material. The overall studies conducted by various researches shown that the fly ash aggregate produced by pelletization can be an effective aggregate in concrete production. Also, the efficiency of pelletization depends on the speed of the pelletizer, angle of the pelletizer and the type of binder added along with the fly ash. The cost effective and simplified production techniques for manufacturing fly ash aggregate can lead to mass production and can be an ideal substitute for the utilization in many infrastructural projects. In the near future the depletion of the nature resources for aggregate can be suitably compensated from the fly ash aggregate.

REFERENCES

Ahmaruzzaman M (2010). A review on the utilization of fly ash. Prog. Energy Combustion Sci., 36: 327-363.

Amir Elsharief, Menashi D Cohen, Jan Olek (2005). Influence of Lightweight Aggregate on the Microstructure and Durability of Mortar. Cement Concrete Res., 35: 1368-1376.

Amor Ben Fraj, Mohamed Kismi, Pierre Mounanga (2010). Valorization of coarse Rigid Polyurthane Foam Waste in Lightweight Aggregate Concrete. Constr. Build. Mater., 24: 1069-1077.

Behera JP, Nayak BD, Ray HS, Sarangi B (2004). Lightweight Weight Concrete with Sintered Fly ash Aggregate: A Study on Partial Replacement to Normal Granite Aggregate. IE (I) Journal-CV., 85: 84-87.

Bekir Ilker Topcu, Tayfun Uygunoglu (2007). Properties of Autoclaved Lightweight Aggregate Concrete. Build. Environ., 42: 4108-4116.

Bijen JMJM (1986). Manufacturing processes of artificial lightweight aggregates from fly ash. Int. J. Cement Composit. Lightweight concrete. 8(3): 191-9.

Byung-Wan Jo, Seung-Kook park, Jong-bin Park (2007). Properties of concrete made with alkali-activated fly ash lightweight aggregate (AFLA). Cement Concrete Composit., 29: 128-135.

Chao-Lung Hwang, Meng-Feng Hung (2005). Durability design and performance of self-consolidating lightweight concrete. Constr. Build. Mater., 19: 619-626.

Chi JM, Huang R, Yang CC, Chang JJ (2003). Effect of Aggregate Properties on the Strength and Stiffness of Lightweight Concrete. Cement Concrete Composit., 25: 197-205.

Chung-Chia Yang, Ran Huang (1998). Approximate Strength of Lightweight Aggregate using Micromechanics Method. Adv. Cement Based Mater., 7; 133-138.

Geetha S, Ramamurthy K (2011). Properties of Sintered Low Calcium Bottom Ash Aggregate with Clay Binders. Constr. Build. Mater., 25: 2002-2013.

Grubl P (1979). Mix Design of Lightweight Aggregate Concrete for Structural Purposes. Int. J. Lightweight Concrete, 1(2): 63-69.

Harikrishnan KI, Ramamurthy (2006). Influence of Pelletization Process on the Properties of Fly Ash Aggregates. Waste Manag., 26: 846-852.

Kayali O (2008). Fly ash lightweight aggregates in high performance concrete. Constr. Building Mater., 22: 2393-2399.

Kayali O, Haque MN, Zhu B (1999). Drying shrinkage of Fibre reinforced lightweight aggregate concrete containing fl ash. Cement Concrete Res., 29: 1835-1840.

Kayali O, Haque MN, Zhu B (2003). Some characteristics of high strength fiber reinforced lightweight aggregate concrete. Cement Concrete Composit., 25: 207-213.

Lo Tommy Y, Cui HZ, Tang WC, Leung WM (2006). The effect of aggregate absorption on pore area at interfacial zone of lightweight concrete. Constr. Build. Mater., 22: 623-628.

Lo Y, Gao XF, Jeary AP (1999). Microstructure of pre-wetted aggregate on lightweight concrete. Build. Environ., 34: 759-764.

Manikandan R, Ramamurthy K (2007). Influence of fineness of fly ash on the aggregate pelletization process. Cement Concrete Composites, 29: 456-464.

Manikandan R, Ramamurthy K (2008). Effect of Curing Method on Characteristics of Cold Bonded Fly Ash Aggregate. Cement Concrete Composites, 30: 848-853.

Mehmet Gesoglu, Turan Ozturan, Erhan Huneyisi (2004). Shinkage Cracking of Lightweight Concrete made with Cold-bonded Fly ash Aggregates. Cement Concrete Res., 34: 1121-1130.

Min-Hong Zhang, Gjorv Odd E (1990). Microstructure of the interfacial zone between lightweight aggregate and cement paste. Cement Concrete Res., 20: 610-618.

Nambu Masateru, Kato Masahiro, Anosaki Takao, Nozaki Kenji, Ishikawa Yoshitaka (2007). Pozzolanic Reactions between natural and Artificial Aggregate and the Concrete matrix. World of Coal ash.

Niyazi Ugur Kockal, Turan Ozturan (2010). Effects of lightweight fly ash aggregate properties on the behavior of lightweight concretes. J. Hazard. Mater., 179: 954-965.

Niyazi Ugur Kockal, Turan Ozturan (2011). Charecteristics of lightweight fly ash aggregates produced with different binders and heat treatments. Cement Concrete Composites, 33: 61-67.

Niyazi Ugur Kockal, Turan Ozturan (2011). Durability of lightweight concretes with lightweight fly ash aggregates. Constr. Build. Mater., 25: 1430-1438.

Niyazi Ugur Kockal, Turan Ozturan (2011). Optimization of Properties of Fly ash Aggregates for High-strength Lightweight Concrete Production. Mater. Des., 32: 3586 - 3593.

Niyazi Ugur Kockal, Turan Ozturan (2011). Strength and Elastic Properties of Structural Lightweight Concretes. Mater. Des., 32: 2396-2403.

Okafor OF (1988). Palm Kernel Shell as a Lightweight Aggregate for Concrete. Cement Concrete Res., 18: 901-910.

Ramamurthy K, Harikrishnan KI (2006). Influence of binders on properties of sintered fly ash aggregate. Cement Concrete Composit., 28: 33-38.

Santish Chandra, Leif Berntsson (1983). Technical notes: Influence of polymer microparticles on acid resistance of structural lightweight aggregate concrete. Inter. J. Cement Composit. Lightweight Concrete, 5(2): 127-131.

Sari D, Pasamehmetoglu AG (2005). The effects of gradation and admixture on the pumice lightweight aggregate concrete. Cement Concrete Res., 35: 936-942.

Shannag MJ (2011). Characteristics of Lightweight Concrete Containing Mineral Admixtures. Constr. Build. Mater., 25: 658-662.

Shondeep L Sakar, Satish Chandra, Leif Berntsson (1992). Interdepence of Microstructure and strength of Structural Lightweight Aggregate Concrete. Cement Concrete Composites, 14: 239-248.

Swamy RN, Lambert GH (1981). The microstructure of Lytag aggregate. Int. J. Cement Composit. Lightweight Concrete, 3(4): 273-282.

Verma CL, Handa SK, Jain SK, Yadaw RK (1998). Techno-commercial perspective study for sintered fly ash light-weight aggregates in india. Constr. Build. Mater., 12: 341-346.

Wang Lijiu, Zhang Shuzhong, Zhao Guofan (2005). Investigation of the Mix Ratio Design of Lightweight Aggregate Concrete. Cement Concrete Res., 35: 931-935.

Wasserman R, Bentur A (1997). Effect of Lightweight Fly ash Aggregate Microstructure on the Strength of Concretes. Cement Concrete Res., 27(4): 525-537.

Yun Bai, Ratiyah Ibrahim, Muhammed Basheer PA(2004). Properties of Lightweight Concrete Manufactured with Fly ash, Furnace Bottom ash and Lytag. International Workshop on Sustainable Development and Concrete Technology. pp. 77-87.

Experimental research and theoretical study on bending capacity of tube-gusset K-joint connection

Menghong Wang and Xiaodong Guo

School of Civil and Transportation Engineering, Beijing University of Civil Engineering and Architecture, Beijing 100044, China.

This study investigates bending behavior of tube-gusset K-joint using static experimental research and finite element analysis (FEA). Firstly, five groups of full scale tube-gusset K-joint models with different parameters which mainly refer to diameter and thickness of main tube were carried out to study the response and bearing condition of nodes in loading process. Then, finite element model of the joint was established, and the influence of main parameters including diameter and thickness of main tube, and length of gusset on the node mechanical performance was studied. Test and FEA results show that node bending capacity decreases with the increment of main tube diameter and increases with the increment of the main tube thickness and gusset length. On the basis of experimental and theoretical analysis, bending capacity calculation formulas of tube-gusset k-joint were proposed with numerical method and its applicability is verified.

Key words: Tube-gusset K-joint, static experiment, finite element analysis, bending capacity.

INTRODUCTION

Steel component with pipe cross section has many advantages, for instance, larger in radius of gyration and torsional stiffness, no weak axis under bending moment, high bearing capacity and good corrosion resistance after port closed, etc (Wang, 2011). Tube-gusset joint is mainly made up of steel tube and has above advantages, which avoid complex process of tubular welding joint and has better bearing performance; hence it is widely applied in hollow section structures. At present, tube-gusset joint is one of main node forms in industrial and civil constructions (Luo, 2010), which is being applied in practical engineering as shown in Figure 1.

Throughout specifications of hollow section structure at home and abroad, theoretical study and experimental research have rarely been done; mature design formula for the node has not been proposed yet. Engineers always refer to tubular node and consider certain safety factors when designing this type of node. Theoretical study and experimental research in-depth are needed.

Access to the data referred by the author and relevant research to this paper mainly includes: reduced scale experiment of tubular node which forced simultaneous on main tube and branch tube has been carried by Kim (2001); and established finite element model which replace axial force of branch tube by equivalent loads, on the basis of preceding work, calculation method of branch tube axial force and node moment was proposed; deduced dimensionless interactional relation of main tube

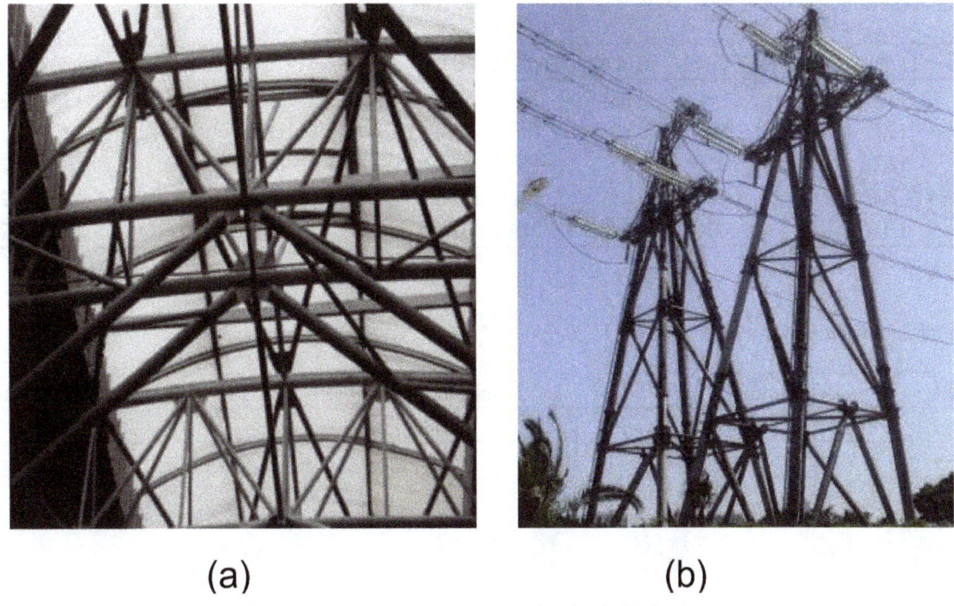

(a) (b)

Figure 1. Practical use of the node in engineering (a) truss (b) Transmission line tower.

axial force, component of branch tube axial force along main tube direction and node moment. Wang et al. (2000) studied stress distribution of tubular K-joint by FEA method, analyssed parameters which influence the node bearing capacity and concluded infinite element calculation formula of tubular K-joint. All above researches focus on experiment research and FEA of tubular welding node; experimental research and theoretical analysis of full scale tube-gusset joint has not been found in references; and practical design formula of tube-gusset node is still in blank.

The purpose of this research is to acquire the node mechanical performance under static load and calculation formula. Firstly, finite experiment of full scale tube-gusset K-joints have been done; thereafter FEA models of the nodes were founded and calculated; and finally, practical bending capacity calculation formula of tube-gusset K-joint through numerical regression analysis method concluded the study.

SUMMARY OF TEST

Specimens design

Five groups (K1 to K5) specimens have been designed. To avoid coincidence from a single result, each group has two identical specimens in same test process, and average value of each group was adopted for the following analysis.

The mainly relevant parameters in experiment are diameter and thickness of main tube. To study the failure mechanization of node area, stiffness of gusset and branch tube should be enough to the extent that would not damage before main tube. The specimens have three different main tube thicknesses: t (6 , 8 and 10 mm), main tube diameter, D (152, 168 and 219 mm). All gussets in specimens are unified 380 × 140 × 12 mm; the size of branch tube $d_z \times t_z$ are all 95 × 10 mm; angle between main and branch tube is maintained at 60°. Basic parameters of the specimens are given in Table 1. The node dimension is as shown in Figure 2.

Main tube and gusset were connected through fillet weld, while branch tube and branch plate were connected by open welding. Geometric size and structure of K1 is shown in Figure 2.

Property of the material

All specimens were made of Q235. Four standard test samples were fabricated by scrap reserved in baiting process and average test results were taken as the specimens mechanical property as shown in Table 2.

Load application

The test was carried out in Beijing University of Civil Engineering. The adopted load application facility was 500 KN hydraulic jack and self-balanced loading frame. In order to imitate the model which was destroyed under bending moment in reality, load was applied to the bottom of main tube horizontally by jack, which is one-way loading by steps (studied node mechanical property under bending). Loading devices are shown in Figure 3.

Ultimate load was obtained when load-displacement curve of main tube skin point to where transformed maximum emerged, with the decreased part or deformation value in this paper surpassing ultimate deformation (3% diameter of main tube) (Zhao, 1995; Van der Vegte, 1995). As location of decrease point was hardly determined, the later criterion was used as basis for judgment.

Two control modes including force control and displacement control were employed in load application process; the specific steps taken are as follows: 10 KN applied in first loading stage, increment of following stages are 5 KN, and displacement jumped when load was applied up to 60 KN; thereafter, there was a reverse to adopt displacement control with 10 mm added to original deformation in the first stage, the following increments 5 mm (the strain of measuring point remaining unchanged); every stage was

Table 1. Basic parameters of test specimens.

Specimen	L (mm)	t (mm)	l_z (mm)	d_z (mm)	D (mm)	θ (°)	Boundary of non-loading end	Amount of specimens
K1	1200	6	304	95	168	60	fixed	2
K2	1200	8	304	95	168	60	fixed	2
K3	1200	10	304	95	168	60	fixed	2
K4	1200	6	304	95	152	60	fixed	2
K5	1200	6	304	95	219	60	fixed	2

Table 2. Mechanical property of material.

Type of steel	Yield strength (f_y/MP)	Tensile strength (f_t/MP)	Elongation percentage (ε/%)
Q235	284.3	470.8	29.4

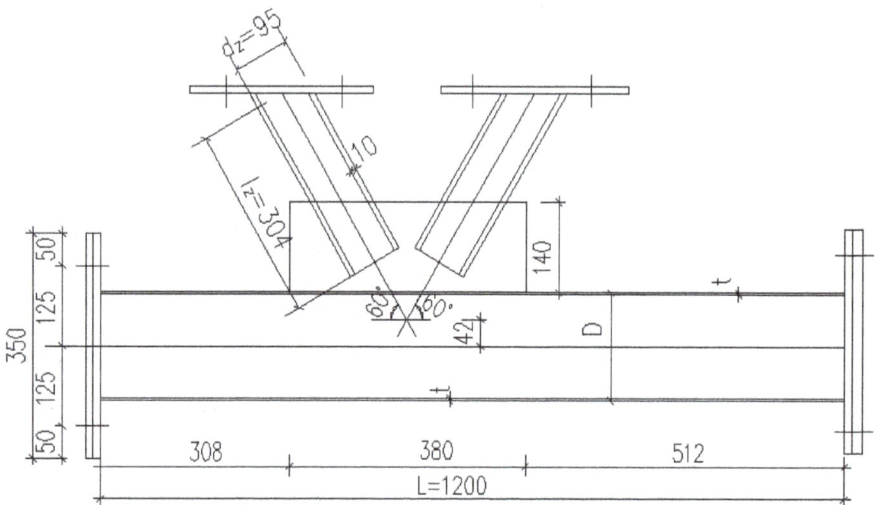

Figure 2. Size of node.

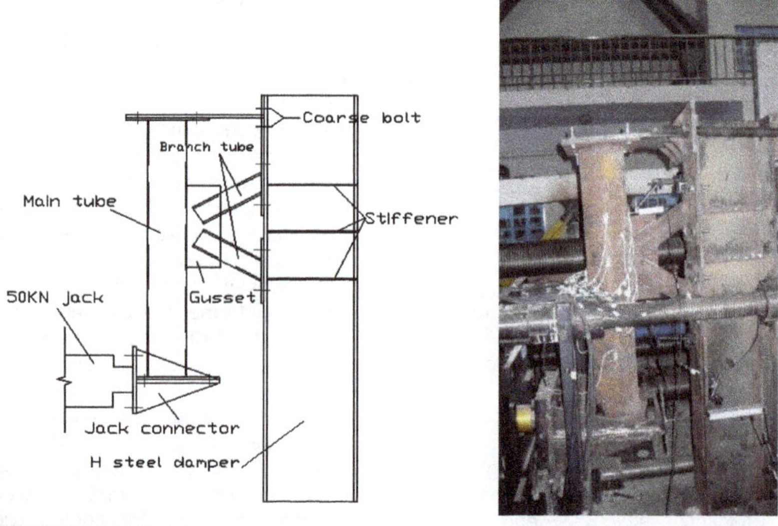

Figure 3. Test loading mode.

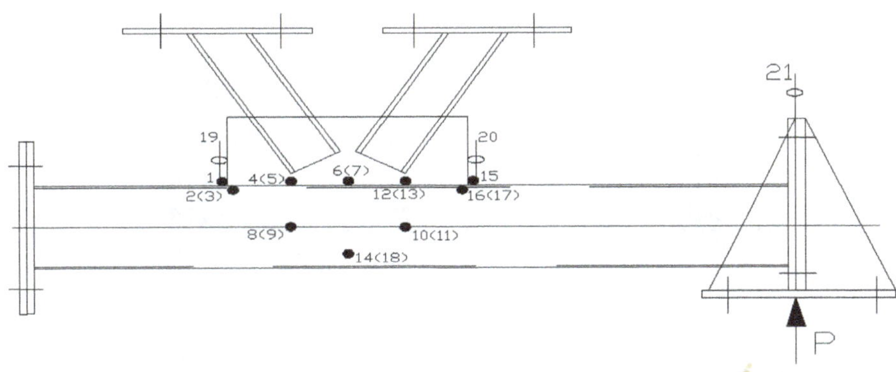

Figure 4. Layout of measuring points.

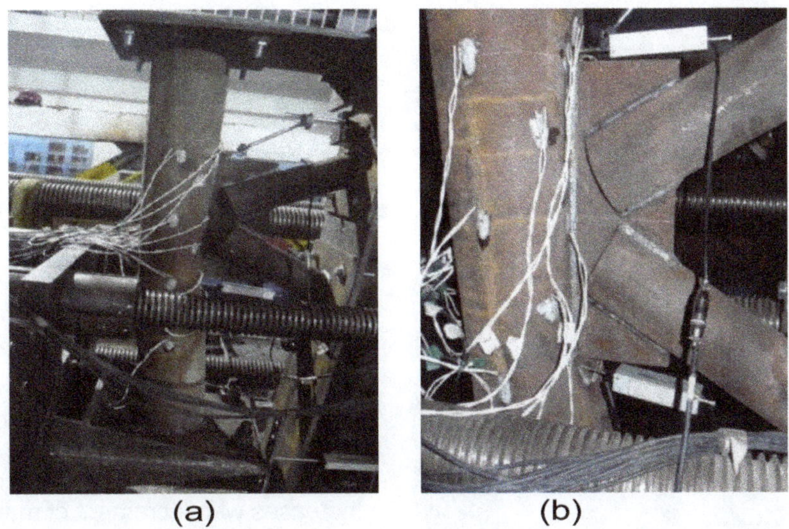

Figure 5. Deformation graph of K2 after failure (a) Global deformation (b) Local deformation.

kept static for two minutes before being entered into next stage. The entire process lasted about 2 h for one specimen.

Measuring point arrangement

Strain gauges (1~3, 15~17) were glued at connection part measuring change of strain, gauges (4-14, 18) were glued around the joint and in scope of 0.5 length of branch tube. Displacement meters (19, 20) was arranged at the connection, 19 measures local deflation deformation of the joint, 20 measures depression deformation of the joint, 21 measures displacement of loading end. Measuring points distributed shown in Figure 4, revealed 1~18 and 19 are strain gauges, and 21 represent displacement meter.

TEST RESULTS

Deflation deformation

The largest bending moment emerges in connection between main tube and gusset when horizontal force was applied on one end of main tube; because of small area in connection, stress concentration occurs in this place. Taking K2 as an example, the yield load is about 65 KN. This does not mean that further load increase will result to destruction as plastic stress redistribution take place in this region; though there will be node failure when local plastic deformation is too large. The failure load is about 103 KN. Local denting can be seen in compression side after joint failure, meanwhile, deflation deformation can be seen in tension side, specimen K2 after failure as shown in Figure 5.

Local denting

The test result is as shown in Table 3. Among them, refers to the replacement of 21, *M* refers to the moment of measuring point, and *s* refers to the deformation of 20.

Table 3. Measurement result in loading stages of K2.

P (kN)	Δ (mm)	M (kN•m)	s (mm)
4.95	0.43	2.534	0.086
66.5	7.91	34.048	1.511
102.27	26.93	52.364	5.038
140.7	41.84	72.038	8.224

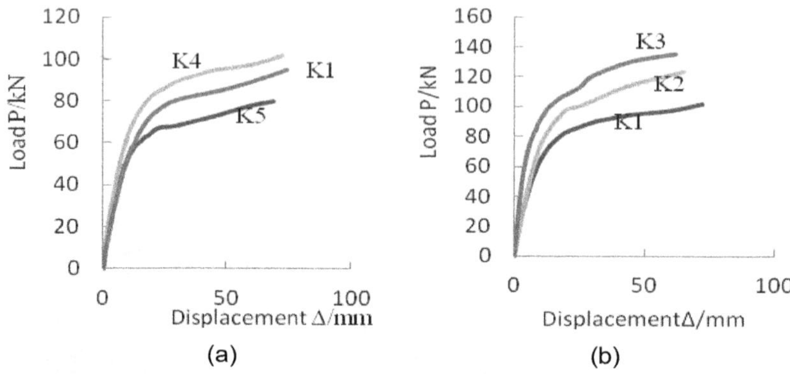

Figure 6. Stress-strain curves (a) Alter diameter of main tube (b) Alter thickness of main tube.

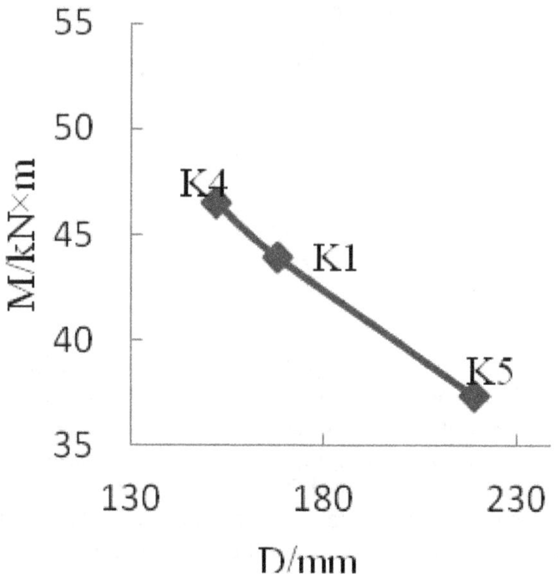

Figure 7. Variation of ultimate strength main tube with diameter.

Load-Displacement curve

The load-displacement curves of specimens under loading are as shown in Figure 6; force and displacement are in linear relationship at the beginning of load application, which indicate that specimens are in elastic stage, transferred to plastic stage with larger load application, faster increasing deformation of specimens, but small increment of bearing capacity of specimens.

Figure 6(a) shows bearing capacity decreases with increment of main tube diameter while 6(b) shows increases with increment of main tube thickness.

Parameters analysis of test result

The influence of main tube diameter on bearing capacity shown in Figure 7 reveals joints bearing capacity decrease with the increment of main tube diameter.

The influence of main tube thickness on bearing capacity shown in Figure 8 reveals joints bearing capacity increase with the increment of main tube thickness. Specimens were yield successive as load increased instantly, promoting bending rigidity as the increment of main tube thickness; therefore resulting in larger bearing capacity.

INFINITE ELEMENT ANALYSES

Established infinite element analytical models

Infinite element models adopted unit-solid 45 to imitate this type of unite combined with 8 nodes; each node has three translational freedom degrees along coordinate direction of *x*, *y*, *z*. Q235 was selected as material of

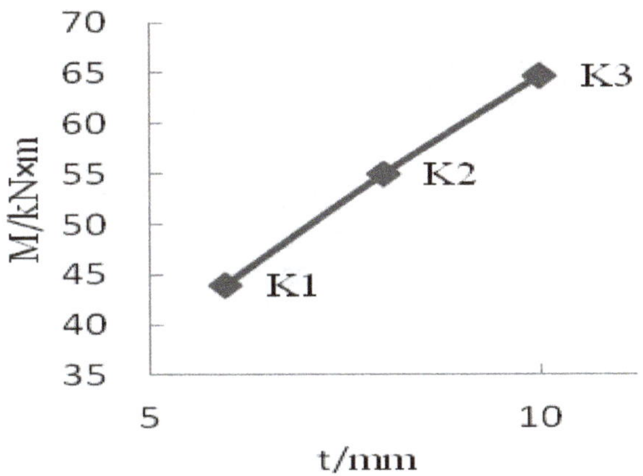

Figure 8. Variation of ultimate strength with main tube thickness.

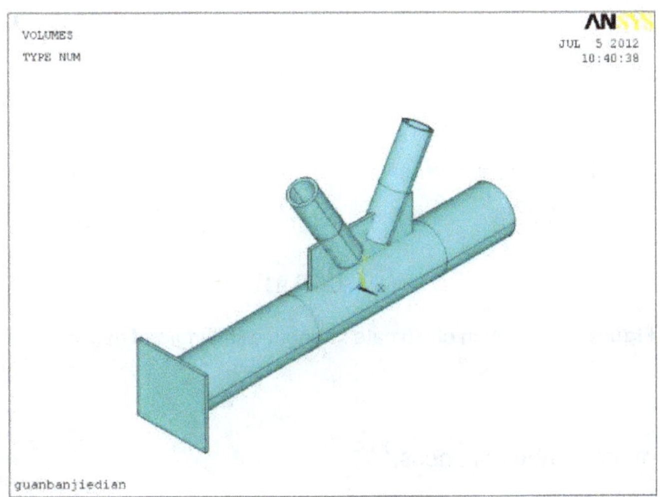

Figure 9. Finite element analysis model.

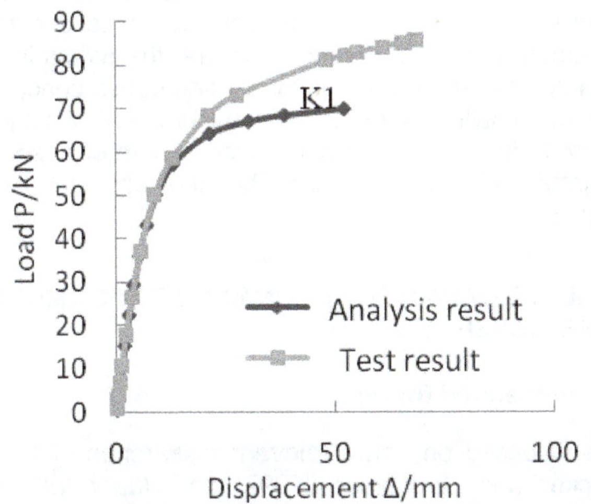

Figure 10. Stress-strain curves of specimen K1.

models, referring to the test result, valued at E=2.05×10^5 N/mm^2, v=0.29, abide by Von-Mises yield criterion and associated flow rules, plastic models sizes, boundary conditions and means of applying load was kept the same as the tests have been done. Main tube node area was subdivided in the process of meshing; branch tube and gusset adopted intelligent mesh; number of main tube element is 3344, while total element of the model is 6762. Weld has not been simulated, as benefit effect and bad effect of residual stress on bearing capacity are basic equivalent. Model is shown in Figure 9. Bearing capacity was judged by ultimate deformation criterion, namely deformation of hot point reached 3% main tube diameter.

Comparison study of FEA to test result

In order to verify reliability of FEA result, comparison of calculation result to results of FEA and test is shown in Table 5. Result of FEA was a bit smaller than that of test in certain range. FEA result and test result of K1 are shown in Figure 10; stress and strain in FEA models well imitated actual process.

Major influential parameters

There are plenty of influential parameters to tube-gusset K-joint, including length, thickness and diameter of main tube; length and thickness of gusset; diameter, thickness and angle of branch tube, etc. Main work in this paper is bearing capacity of the joint under bending, considering major influential parameters which include: main tube diameter (D), main tube thickness (t), length of gusset (l). The selected value of parameters in analysis are listed in Table 4.

Main tube diameter (*D*) effect on node bearing capacity

Bearing capacity of FEA result with t=6 mm, l=380 mm and main tube diameter which varies from 152 to 219 is shown in Figure 11. The joint bending capacity decreases with the increment of main tube diameter. Bending stiffness decrease as a result of diameter-thickness ratio increase due to t remains unchanged with D increases; stress concentrate become more significant; specimens failure in a lower load is observed when larger plastic deformation occur in connection of gusset and main tube.

Main tube thickness (*t*) effect on node bearing capacity

Bearing capacity of FEA result with D=168, l=380 and main tube thickness varies from 6 to 16 as shown in Figure 12. Bearing capacity increases significantly with

Table 4. Parameters value of model.

Main tube diameter (*D*/mm)	Main tube thickness (*t*/mm)	Length of gusset (*l*/mm)
152	6	340
160	7	350
164	8	360
168	10	370
170	12	380
219	16	

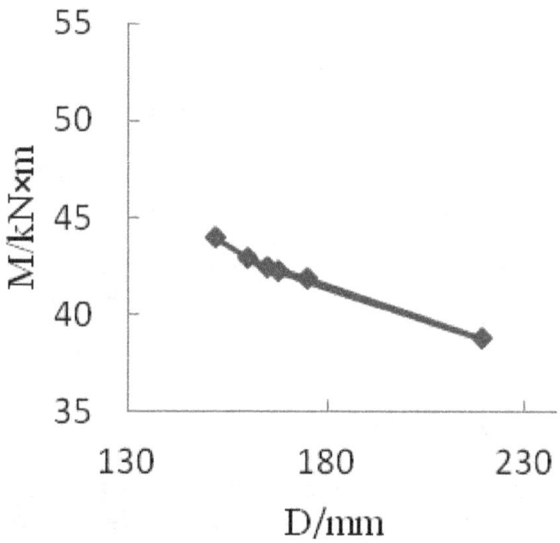

Figure 11. Variation of ultimate strength with main tube diameter.

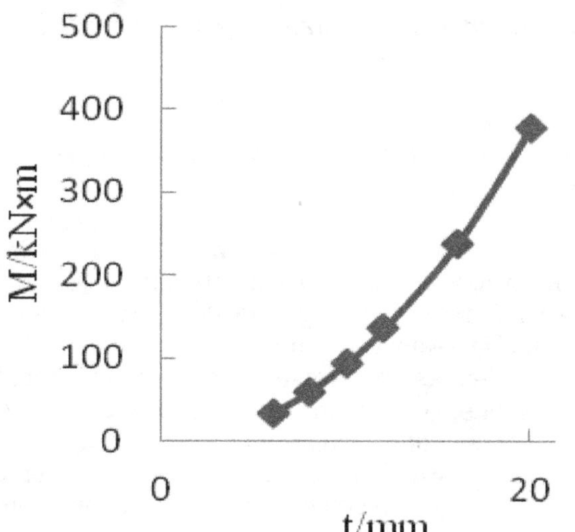

Figure 12. Variation of ultimate strength with main tube thickness.

the increment of main tube thickness; curve of bearing capacity present exponential increase with the increment

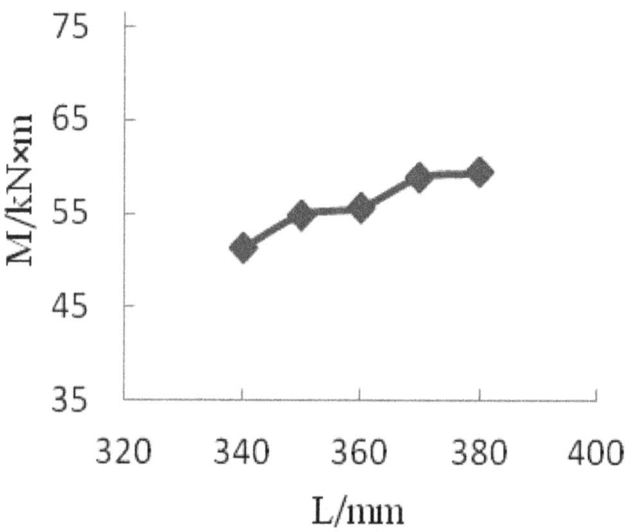

Figure 13. Variation of ultimate strength with length of gusset.

of main tube thickness.

Gusset length (*l*) effect on node bearing capacity

Bearing capacity of FEA result with *D*=168, *t*=8 and branch tube length varies from 340 to 380 as shown in Figure 13. Stiffness of node area increases due to the increment of gusset length; therefore, the stress in node area is distributed more uniformly and stress concentrate become smaller as a result of the increment of subjected length to load. Selected appropriate length of gusset has a great influence on node bearing capacity in real projects.

BENDING CAPACITY RECOMMENDED FORMULA OF TUBE-GUSSET K-JOINT

Recommended formula

This is based on some relevant researches of Canada (Packer and Henderson, 1992) and Japan (AIJ-SRC, 2001) in combination with the test and FEA result in this paper, and considering major influential parameters such

Table 5. Comparison between calculation formula and result of FEA and test of tube-gusset K-joint bending capacity.

Size of models (mm) [$D \times t \times L$]	Value of recommended formula (M_u / KN ×m)	Test result [M_{ut} / KN ×m]	Value of FEA [M_{ua}/ KN ×m]	(M_{ut}-M_u)/M_{ut} (%)	(M_{ua}-M_u)/M_{ua} (%)
152×6×380	27.51	30.68	28.74	10.33	4.28
168×6×380	26.51	29.37	27.15	9.73	2.36
219×6×380	24.31	24.31	25.96	10.59	10.59
168×8×380	47.14	52.36	49.47	9.96	4.70
168×10×380	73.66	80.69	76.92	8.71	4.23
168×8×340	40.59	/	42.03	/	3.42
168×8×360	43.82	/	46.27	/	5.29

as diameter and thickness of main tube and length of branch tube. Bending capacity calculation formulas of tube-gusset K-joint proposed according to sample and practical principle as follow:

$$M = A\left[1 + B\frac{l}{D}\right]t^2 l f_y \quad (1)$$

Among them, A and B are undetermined coefficients.

According to test and FEA results, fitting parameters (D, t, l in Table 3) and value of FEA result (M in Table 5) to formula (1), results in A and B; with formula (1) divided by 1.25 safety margin, then, bending capacity calculation formula of tube-gusset K-joint is attained as follow:

$$M = 5.3[1 + 0.246\frac{l}{D}]t^2 l f_y \quad (2)$$

Applicability analysis of the formula

In order to verify reliability of the formula proposed above, comparison of calculation result to results of FEA and test is shown in Table 5. Test and FEA data in Table 4 considering safety margin (1.25) due to it also considering formula (1), thus, unified the standard adopted in the process of comparison.

It can be seen in Table 5 that value of formula proposed above is closed, but a little smaller to result of FEA and test, which shows that formula (2) well reflects bending capacity of K-joint, and emphasis on safety. Recommend calculation formula has been adopted (Zhang, 2013) to contrast with FEA result of tube-gusset K-joint models (D=150 mm ~ 400 mm, t=6 mm ~ 40 mm, l=250 mm ~ 800 mm); the error is controlled within 5%; the applicability and reliability are verified again.

Conclusions

(1) FEA and test results show that the node bending capacity decreases with the increment of main tube diameter and increases exponentially with the increment of main tube thickness. To increase the thickness of main tube is a good means to improve capacity of this joint in reality. The article recommend ratio of D/t within 16 to 25 is reasonable, satisfy members lighter and not easily result in local bulking.

(2) The influence of gusset length to joint bending capacity is linear, and the impact on bearing capacity is small compared to other factors, simply making the length of gusset satisfying branch tube layout enough.

(3) Bending capacity of tube-gusset K-joint calculation formula proposed in this paper with universal applicability which can be used in strength-checking calculation in designing.

Conflict of Interests

The author(s) have not declared any conflict of interests.

ACKNOWLEDGEMENT

The authors of this paper acknowledge the financial support from the National Natural Science Foundation of China (Grant No. 51078016), Beijing Natural Foundation (Grant No. 8132023).

REFERENCES

Wang M (2011). Nonlinear analysis and research on dynamic stability of steel structure [M]. China Architect. Build. Press pp. 26-46.

Luo Y (2010). Ultimate strength research of the new tube-plate connection [D]. Beijing University of civil engineering and architecture,

Kim WB (2001). Ultimate strength of tube-gusset plate connections considering eccentricity J. Eng. Struct. 23(11):1418-1426.http://dx.doi.org/10.1016/S0141-0296(01)00050-5

Wang B, Hu N, Kurobane Y, Makinob Y, Lie ST (2000). Damage criterion and safety assessment approach to tubular joints. J. Eng. Struct. 22(5):424-434. http://dx.doi.org/10.1016/S0141-0296(98)00134-5

Van der Vegte GJ (1995). The static strength of uniplanar and multiplanar tubular T- and X- joints [D]. Delft University Press, pp. 48-59.

Packer JA, Henderson JE (1992). Design guide for hollow structural.

section connection [M]. Canadian Instit. Steel Constr. pp. 52-58
AIJ-SRC (2001). Architectural Institute of Japan. Recommendations for the design and fabrication of tubular structures in steel [s], Architectural Institute of Japan.
Zhang Y (2013). Experimental research and theoretical study of ultimate strength at node region of the tube-gusset plate connection. [D]. Beijing university of civil engineering and architecture, pp. 40-60.

7

Experimental study on axial compressive strength and elastic modulus of the clay and fly ash brick masonry

Freeda Christy C , Tensing D and Mercy Shanthi R

School of Civil Engineering, Karunya University, Coimbatore, TamilNadu, India, 641 114.

Brickwork is a composite material with bricks as the building units and the mortar as the jointing material. When this two element combined to form a brickwork unit, the properties of the materials influences the strength of the brickwork. Short prisms have been tested under axial compressive load using two types of masonry units: clay brick and flyash brick using flyash cement mortar. The brick masonry is reinforced with woven wire mesh at the alternate bed joint and tested for its axial strength and elastic modulus of the prisms specimens. They confirm that masonry prisms may be used for determining the basic compressive strength. Areas needing further investigation include the effect of moisture on the strength of brick masonry and the strength of eccentrically loaded brick work. In the present research, design strength was determined.

Key words: Prism, flyash, mortar, brick masonry, elastic modulus.

INTRODUCTION

Buildings that are constructed by using bricks have high compressive strength and durability against foreign disturbances. Structural components of the buildings that are built out of bricks also have multiple resistances against heat and sound. Due to the resistances, the masonry components also act as insulator within certain part of the building. Bricks also provide aesthetic surfacing to the brick work. In term of workability and economy, the usage of brick masonry makes the whole building construction easier, faster and cheaper. Masonry is a non-homogeneous material with two constitutive elements: bricks and mortar. The mortar has different functions inside the masonry, that is, it forms a layer to assemble the bricks and permits a uniform transmission of the internal forces. It is important that the mechanical properties of the masonry depend on the mechanical properties of the constitutive materials, as well as depend on the arrangement of the bricks inside the masonry.

REVIEW OF THE LITERATURE

The present research included a study on compressive strength of brick masonry subjected to axial loading. The study focuses on the effect of the masonry components with different types of bonding on compressive strength. Mohamad et al. (2005) carried out experimental tests on masonry prisms subjected to compression. The failure mechanism of masonry depends on the difference of elastic modulus between brick unit and mortar. The mortar governed the non-linear behavior of masonry. Oliveira et al. (2000) carried out the tests on prisms under cyclic loading and the stress-strain behaviour of the brick prisms showed a bilinear pre-peak behaviour. Gumaste et al. (2007) studied the properties of brick masonry using table moulded bricks and wire-cut bricks from India with various types of mortars. The table moulded brick masonry using lean mortar failed due to loss of bond between brick and mortar. The wire-cut brick masonry exhibited a better correlation between mortar strength and masonry strength. Mosalam et al. (2009) investigated the mechanical properties of masonry which was a heterogeneous composite in which brick units made from clay, compressed earth, stone or concrete were held together by mortar. Maurenbrecher (1980) described the effects of various factors on prism strength. The Canadian masonry design standard for buildings allow two methods of determining compressive strength of masonry, (i) tabular values based on unit strength and mortar type, (ii) axially loaded prisms such as two-course block-work stacks. Elizabeth and Eleni-Eva (2001) investigated the effect of deep rejointing behaviour of brick masonry subjected to axial compression. In all specimens, typical vertical cracks due to compression appeared both along the length and the width of prisms. In addition to those cracks, spalling of bricks was observed in prisms to which deep rejointing was applied. Hemant et al. (2007) developed a simple analytical equation by regression analysis of the experimental data to estimate the modulus of elasticity and to plot the stress–strain curves for masonry. A significant improvement in ductility of masonry was observed because of the presence of lime in the mortar without any considerable reduction in its compressive strength. This showed that lime in the mortar offered distinct structural advantages. The compressive strength of masonry was found to increase with the compressive strength of bricks and mortar. The trend was more prominent in case of masonry constructed with weaker mortar. Mojsilović (2005) derived masonry characteristics from compression tests. The masonry behaved more or less as linear-elastic material, in particular for working loads (loads up to 30% of the failure load); for higher loads, concrete and calcium-silicate block masonry exhibited nonlinear behaviour, while clay brick masonry remained linear-elastic up-to failure. Bryan and Mervyn (2004) captured the stress-strain characteristics of unconfined and confined clay brick masonry. Confinement plates dramatically improved the compressive strength of clay brick masonry. It was noted that confinement plates placed within the mortar bed joints restricted the lateral expansion of the joint and the differential expansion between the clay brick unit and the joint. The plates increased the ultimate strength by 40%. Jagadish et al. (2002) examined an additional feature known as containment reinforcement which controlled the post-cracking deflections and impart flexural ductility of masonry walls. Masonry buildings in mud mortar or lime mortar are prone to severe damage due to lack of bond strength. Masonry with cement mortar (which has higher bond strength) generally behaved better. Since the brittle nature of masonry building is the major cause for collapse of buildings and loss of lives, there is a need to introduce remedial measures in the construction of such buildings. In the construction industry, it is believed that the strength and durability of the structure mostly depend on the quality of bricks. But, the mortar joint also contribute great effect on the compressive strength and durability of the entire structure.

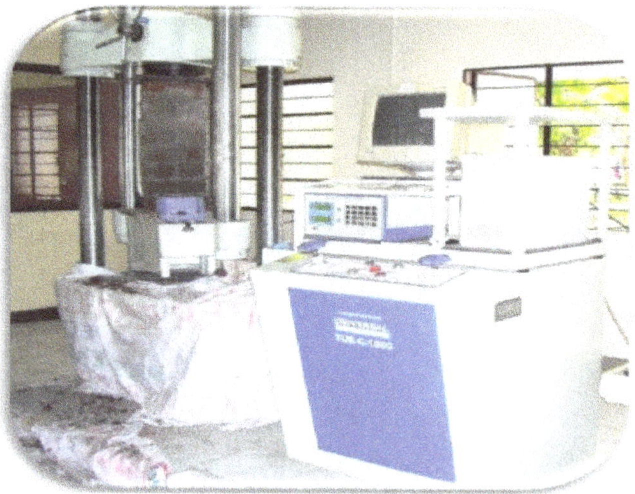

Figure 1. Axial load test setup with the data acquisition system.

EXPERIMENTAL INVESTIGATIONS

In engineered masonry, the compressive strength f_{pm} and the modulus of elasticity E_{pm} of the material are the two main components of the element. Compressive strength is important because it determines the bearing capacity of the element; the modulus of elasticity is important because it provides the estimate of deformation of the element under loading. The compressive strengths of masonry unit and mortars are two of the most tested properties for typical projects simply because the specimens are relatively easy and inexpensive to prepare when compared to testing for other properties. Axial compression tests of brick masonry prisms are used to determine the specified axial strength of the brick masonry f_a. The Bureau of Indian Standards IS: 1905 (1987) suggest to use brick masonry prisms having the dimensions of minimum 40 cm height with aspect ratios (h/t) between 2 to 5 in order to determine the axial strength of the brick masonry. Axial compression test was conducted on the brick masonry prisms with the aspect ratio (h/t) of 3.63 in 1:6 cement mortar with 0, 10 and 20% replacement of fine aggregate with fly ash. The plywood sheet having the thickness of 3 mm was placed on the top and the bottom of the masonry prism specimen which is loaded in-between the steel plate having the thickness of 25 mm. The above specimen was placed on the computerized universal testing machine and the axial compressive load was applied and the deformation was recorded by a sensor available in the computer based data acquisition system as shown in Figure 1.

Codes of practice on masonry design give the guidelines to assess the compressive strength of the brick masonry by considering compressive strength of the masonry unit, height of the masonry unit and the type of the mortar (cement (C): fly ash(F): fine aggregate(FA)). Five brick stack bonded masonry prism tests were performed under axial compression tests to obtain the basic compressive strength of the brick masonry. The brick masonry is also reinforced with the locally available galvanized hexagonal woven wire mesh (chicken wire mesh) at the alternate bed joint as shown in Figure 2.

The prism tests were conducted with clay brick and fly ash brick assemblages with different combinations of mortars (cement, flyash and fine aggregate) as indicated in Table 1.

Table 1. Specimen details for axial compressive strength of the brick masonry.

S. No	Designation of the prism	Types of brick	Mortar			Details of the reinforcement in the specimen
			C:	F:	FA	
1	CBP	Clay brick	1:	0:	6	Unreinforced clay brick prism
2	CBP10	Clay brick	1:	0.6:	5.4	Unreinforced clay brick prism
3	CBP20	Clay brick	1:	1.2:	4.8	Unreinforced clay brick prism
4	CBPR	Clay brick	1:	0:	6	Reinforced clay brick prism
5	CBP10R	Clay brick	1:	0.6:	5.4	Reinforced clay brick prism
6	CBP20R	Clay brick	1:	1.2:	4.8	Reinforced clay brick prism
7	FBP	Fly ash brick	1:	0:	6	Unreinforced fly ash brick prism
8	FBP10	Fly ash brick	1:	0.6:	5.4	Unreinforced fly ash brick prism
9	FBP20	Fly ash brick	1:	1.2:	4.8	Unreinforced fly ash brick prism
10	FBPR	Fly ash brick	1:	0:	6	Reinforced fly ash brick prism
11	FBP10R	Fly ash brick	1:	0.6:	5.4	Reinforced fly ash brick prism
12	FBP20R	Fly ash brick	1:	1.2:	4.8	Reinforced fly ash brick prism

Figure 2. Mesh at alternate bed course.

Stack bonded unreinforced clay brick prism (CBP) and fly ash brick prism (FBP) of size 230 × 110 × 420 mm were prepared using clay brick and fly ash brick of size 230 × 110 × 70 mm in 1:6 cement mortars with 0, 10 and 20% replacement of fine aggregate with fly ash (CBP10, CBP20, FBP10 and FBP20). The clay brick prism (CBPR, CBP10R and CBP20R) and fly ash brick prism (FBPR, FBP10R and FBP20R) were reinforced with hexagonal woven wire mesh at the alternate bed course as shown in Figure 2 and tested under compression. Mortar joint thickness of 10 to 12 mm was used for all the prism specimens. The specimens were subjected to an axial load up to failure of the test specimen.

The nature of the stresses developed in the masonry unit and the mortar when the brick masonry is subjected to compression greatly depends upon its relative elastic modulus (E). During compression of brick masonry prisms constructed with stiffer bricks, mortar of the bed joint may have a tendency to expand laterally more than the bricks because of lesser stiffness of mortar, Hemant (2007). However, the mortar is confined laterally at the brick mortar interface by the bricks because of the bond between them; therefore, shear stresses at the brick mortar interface result in an internal state of stress consisting of tri-axial compression in mortar and bilateral tension coupled with axial compression in brick as shown in Figure 3. Failure in brickwork occurs when the tensile stress in the brick reaches its ultimate tensile strength, Lenczer (1972). Under uni-axial compression, stack bonded brick masonry prism expands laterally in the plane perpendicular to the direction of loading causes vertical splitting as shown in Figure 4.

The compressive strength of the brick masonry with clay brick prism and fly ash brick prism in 1:6 cement mortar with 0, 10 and 20% replacement of fine aggregate with fly ash were shown in Figure 5. It was found that, in clay brick masonry prism in 1:6 cement mortars with partial replacement of fine aggregate with the fly ash resulted in increase in axial strength of the brick masonry.

From Figure 5, it was also found that the fly ash brick masonry in 1:6 cement mortar with 10% replacement of fine aggregate with fly ash resulted in higher load carrying capacity. From this, it was understood that the fly ash content in the mortar improves the interfacial zone microstructure as reported by Rafat (2003) and Chaid et al. (2004). Also the fly ash brick masonry has higher compressive strength than clay brick masonry.

RESULTS AND DISCUSSION

Elastic properties of clay brick masonry and fly ash brick masonry for unreinforced (CBP, CBP10 and CBP20) and reinforced (with wire mesh) brick masonry (RCBP, RCBP10 and RCBP20) were studied. Stress-strain characteristics of brick masonry were examined through prism test as per IS 1905 (1987) and ASTM C 67 (2009). The stress-strain behaviour of both unreinforced and reinforced clay brick masonry in 1:6 cement mortar with partial replacement of fine aggregate with flyash is indicated in Figure 6. From the stress – strain behaviour, the compressive strength of reinforced clay brick masonry in 1:6 cement mortar with 20% replacement of fine aggregate with fly ash exhibited higher compressive strength and the reinforced brick masonry yielded for more deformation.

The stress-strain curve was found to be linear until 1/3rd of the ultimate stress (f_a) after which cracks began to form in the mortar introducing the non-linearity as shown in Figure 6. The stress-strain curve of both unreinforced and reinforced fly ash brick masonry with three types of mortar is shown in Figure 7. Secant modulus of elasticity at 60% of the ultimate strength of the specimen is calculated from stress-strain curves.

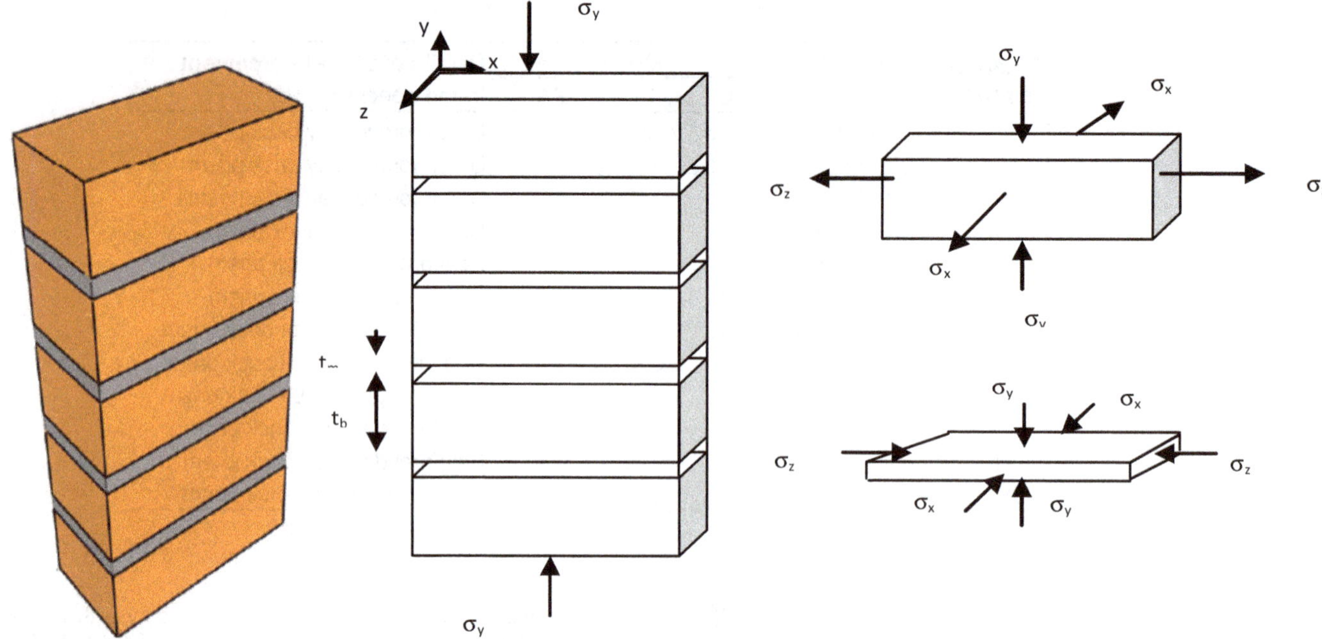

Figure 3. Stress distributions in the composite masonry.

Figure 4. Failure of clay brick prism and fly ash brick prism.

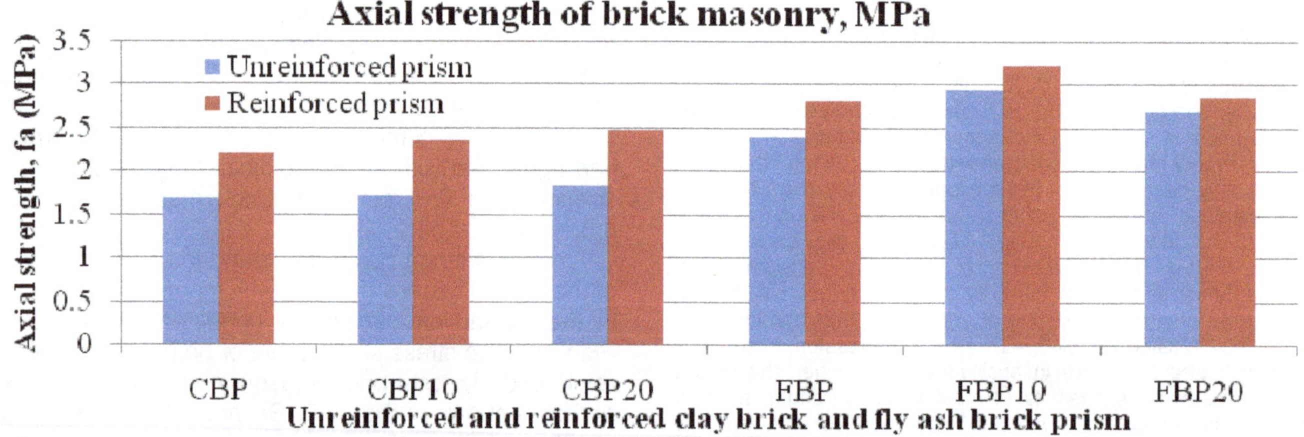

Figure 5. Comparison of axial strength of brick masonry.

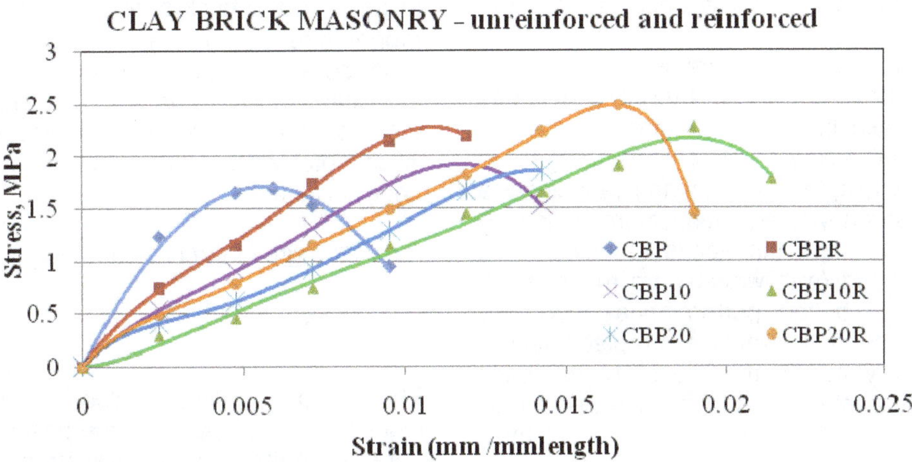

Figure 6. Stress-strain curve of clay brick masonry.

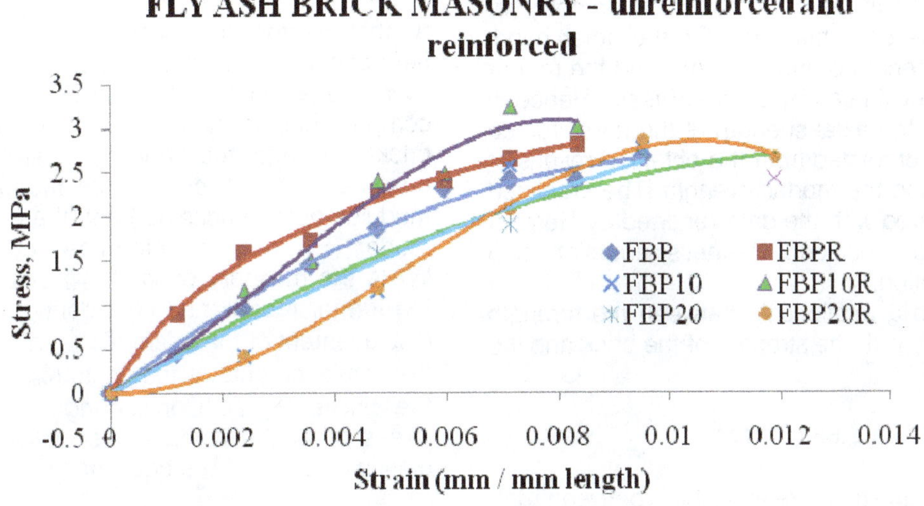

Figure 7. Stress-strain curve of fly ash brick masonry.

Figure 8. Splitting of brick masonry.

From the stress - strain curve, it was found that the compressive strength of reinforced fly ash brick masonry in 1:6 cement mortar with 10% replacement of fine aggregate with fly ash exhibited higher strength. Reda Taha and Shrive (2002) reported that the effect of fly ash on brick masonry attributed its pozzolanic activity, by which the pozzolans chemically convert the weak CH crystals to strong CSH fibrous gel. The pozzolanic activity depends mainly on the chemical composition and the fineness of the pozzolans. The pozzolanic reaction of fly ash was reported to have a significant effect on long-term strength development. The fly ash brick masonry prisms were damaged with visible vertical cracks (macro cracking) along the entire surface as shown in Figure 8. Lenczer (1972) and Mosalam (2009) reported that the mortar joints can develop lateral compression while brick develops lateral tension in brick masonry. However, the stress-strain curve of fly ash brick masonry was found to be non-linear.

The compressive strength of the unreinforced clay brick prism varies in the range of 1.69 to 1.85 MPa whereas the unreinforced fly ash brick prism varies from 2.4 to 2.68 MPa. With partial replacement of fine aggregate in the mortar with the fly ash, the load carrying capacity was increased and the strain yielded much more indicating ductility in the mortar. From the above results, it was found that the reinforced (with woven wire mesh) brick masonry resulted in better performance than the unreinforced brick masonry. However, the replacement of fine aggregate with fly ash in the mortar of the brick masonry reduces the cost of the construction in addition to the enhancement of load carrying capacity of the brick masonry.

Analysis and design of the masonry buildings with masonry require material properties like axial strength of the brick masonry. It is not always feasible to conduct the compression test on masonry prisms to get the actual prism strength, which is the basic structural property for the designing of the brick masonry. On the other hand, the compressive strength of the brick (fb) and the mortar (fm) can easily be evaluated by standard tests. Hence in this research work, the axial strength of the unreinforced brick masonry was predicted from the obtained results of the brick strength and the mortar strength. The predicted values were compared with the data reported by Hemant et al. (2007) which included Bennet's equation and Dayaratnam's equation.

Bennet has given a relationship between the strength of the brick masonry with the strength of the brick and the mortar as,

Masonry strength, $f_{m'} = 0.63\, f_b^{0.49}\, f_m^{\ 0.32}$ (1)

Dayaratnam has given a relationship between the strength of brick masonry with the strength of the brick and the mortar as,

Masonry strength, $f_{pm} = 0.275\, f_b^{0.5}\, f_m^{\ 0.5}$ (2)

The equation proposed by Bennet and Dayaratnam gives almost equal weight age to the compressive strength of the brick and the mortar. Hemant et al. (2007) reported that, in such cases, the errors in the estimation of masonry compressive strength may be higher.

The generalized equation is proposed for estimating the axial strength of the brick masonry as,

Axial strength of the brick masonry, $f_a = k\, f_b^{\alpha} f_m \beta$ (3)

Where,

k, α and β = Constants

f_b = Strength of brick in MPa

f_m = Strength of mortar in MPa

In general, the brick strength is usually greater than the mortar strength, hence 'α' must be greater than 'β' as reported as reported by Hemant et al. (2007). However, the axial strength of brick masonry is calculated based on the experimental results of the present study, the value of 'k' is obtained by least square methods of regression analysis as 0.35.

Axial strength of the brick masonry,

$$f_a = 0.35 \times f_b^{\ 0.65} \times f_m^{\ 0.25} \quad (4)$$

The average strength value of the mortar was much higher when compared to the prism masonry specimens, but near to the average strength of the bricks. Based on Eurocode 6, modulus of elasticity of masonry is derived as,

Elastic modulus of brick prism, $E_{pm} = K_E f_{bc}$ (5)

At the macroscopic scale, the assumption is that the heterogeneous masonry material can be represented as a homogeneous material. For the masonry under compression, the nature of the stresses developed in the brick unit and the mortar depend upon the relative modulus of the brick and the mortar. Thus, the elastic modulus of the bricks (E_b) and the elastic modulus of the mortar (E_m) can be determined by the standard tests. While the masonry prism (E_{pm}) can be calculated using the former moduli and considering that the total vertical displacement of the prism (δ_{prism}) with mesh is the sum of the displacements of the joints (δ_{mortar}) with mesh and of the bricks (δ_{brick}). Considering the same compressive stress in all the components of the brick masonry, the elastic modulus of the brick masonry was derived. Hence,

Elastic modulus of brick masonry,

$$E_{pm} = p\left[\frac{1+\gamma_t}{1+\dfrac{\gamma_t}{\gamma_{mb}}}\right]E_b \quad (6)$$

$$\gamma_t = \frac{t_m}{t_b}\ ; \qquad \gamma_{mb} = \frac{E_m}{E_b}\ ;$$

$$p = \frac{E_b}{E_{mesh}}$$

γ_t – Thickness ratio between the mortar and the brick

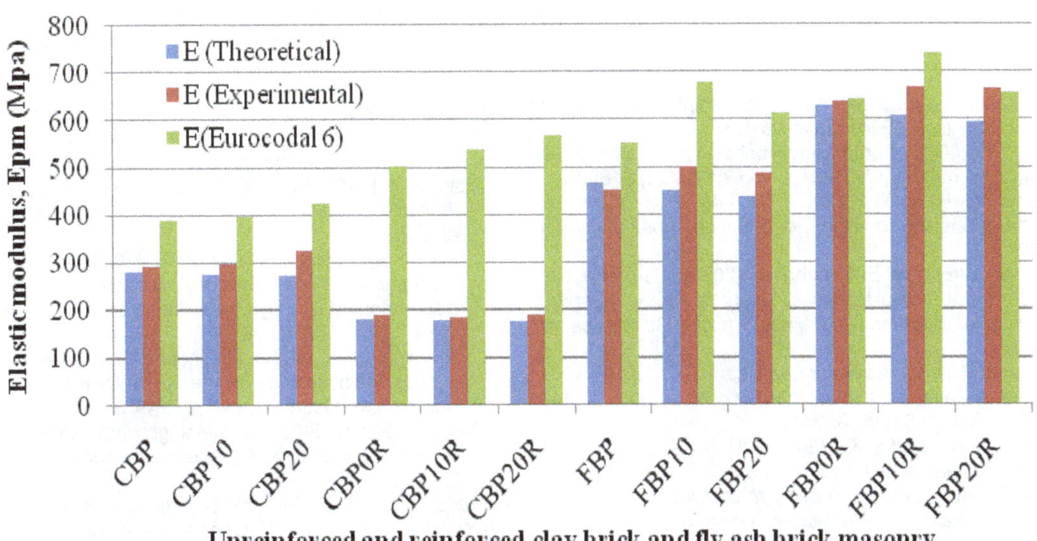

Figure 9. Equivalent homogenized elastic modulus of brick masonry.

γ_{mb} – Elastic modulus ratio between the mortar and the brick
p – Reinforcement constant as elastic modulus ratio between the brick and the mesh
For, Unreinforced clay brick masonry, CBP, p = 1.0
Reinforced clay brick masonry, CBPR, p = 0.65
Unreinforced fly ash brick masonry, FBP, p = 1.0
Reinforced fly ash brick masonry, FBPR, p = 1.35
t_m – Thickness of the mortar in mm
t_b – Thickness of the brick in mm
E_m – Elastic modulus of the mortar in MPa
E_b – Elastic modulus of the brick in MPa
E_{mesh} – Elastic modulus of the woven wire mesh in MPa

The comparison of the equivalent homogenized elastic modulus of brick masonry with the observed experimental elastic modulus of the brick masonry is shown in Figure 9.

From the Figure 9, it was understood that the existence of horizontal mesh reinforcement distributed the strain at the region of reinforced clay brick masonry which resulted in reduction of the elastic modulus of the reinforced clay brick masonry. Further, the mesh reinforcement effectively influenced the distribution of the total strain through the clay brick masonry. It was noted that the effect of mesh reinforcement on the strain distribution of the fly ash brick masonry was found to be less. After observing the failure of the prism in the case of clay brick masonry with mesh reinforcement, the composite action was found to be less effective whereas in the case of fly ash brick masonry with mesh reinforcement, the composite action was very effective. The observed experimental elastic modulus of the brick masonry was found to be with the average variation of 5.9% with the theoretical elastic modulus of the brick masonry. Areas needing further investigation include the effect of moisture on the strength of brick masonry and the strength of eccentrically loaded brick work.

Conclusions

(i) The mortar with the ratio of 1:6 cement mortar with 20% replacement of fine aggregate with fly ash exhibited a higher compressive strength than the control mix after 28 days of curing.
(ii) The compressive strength of unreinforced fly ash brick masonry was 34% more than the unreinforced clay brick masonry. The reinforced fly ash brick masonry was 20.7% more than the reinforced clay brick masonry.
(iii) The introduction of wire mesh in the clay brick masonry resulted in an increase of load carrying capacity by 25%, while the introduction of mesh in fly ash brick masonry resulted in an increase of load carrying capacity by 10% as the strength of the fly ash brick contributed more in the brick masonry strength.
(iv) Incorporation of fly ash in the brick masonry results in the reaction of pozzolanas with the calcium hydrate which forms produced strong calcium silicate hydrates, thus enhancing the bond strength of the brick masonry with the modification of the microstructure of the mortar-brick unit interface.
(v) The elastic modulus of the brick masonry (Epm) was determined with the prism strength (fpm).
(vi) The equivalent homogenized elastic property of the masonry was derived with the elastic properties of brick,

mortar and the reinforcement.

REFERENCES

ASTM C- 67-09 (2009). "Standard test method of sampling and testing brick and structural clay tile." ASTM Standard, USA.

Bryan DE, Mervyn JK (2004). "Compressive behaviour of unconfined and confined clay brick masonry." J. Struct. Eng. @ ASCE, p. 650.

Chaid R, Jauberthie R, Rendell F (2004). "Influence of a natural pozzolana on the properties of high performance mortar." Indian Concrete J. p. 22.

Elizabeth N Vintzileou, Eleni-Eva E Toumbakari (2001). "The effect of deep rejointing on the compressive strength of brick masonry historical constructions." Lourenço P B, Roca P (Eds), Guimarães, p. 995.

Gumaste KS, Nanjunda Rao KS, Venkatarama Reddy BV, Jagadish KS (2007). "Strength and elasticity of brick masonry prisms and wallettes under compression." Mater. Struct. 40:241.

Hemant BK, Durgesh CR, Sudhir KJ (2007). "Uniaxial compressive stress–strain model for clay brick masonry." Curr. Sci. 92(4):25.

IS 1905- 1987 (1987). "Indian standard code of practice for structural use of un-reinforced masonry." Bureau of Indian Standards, New Delhi, India.

Jagadish KS, Raghunath S, Nanjunda RKS (2002). "Shock table studies on masonry building model with containment reinforcement." J. Struct. Eng. 29:9.

Lenczer D (1972). "Elements of load bearing brickwork." Pergamon Press, Oxford.

Maurenbrecher AHP (1980). "Effect of test procedures on compressive strength of masonry prisms." Proceedings of the Second Canadian Masonry Symposium, held in Ottawa 9 -11, p. 119.

Mohamad G, Lourenço PB, Roman HR (2005). "Mechanical behavior assessment of concrete block masonry prisms under compression", proceedings of International Conference on Concrete for Structures (INCOS 05), Coimbra, p. 261.

Mojsilović N (2005). "A discussion of masonry characteristics derived from compression tests." 10th Canadian Masonry Symposium, Banff, Alberta.

Mosalam K, Glascoe L, Bernier J (2009). "Mechanical properties of unreinforced brick masonry section -1." Documented to U.S. Department of Energy by Lawrence Livermore National Laboratory.

Oliveira DV, Lourenço PB, Roca P (2000). "Experimental characterization of the behaviour of brick masonry subjected to cyclic loading." Proceedings of the 12th International Brick/Block Masonry Conference, Madrid, Spain, p. 2119.

Rafat S (2003). "Effect of fine aggregate replacement with class F fly ash on the mechanical properties of concrete." Cem. Concrete Res. 33:539.

Reda TMM, Shrive NG (2002). "The use of pozzolans to improve bond and bond strength." 9th Canadian Masonry Symposium.

8

Effect of heavy metal and magnesium sulfate on properties of high strength cement mortar

B. Madhusudana Reddy* and I. V. Ramana Reddy

Department of Civil Engineering, Sri Venkateswara University (S. V. U.) College of Engineering, Sri Venkateswara (S. V.) University, Tirupati -517502, Andhra Pradesh, India.

The effect of lead (Pb) present in mixing water on compressive strength, setting times, soundness and magnesium sulfate attack on high strength cement mortar was experimentally evaluated. Cement mortar specimens were cast using deionised water and lead (Pb) spiked deionized water for reference and the test specimens as mixing waters respectively. On comparison with reference specimens, at higher concentrations of lead in mixing water, test samples had shown considerable loss of strength, and also their setting times had significantly increased. However, at 2000 mg/L concentration of lead (Pb), the compressive strength marginally increased. Apart from that when reference specimens and test specimens were immersed in various concentrations of magnesium sulfate solution at different immersed ages, the loss in compressive strength was found to have been slightly less in test specimens than that in reference specimens. X-ray diffraction (XRD) technique was employed to find out main compounds.

Key words: Cement mortar, lead (Pb), silica fume (SF), magnesium sulfate, superplasticiser (SP).

INTRODUCTION

It is a well known fact that quality and quantity of mixing water in fresh cement mortar and concrete are important in determining properties of cement mortar and concrete. Water has both beneficial and detrimental effect on concrete (Neville, 2000). Generally, if water is potable, it is also suitable as mixing water for concrete. However, non-potable water, such as treated industrial wastewater, which contains heavy metals (Hg, Cu, Ni, Zn, Cr, Pb, Cd, and Fe), was satisfactorily used in making cement mortar (Babu et al., 2009). The water quality with respective to impurities could be made less stringent for curing if no chemicals that harm concrete remain on the surface after evaporation. Even greater amounts of impurities could be permitted in water if it was used for washing concrete equipment (Hooton, 1993; Steinour and Harold, 1990). Even though a huge volume of research was carried out to understand the interaction of different ingredients of concrete such as cement, aggregate, chemical and mineral admixtures, considerable research work was not carried out in the role of mixing water on concrete.

However, few researchers (Babu, 2009) had worked on use of treated and partially treated wastewaters, but a particular constituent effect and its maximum permissible limit in mixing water was not reported. For this reason, a guideline based on careful scrutiny on tolerable limit of a specific constituent in mixing water is highly needed.

RESEARCH SIGNIFICANCE

This paper examines the maximum permissible limit of lead (Pb) present in mixing water for cement mortar and effect of magnesium sulfate in various concentrations on the same specimens. Babu et al. (2007) reported Pb, Zn, hg, Cu, Ni, Fe and Cr were friendly with cement mortar up to 600 mg/L. Mindess and Young (1981) reported the tolerable limit of Cu, Pb, Zn, Mn was 500 mg/L. Tay and Yip (1987) showed that the use of reclaimed wastewater for concrete mixing did not have any adverse effect on concrete. Ramana et al. (2006) reported inexplicit results both positive and negative with biologically contaminated water. Cebeci and Saatci found that biologically treated domestic wastewater was indistinguishable from distilled water when used as mixing water in concrete (Cebeci

*Corresponding author. E-mail: srinamasa@gmail.com.

Table 1. Physical properties of cement.

S/N	Property	Result
1.	Specific gravity	3.17
2.	Fineness	225 m^2/kg
3.	Initial setting time	114 min
4.	Final setting time	224 min
5.	Compressive strength	MPa
a)	3 days	33
b)	7days	43
c)	28 days	54
6.	Soundness	0.5 mm

Table 2. Chemical composition of cement.

S/N	Oxide composition	Percent
1.	CaO	64.59
2.	SiO_2	23.95
3.	Al_2O_3	6.89
4.	Fe_2O_3	3.85
5.	MgO	0.78
6.	SO_3	1.06
7.	K_2O	0.46
8.	N_2O	0.12
9.	Loss on ignition	1.2
10.	Insoluble residue	0.35

and Saatci, 1989). However, heavy metals such as Cu, Zn, Pb, caused a retardation of the early hydration and strength development of cement mortar (Tashiro, 1980). These metals delayed setting and early strength development (Barth, 1990).

Even though biologically treated sewage and reclaimed wastewater are reported to be usable in concrete for mixing, there is very little information on the maximum permissible limit of heavy metals in mixing water and cement mortar made with metal spiked deionised water is exposed to sulfate environment. Hence, this investigation was carried out to understand the effect of lead in mixing water on compressive strength, setting times, soundness of high strength cement mortar and to evaluate magnesium sulfate attack on the same cement mortar.

METHODOLOGY

Materials

Cement

53-grade ordinary Portland cement conforming to IS: 12269-1987 was used. The physical properties and chemical composition of major compounds of cement are given in Tables 1 and 2 respectively.

Sand

Ennore sand conforming to IS: 650-1966 was used. Physical properties are given in Table 3. The cement to fine aggregate ratio was maintained at 1:3 (by weight) in the mortar mixes.

Superplasticiser

Commercial superplasticiser was used. Based on a number of trials, 0.8% (by weight of cement) was arrived.

Water

Deionised water was used in reference specimens and lead spiked deionised water in different concentrations was used in test specimens.

Silica fume

Silica fume was used in the present investigation. 9% of the cement was replaced by silica fume, where maximum compressive strength was achieved. The chemical composition is given in Table 4.

Sulfate

Magnesium sulfate was used in different concentrations in order to

Table 3. Physical properties of sand.

S/N	Property	Result
1.	Specific gravity	2.65
2.	Bulk density	15.84 kN/m^3
3.	Grading	percent
4.	Passing 2 mm sieve	100%
5.	Passing 90 µ sieve	100%
6.	Particle passing 2 mm and retained 1 mm	33.33%
7.	Particle passing 1 mm and retained 500 µ	33.33%
8.	Particle passing 500 µ and retained 90 µ	33.33%

Table 4. Chemical composition of silica fume.

S/N	Oxide composition	Percent
1.	CaO	0.5
2.	SiO_2	92.3
3.	Al_2O_3	2.7
4.	Fe_2O_3	1.4
5.	MgO	0.3
6.	SO_3	0.1
7.	K_2O	0.1
8.	N_2O	0.1
9.	Loss on ignition	1.8

Table 5. Physical properties of reference cement mortar.

S/N	Property	Result
1.	Initial setting time	160 min
2.	Final setting time	272 min
3.	Compressive strength	MPa
a)	3 days	49
b)	7 days	59
c)	28 days	75
d)	90 days	77
e)	180 days	81
f)	365 days	82
4.	Soundness	0.7 mm

study the magnesium sulfate attack on reference and test specimens.

Procedure

Lead was introduced into the deionised water in predetermined concentrations such as 10, 50, 100, 500, 1000, 2000, 3000, 4000, 5000 mg/L. The concentrations were arrived based on the literature. After a number of combinations tried, a combination (cement + 9% SF + 0.8% SP) was fixed for reference specimens where maximum compressive strength was attained. The physical properties of reference specimens are given in Table 5.

Nine series of specimens were cast for test. The test specimens were cast with (cement + 9% SF + 0.8% SP + lead). Lead concentrations of 10, 50, 100, 500, 1000, 2000, 3000, 4000, and 5000 mg/L were introduced into the deionised water used as mixing water for test specimens. The quantities of cement, Ennore sand and mixing waters for each specimen were 200, 600 g and (P/4) + 3, where P denotes the percentage of mixing water required to produce a paste of standard consistence. Initial and final setting times were found out by Vicat's apparatus. Le-Chatelier equipment was used to find soundness of reference and test specimens. The reference and test specimens were prepared using standard metallic cube mould of size 7.06 x 7.06 x 7.06 cm for compressive strength of mortar. The blended cement to sand ratio was 1: 3 by weight throughout the tests. The compressive strength of reference and test specimens was studied at different ages, that is, 3, 7, 28, 90, 180 and 365 days. The compacted specimens in mould were maintained at a controlled temperature of 27 ± 2° and 90% relative humidity for 24 h by keeping the moulds under gunny bags wetted by the deioned water and then demolded. After demolding, the specimens were cured in deionised water for 27 days. From the experiments of setting and soundness tests, an average of three values was used to compare the results of the reference specimens. In the case of compressive strength tests, three test specimens were compared with three reference specimens.

In order to study magnesium sulfate attack, After 28 days curing, the reference mortar specimens were immersed in five plastic tanks. Magnesium sulfate concentrations maintained in the tanks were 1, 1.5, 2, 2.5 and 4% respectively. These concentrations represent very severe sulfate exposure conditions according to ACI 318-99, that are widely prevalent in many parts of the world Al-Amoudi et al., 1992, 1994).

The exposure magnesium sulfate solutions were prepared by dissolving magnesium sulfate in deionised water. Fifteen specimens were immersed in each concentration for up to 12 months. The concentration of the solution was checked periodically and the solution was changed every 4 months. The previous procedure was adopted for test specimens. Three mortar specimens representing similar compositions were retrieved from the magnesium sulfate solutions after 1, 3, 6, 9, 12 months of immersion. The effect of magnesium sulfate concentrations on the performance of reference and test specimens was evaluated by measuring the reduction in compressive strength. The reduction in compressive strength of reference and test specimens immersed in magnesium sulfate solutions were compared with that of reference specimens cured in deionised water.

Powdered X- ray diffraction studies

Powder X- ray diffraction (XRD) is one of the commonly used techniques for investigation of crystalline compounds in hydrated

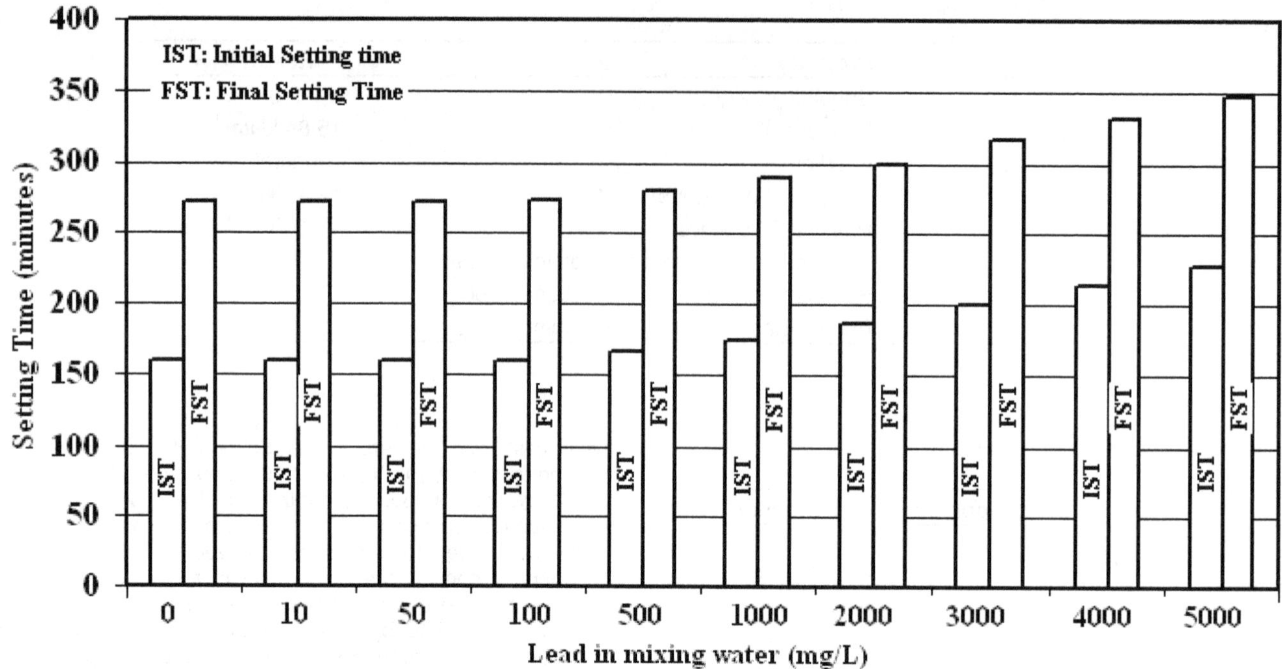

Figure 1. Effect of lead on setting times of blended cement.

cement paste (Knudsen, 1976). The reference sample (Cement + 9% SF + 0.8% SP + deionised water) and test sample (Cement + 9% SF + 0.8% SP + lead spiked (2000 mg/L) deionised water) for XRD were ground to a fine powder and a flat specimen was prepared on a glass surface using an adhesive. The diffracted intensities were recorded with powdered diffractometer using monochromatic copper Kα radiation.

RESULTS AND DISCUSSION

Setting times

Figure 1 shows the effect of deioniesd water (Reference) and lead spiked deionised water (Test) on initial and final setting times. The initial and final setting times increased as the concentration of lead increased.

At a maximum concentration of 5000 mg/L, the test samples had 67 min increase in the initial setting time and 74 min increase in the final setting time, compared to the reference specimens. At the opted concentrations (10, 50, 100, 500, 1000, 2000, 3000, 4000, 5000 mg/L), the increases in initial setting times observed were 0, 0, 1, 7, 14, 26, 40, 54, and 67 min respectively. The corresponding increases in the final setting times were 0, 0, 1, 8, 16, 28, 44, 59, and 74 min.

Compressive strength

Figure 2 shows the change in compressive strength of test samples due to the use of lead spiked deionised water. The strength developments in reference and test specimens were the same for concentration of up to 100 mg/L. For the concentration of 500 mg/L, the observed decrease in compressive strength at 3 days was 2.4%, compared to reference specimens. After 3 days, compressive strength developments in reference and test samples were the same. For the concentration of 1000 mg/L the decrease in compressive strength at 3 days was 3.06%, but at 7, 28, 90, 180, and 365 days a slight increase in compressive strength was observed by 0, 0.66, 1.03, 0.61, and 0.61%, respectively, compared with reference specimens.

For the concentration of 2000 mg/L the decrease in compressive strength at an early age (3 and 7 days) was by 8.16 and 3.38%, but from 28, 90, 180, 365 days a marginal increase in compressive strength was noticed of 2.0, 2.33, 1.85, and 1.82% respectively, compared with reference specimens. However, the rate of decrease in compressive strength increased from 3000 to 5000 mg/L. At 5000 mg/L, the decrease in compressive strength was by 32.65, 28.81, 22.00, 23.37, 25.92, 29.97% for 3, 7, 28, 90, 180, 365 days respectively. Eventually, compressive strength results reveal that at 2000 mg/L concentration, maximum increase in compressive strength is observed.

Soundness

The Le-Chatelier's test result for expansion measurement in cement should not be more than 10 mm. The effect of deionised water (Reference) and lead spiked deionised

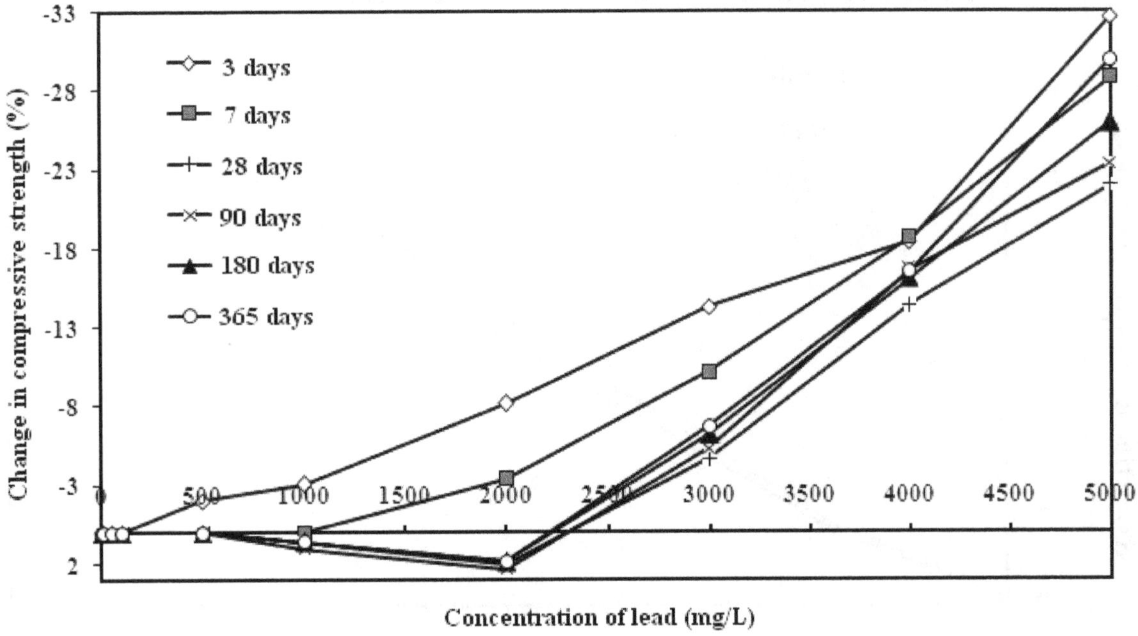

Figure 2. Change in compressive strength of mortal specimens made with various concentrations of lead at different age.

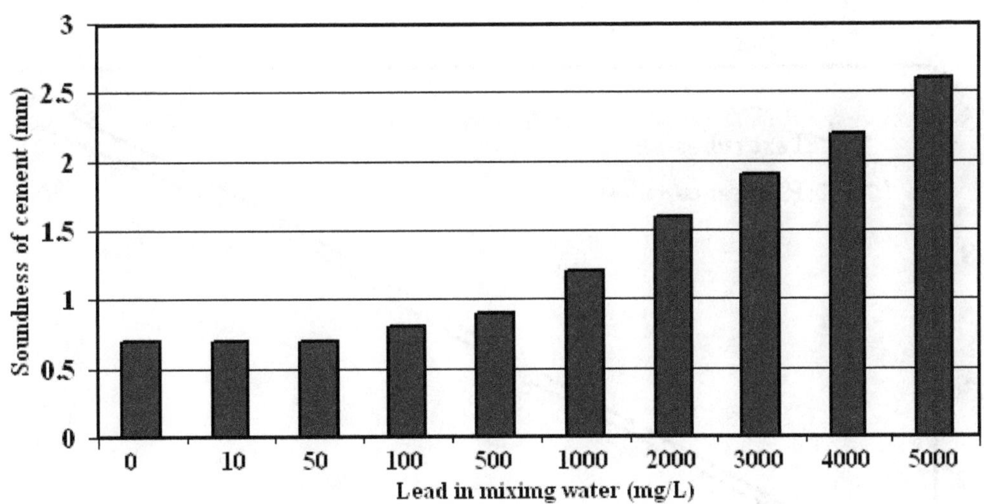

Figure 3. Effect of lead on soundness of blended cement.

water (Test) on soundness is shown in Figure 3. The expansions measured were 0.7, 0.7, 0.7, 0.8, 0.9, 1.2, 1.6, 1.9, 2.2 and 2.6 mm for 0, 10, 50, 100, 500, 1000, 2000, 3000, 4000 and 5000 mg/L concentrations respectively. Since all measured values were less than 10 mm, all the samples are considered sound.

Sulfate attac

Figures 4 to 8 show reference and test specimens immersed in 1, 1.5, 2.0, 2.5 and 4% magnesium sulfate solutions for 12 months. The decrease in compressive strength, with increase in concentration and period of exposure, was noted in reference and test specimens, compared with reference specimens immersed in deionised water. The decrease in compressive strength was similar in reference and test specimens for any concentration of magnesium sulfate solution. The decrease in compressive strength was insignificantly less in test specimens, compared with reference specimens. However, at 12 months, the decrease in compressive

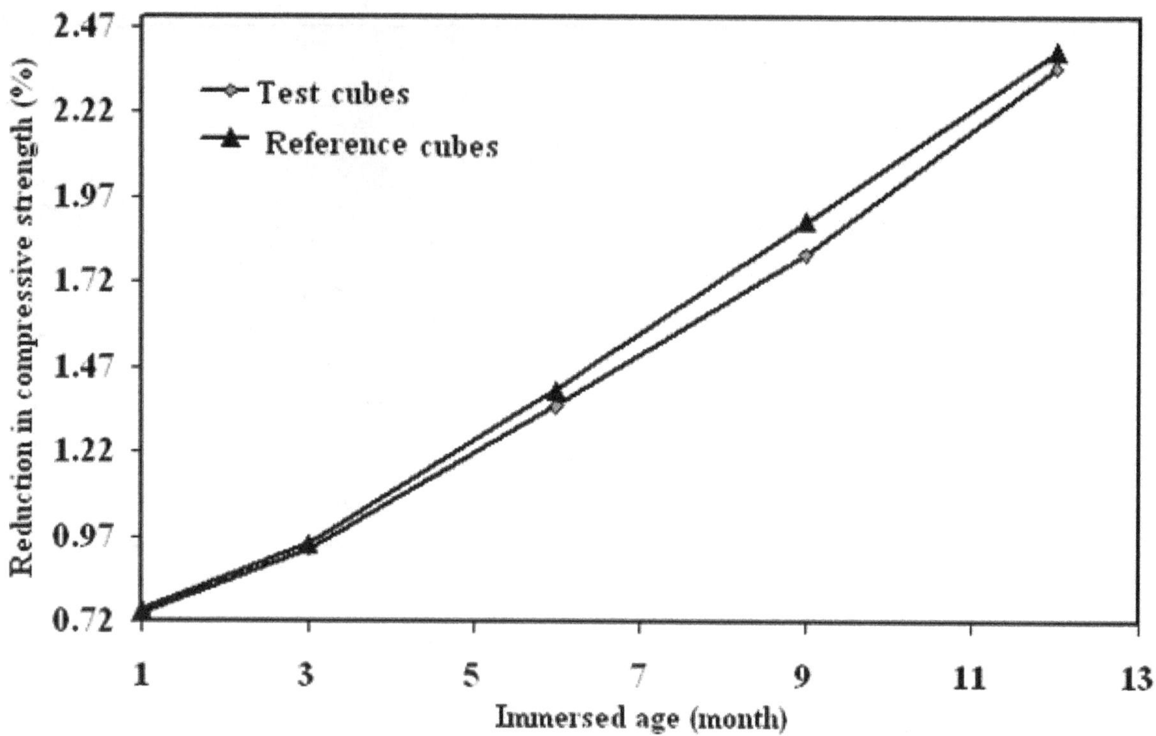

Figure 4. Reduction in compressive strength of mortal specimens immersed in 1% magnesium sulfate solution.

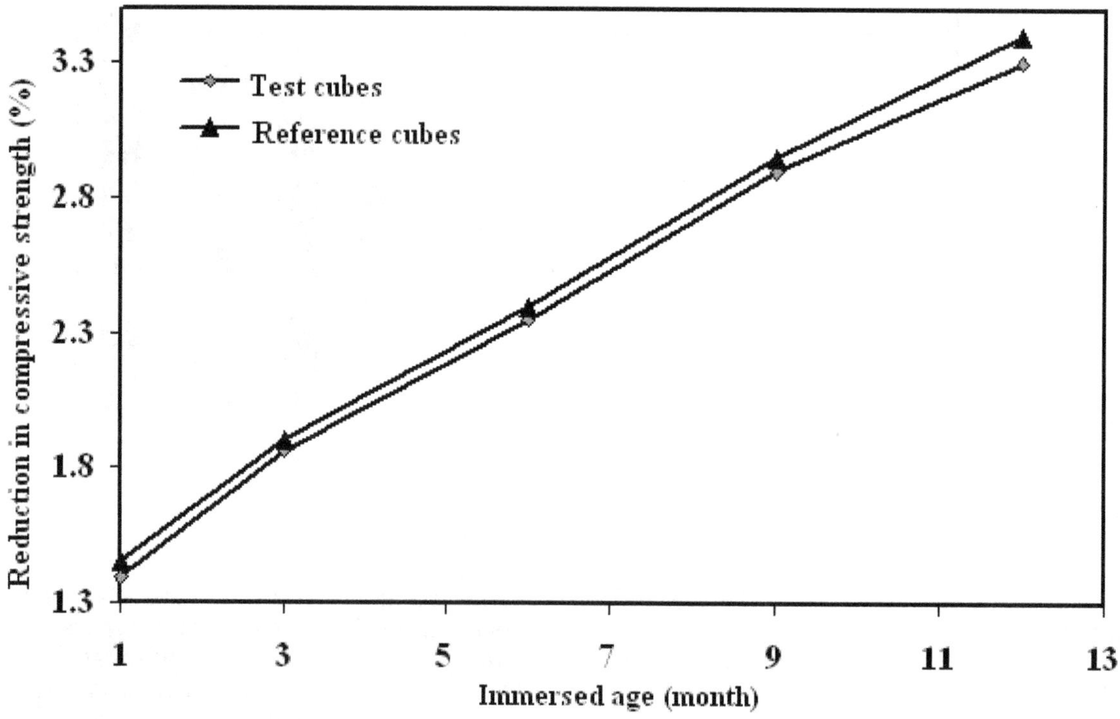

Figure 5. Reduction in compressive strength of mortal specimens immersed in 1.5% magnesium sulfate solution.

strength was by 2.4, 3.4, 4.7, 6.85, 10.5% in reference specimens and 2.35, 3.3, 4.59, 6.8, 10.2% in test specimens for 1, 1.5, 2.0, 2.5, and 4% concentration of magnesium sulfate solution respectively.

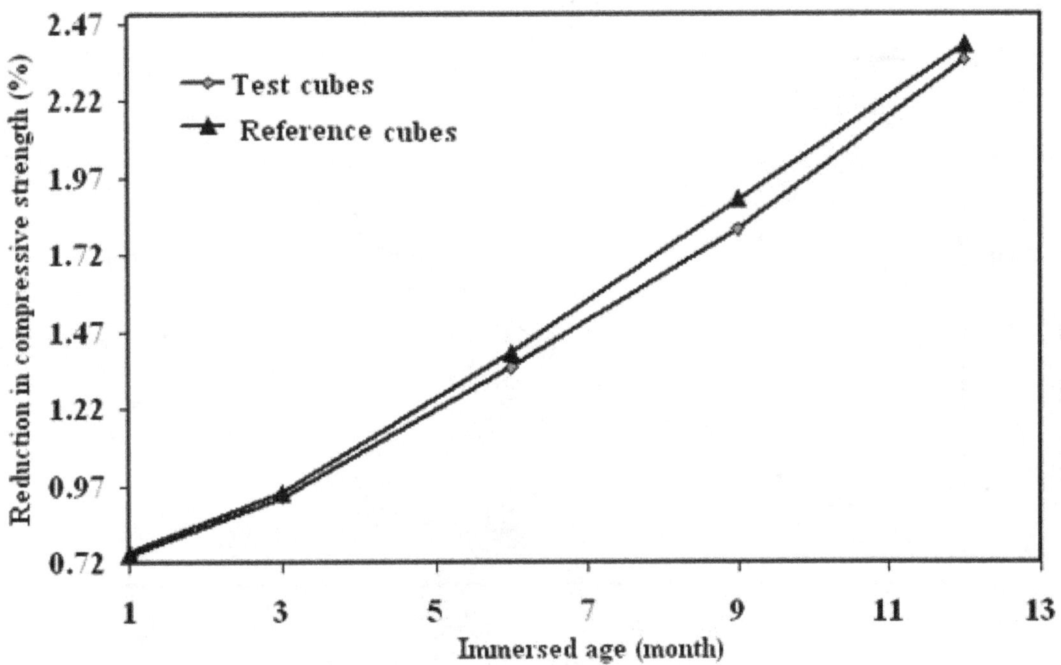

Figure 6. Reduction in compressive strength of mortal specimens immersed in 2.0% magnesium sulfate solution.

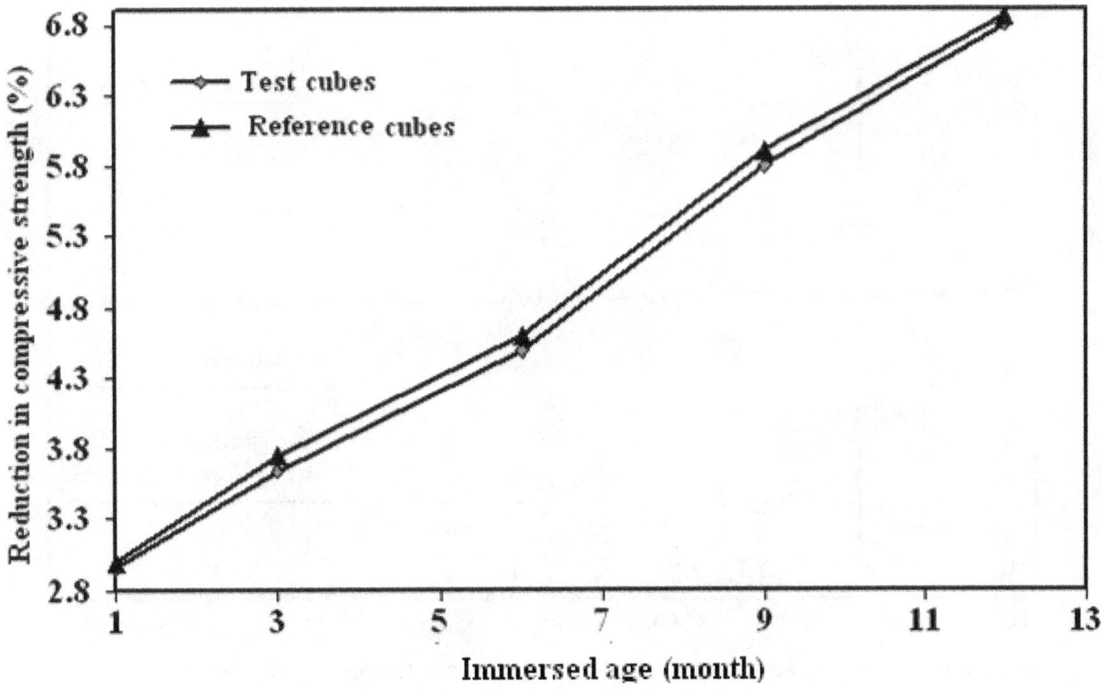

Figure 7. Reduction in compressive strength of mortal specimens immersed in 2.5% magnesium sulfate solution.

XRD analysis of blended cement paste made with deionised water and lead spiked deionised water

Figure 9 shows that powder X- ray diffraction patterns of reference and test samples. Both reference and test sample (2000 mg/L) were cured for 28 days before being subjected to XRD technique. After employing XRD for test sample, some new compounds were found along

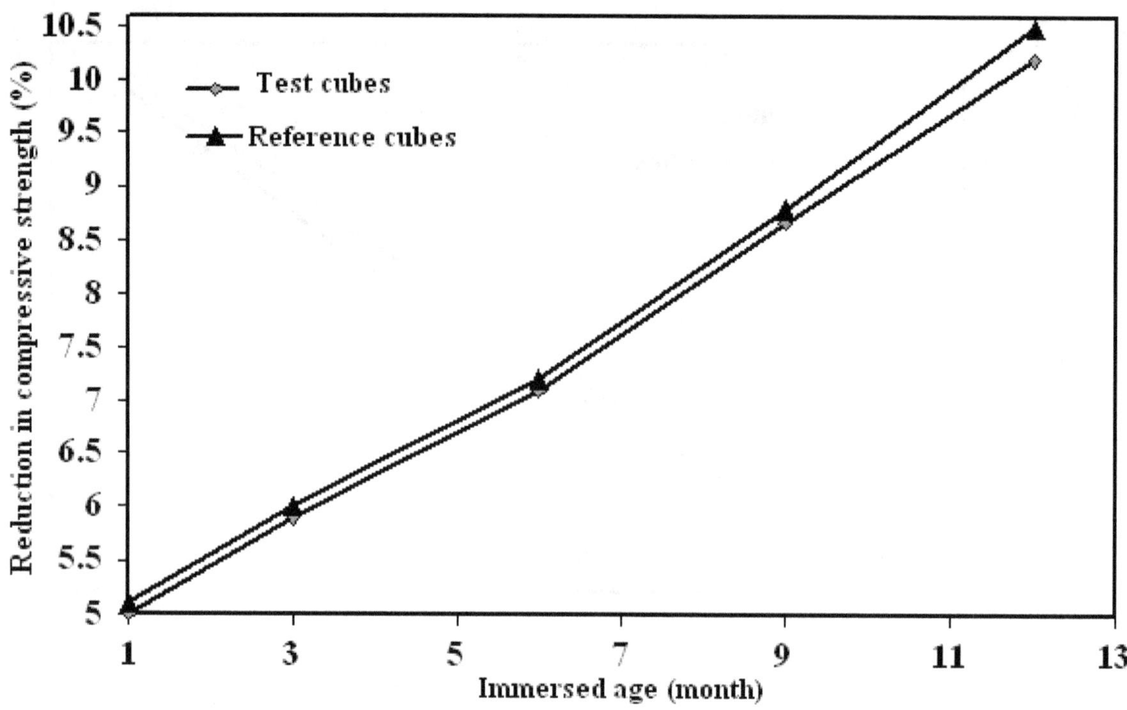

Figure 8. Reduction in compressive strength of mortal specimens immersed in 4% magnesium sulfate solution.

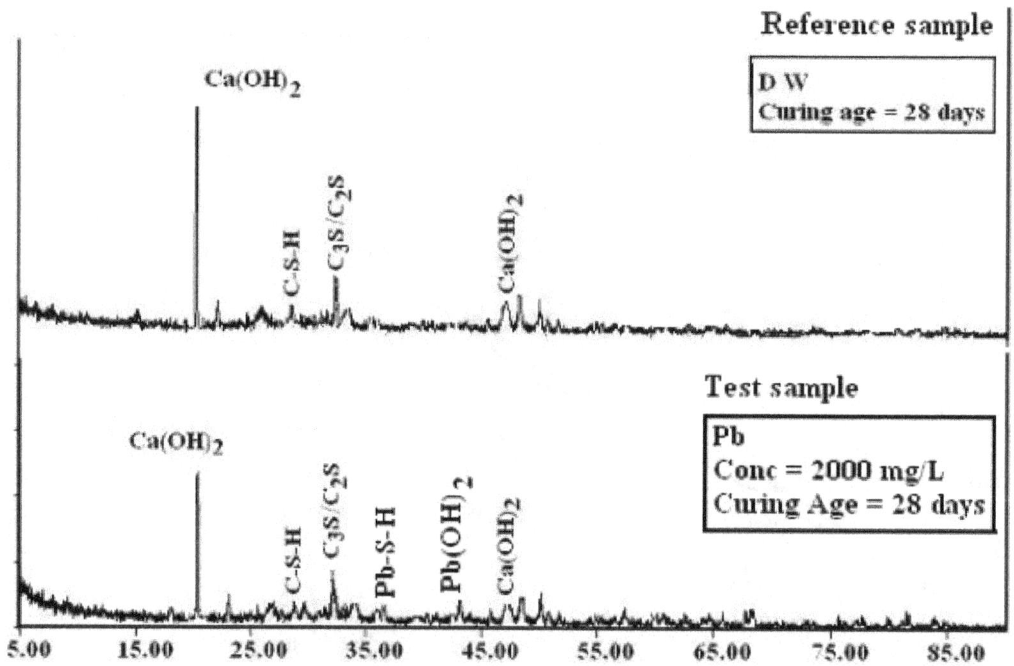

Figure 9. Comparison between XRD patterns of reerence and test samples at 28 days.

with hydrated compounds such as C_3S, C_2S, C-S-H, $Ca(OH)_2$, $Pb(OH)_2$, Pb-S-H, at 32.6, 32.6, 29, 20.6, 42.5 and 31.5° respectively. Lead hydroxide precipitation is expected to form quickly in the high alkaline environment of a cement mix. The precipitated lead hydroxide might be coated on hydrated and anhydrate cement compounds, there by delaying the setting process and slowing early strength development. However, for concentrations of lead above 2000 mg/L, delay in setting time was found to have increased. The compressive strengths of test specimens were considerably decreased compared with reference specimens as concentration of lead was increased. The possible reasons, as concentration of lead increases, lead hydroxide precipitation increases and may increase the replacement of Ca by Pb in hydrated compounds.

Conclusions

Based on the results of this investigation, it was concluded that: Lead spiked deionised water affected setting times. For the concentration of 3000 mg/L and above, setting times were significantly increased. The presence of lead in high concentrations (≥ 3000 mg/L) in deionised water considerably decreased the compressive strengths. For a concentration of 2000 mg/L, at early ages of 3 and 7 days, compressive strength development was slow but for 28 days and onwards, compressive strength development was slightly higher than that of reference specimens. The presence of lead in cement matrix up to 2000 mg/L positively influences engineering properties of mortar.

The compressive strength loss in reference and test specimens was almost the same when they were immersed in magnesium sulfate solutions. The strength loss exhibited was due to dissociation of calcium hydroxide and decalcification of C-S-H.

REFERENCE

Al-Amoudi OSB, Abduljawad SN, Rasheeduzzafar, Maslehuddin M (1992). Effect of chloride and sulphate contamination in soils on corrosion of steel and concrete, Trans. Res. Rec., 1345: 67-73.

Al-Amoudi OSB, Rasheeduzzafar, Maslehuddin M, Abduljawad SN (1994). Influence of chloride ions on sulphate deterioration in plain and blended cements, Mag. Concr. Res., 46(167): 113-123.

Babu RG (2009). Effect of metal ions in industrial wastewater on setting, compressive strength, hardening and soundness of cement, Ph D, Thesis submitted to J N T University Anantapur, pp. 111-118.

Babu RG, Sudarsana HR, Ramana IVR (2007). Use of Treated Industrial Wastewater as Mixing Water in Cement Works. Nat. Environ. Pollut. Technol. J., 6: 595-600.

Babu RG, Sudarsana HR, Ramana IVR (2009). Effect of metal ions in industrial wastewater on cement setting, strength development and hardening, Indian Concrete J., 83: 43-48.

Barth EE (1990). Solidification of hazardous wastes, park Ridge, New York, Noyes data.

Cebeci OZ, Saatci AM (1989). Domestic sewage as mixing water in concrete, ACI J., 86: 503-506.

Hooton RD (1993). Influence of silica fume replacement of cement on physical properties and resistance to sulfate attack, freezing, thawing and alkali silica reactivity, ACI Mater. J., 90: 143-151.

Knudsen T (1976). Quantitative analysis of the compound composition of cement and cement clinker by X – ray diffraction. Am. Ceram. Soc. Bull., 55(12): 1052-1055.

Mindess S, Young JF (1981). Concrete Prentice-Hall, Inc, Engle wood Cliffs, pp. 112-116.

Neville A (2000). Water and concrete-A love-hate relationship, Point view. Concrete Int., 22: 34-38.

Ramana IVR, Prasad NRSR, Reddy GB, Kotaiah B, Chiranjeevi P (2006). Effect of biological contaminated water on cement mortar properties. Indian Concrete J., 80: 13-19.

Steinour, Harold H (1990). Concrete mix water-How impure it can be? P.C.A Research and Development Labs., 2: 32-50.

Tashiro C (1980). Proceedings of the 7th International congress of chemistry of cement, Paris, 11: 11-37.

Tay JH, Yip WK (1987). Use of reclaimed wastewater for concrete mixing ASCE, 113: 1156-1161.

Toughness characterization of steel fibre reinforced concrete – A review on various international standards

A. Sivakumar* and V. M. Sountharajan

Structural Engineering Division, SMBS, VIT University, Vellore, Tamilnadu, India.

Toughness measurements are considered to be an important scale for evaluating the post crack performance of a fibre reinforced concrete. There are various international standards that lay down different testing procedures and the corresponding deflection measurements. This paper presents a complete review on the various flexural testing methods for fibre reinforced concrete (FRC) prescribed by different standards and the methods for characterizing the toughness of FRC. Also reviewed are the significant advantages of these methods, the ways in which the deflections are measured and the practical problems associated with the measurement of deflection. This paper also discusses the various factors such as size of the specimen, stiffness of the testing machine, the rate of loading and type of loading, which influence the test results.

Key words: Toughness, fibre reinforced concrete, flexural loading, size effect, stiffness.

INTRODUCTION

Toughness characterization of fibre reinforced concrete (FRC) becomes more complicated due to erroneous misrepresentation of post peak behavior as a result of the extraneous deflections arising out at testing. The source of error lies either from the machine in which the deflection is recorded or at the point of measurement of deflection. In general the deflection can be measured either from the flexural specimen or outside the specimen. In the former case, the deflection is measured by means of providing notches in the flexural specimen, and the crack mouth opening displacement is measured, and in the latter case, the net deflection is calculated either by measuring the cross head displacement, or by setting up a Japanese yoke at the neutral axis to calculate the net deflection (Gopalaratnam and Gettu, 1995; Barr et al., 1996). Since the deflection of flexural specimens essentially reflects the post cracking behavior of FRC, it becomes vital to calculate it ideally and accurately. Erroneous deflection measurements could lead to overestimation of the resultant toughness of FRC, and cause misconceptions about the composite material property. In the present study, a brief review of various testing methods for FRC and deflection measurement techniques adopted by different standards is presented. There is a need to develop a new set of guidelines which can enhance the experimental techniques in FRC. These guidelines would draw from the experience of the current standards.

*Corresponding author. E-mail: sivakumara@vit.ac.in

Proposed guidelines by different standards for flexural testing

In general, the guidelines proposed by various standards call for similar testing methodology (third point loading) and to some extent differ in the size of specimen adopted. The real adequacy of any tests method lies entirely in preventing the extraneous deflections which can occur either due to the support settlement, lack of stiffness of the testing machine or rigidity of the deflection measuring device (LVDT). Over decades, toughness measurements have been evaluated using an un-notched concrete beam in flexure either by using a four-point loading (or third point loading) or midpoint loading arrangement. Due to the problems associated with support settlement, lifting of beams at supports and sudden drop in load after peak load (lack of stiffness of testing machine) leads to extraneous deflections and

Table 1. Experimental test methods and toughness characterization by various standards.[1]

Name of the standard	Dimensions of the specimen (L*b*h) mm	Rate of loading (mm/min)	Type of loading arrangement	Maximum deflection measured	Toughness measurement
ASTM C-1018[5] (1992)	300*100*100	0.05 to 0.10	Third point loading	Up to the point where there is no resistance on further loading	Determination of toughness indices and residual strength factors
ACI-544 guidelines[6] (1988)	350*100*100	0.05 to 0.10	Third point/Mid-point loading	Up to 1.9 mm	Ratio of energy absorbed by a FRC to that of plain concrete.
JCI specifications[7] (1984)	300*100*100	L/1500 to L/l300	Third point loading	Up to L/150	Energy absorbed up to a deflection of L/150 mm
RILEM draft recommendations[8] (1985)	B>50, d<25, L	0.25	Third point loading	Up to 3 mm	Energy absorbed up to a deflection of 3 mm
EFNARC specification[9] (1993)	450*125*75	0.25 ± 0.05	Third point loading	Up to 25 mm	Residual strength factors up to deflection of 1 and 3 mm

ASTM – American Society for Testing and Materials, ACI – American Concrete Institute, JCI – Japanese Concrete Institute, EFNARC – European Federation of National Association of Specialist Contractors and Material suppliers to construction industry, RILEM - International Union of Testing and Research Laboratories for Materials and Structures.

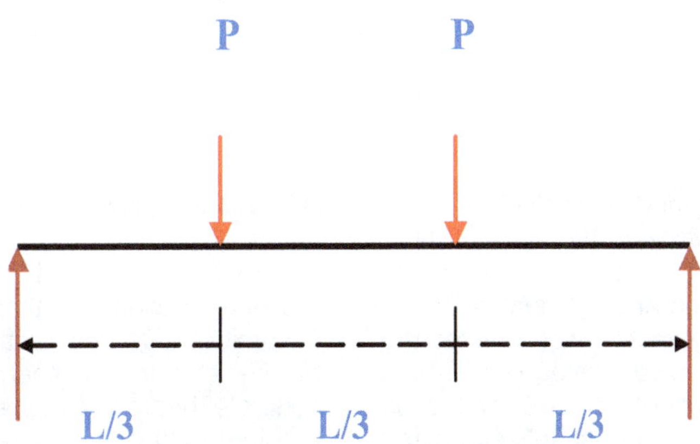

Figure 1. Third point loading test arrangement.

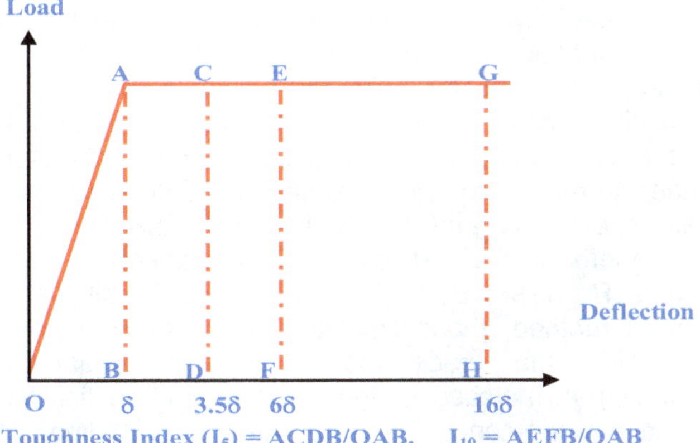

Figure 2. A typical load–deflection plot of FRC.

hence ends up in exaggerated toughness values (Taylor et al., 1997). Recently, it was found suitable to characterize the post peak behavior of the FRC using a centre notched beam wherein, deflections were calculated from the beam specimen and not outside the specimen and gives good characterization of FRC materials. A summary of the various specifications is given in Table 1, and these specifications are discussed in the following part of this work.

ASTM C -1018 Specification (1992)

The ASTM test procedure has been used widely due to its simplicity in testing FRC. A third point loading arrangement as shown in Figure 1 is used for testing the beam specimens. The deflection measurement is done using the cross head displacement and toughness (amount of energy required to deflect and crack an FRC beam) is calculated and reported in terms of indices ((I_5, I_{10}, & I_{20}) and residual strength factors (R_{10} & R_{20}). A typical load-deflection plot and the calculation of toughness indices are shown in Figure 2. The third point loading which is adopted in this method has an

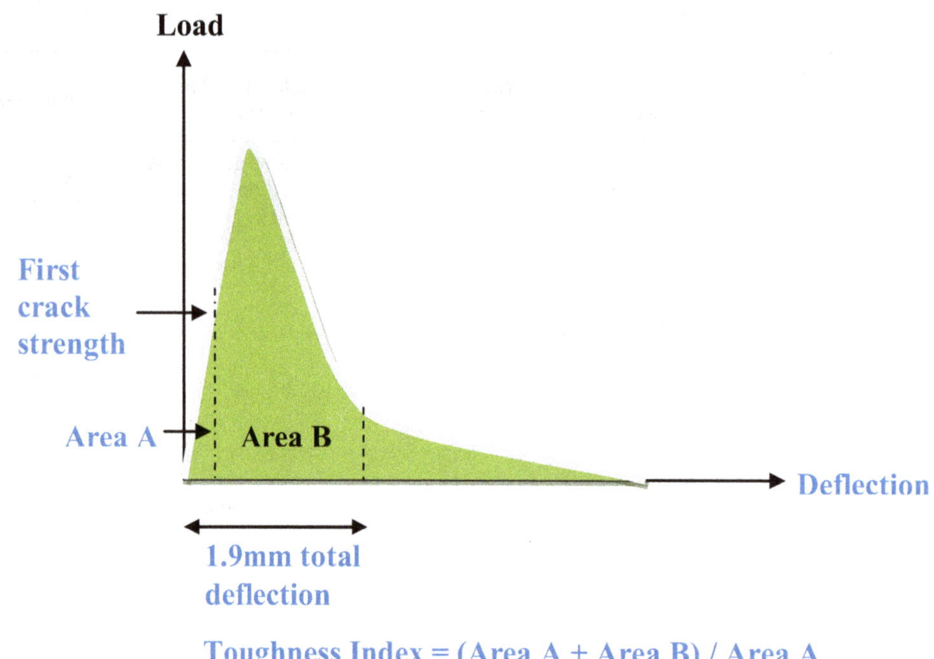

Figure 3. Load deflection curve for fibre reinforced concrete.

advantage, wherein the maximum bending area is increased and subsequently utilizes the efficient fibre action against the bending stresses. On the other hand, the disadvantage with this loading arrangement is due to the failure of beam by shear exactly near one of the loading points. The practical disadvantage with this method lies in the measurement of net deflection (deducting end displacements with central deflection) and first crack deflection which is difficult to locate on the load deflection plot, since the non-linear part of the load deflection curve of FRC is not distinctive. The entire calculation of toughness indices lies in evaluating the exact first crack deflection which is practically impossible to measure. In addition to this, the net deflection is measured against the cross head displacement without taking into account either the support displacement or support settlement which could lead to the calculation of erroneous net deflection. In this standard, toughness indices are evaluated based on multiples of the first-crack deflection which makes it more important to accurately identify the deflection at first crack.

The limitations of this standard are corrected by adding a frame (or 'yoke') around flexural beam specimens that allows direct measurement of the net central deflection of the beam. The use of a yoke eliminate extraneous deflections and results in load deflection curves that are significantly different from those observed by using the traditional cross-head displacement of so-called stiff testing machines. Hence, the practice of measuring displacement directly off the test specimen rather than via the testing machine is preferred by most researchers (Gopalaratnam and Gettu, 1995; Barr et al., 1996).

ACI 544 Specification (1988)

The real application of toughness indices originated with the introduction of the ACI Toughness Index. ACI Committee 544 defines the toughness index as the ratio of the amount of energy required to deflect a fibre concrete beam by a prescribed amount to the energy required to bring the fibre beam to the point of first crack. A sample load –deflection plot and the toughness index calculation is shown in Figure 3. A third point or a four point bend tests are used to characterize toughness. The limitations of this specification include the wide range of parameters that have been used to interpret test results, more variation in the calculated deflections in third-point bend tests compared with three-point bend tests, the difficulty of determining accurately the occurrence of first crack, the extraneous deflections recorded via the testing machine relative to the actual net central deflection of the test specimens and the influence of size effects of specimens on the test results (Gopalaratnam, V.S., and Gettu, R., 1995; Barr, B., *et al.,* 1996). Similar to the ASTM method, these limitations can be overcome by recording the crack mouth opening displacement (CMOD) via notched or the net central deflection via a 'yoke' arrangement subjected to three-point loading.

JCI SF-4 Specification (1984)

The Japanese Concrete Institute (JCI) defines toughness as the area under the load deflection curve up to a limiting deflection of L/150. Identifying the exact occurrence of first crack deflection which is difficult in the ASTM method is not a great concern with this standard. Unlike the ASTM method, the instability in the load-deflection plot right after the first crack is not of major concern in the JCI method, since the end point deflection of span/150 is too far out in the curve to be affected by the instability in the initial portion. However, a limitation of the JCI toughness definition is that the limiting end point deflection is much greater than the acceptable deflection/serviceability limits. The Belgian, Dutch and German specifications have partially overcome this limitation by requiring energy absorption computations also at smaller deflection limits.

EFNARC Specification (1993)

Unlike the other standard methods for the characterization of FRC, this standard recommends the use of a plate test in place of beam test to characterize toughness of FRC. A 600 × 600 mm plate (100 mm thick) is simply supported along all four edges with a 500 × 500 mm span. Load is applied through a 100 × 100 mm punch at a rate of 1.5 mm/min. A plot of the load versus central deflection is used to compute the energy absorbed, until a deflection of 25 mm. The performance of the slab is classified in toughness class a, b or c, for energy absorption capacities of 500, 700 and 1000 J (Nm), respectively. The EFNARC recommendation uses toughness classification identical to that proposed by the Norwegian Concrete Association. However, this approach to characterize toughness was found to be irrelevant for general purpose use.

RILEM Draft Recommendation - 50 FMC (1985)

RILEM recommendation primarily suggests the determination of fracture properties of plain concrete and FRC. This recommendation covers the determination of the critical stress intensity factor and the critical crack tip opening displacement of concrete, using three point bend tests on notched beams. Also there is an advantage of avoiding possible errors due to bending effect by means of reducing the gauge length of LVDT as small as possible and CMOD measured exactly at the centre of beam to avoid eccentricity. This type of testing is unique in that all the material properties can be determined from a single test performed on a notched beam specimen.

FACTORS AFFECTING THE TOUGHNESS RESULTS

Size effects

The size of beam specimens has more direct impact on the test results than the other factors discussed below. It is observed from table 1 that, size of the specimens does not differ greatly for all standards. None of the toughness measurements derived by any standard is size independent. However, for a given size of specimen, the toughness was found to be more sensitive to type of fibre and constituent materials. In reality, even if the energy based indices at small displacements do not exhibit size dependent behavior, the strength and ductility of brittle cementitious composites are inherently size dependent (Gopalaratnam and Gettu, 1995). As a result, none of the toughness measures discussed here and available to date can realistically claim to be truly size-independent.

Type of loading arrangement

It is a general practice of adopting a four point loading test to characterize FRC, since it is easier to conduct and no sophisticated techniques are involved in it. But the real disadvantage with this method is to measure the true deflection at the neutral axis, since bending area is increased. Also, the failure of the beam could occur as a result of shear stress (under the load) rather than bending stress. A mid-point loading configuration is probably more appropriate compared to the four points loading, specifically for notched beam specimens. This setup has numerous advantages in which the stability throughout the test is maintained for both un-reinforced and high strength concretes with low fibre content.

Stiffness of the testing machine

Previously, toughness tests were generally carried out in stiff testing machines that allowed deflection control only. Recently, many research laboratories have carried out tests on closed-loop servo controlled testing machines and achieved stable fracture tests in concrete specimens. The real advantage of such testing machines is in avoiding the sudden drop in load after reaching the peak load. Moreover, one could even control the test by means of the displacement recorded by the opening of the notch. In addition to this, the closed loop testing arrangement allows the crack mouth opening displacement (CMOD) to be used directly to monitor the response of FRC specimen.

Notched versus un-notched beam tests

Compared to an un-notched specimen, deflections in the notched mid-point loaded specimen are always localized at the crack mouth (notch) and the rest of the beam does not undergo any inelastic deformations. This can minimize the energy dissipated over the entire volume of the specimen and, hence, all the energy absorbed can be directed towards the fracture along the notch plane (Gopalaratnam and Gettu, 1995; Barr et al., 1996).

Subsequently, the energy dissipated in these tests can be directly correlated to material response. Also, static tests carried out on centrally notched beam specimen's exhibit the actual deflection of the beam rather than the apparent deflection recorded through the testing machine. Hence, the real advantage of the notched beam test is the possibility of toughness characterization of FRC in terms of CMOD measurements, which are not subjected to any possible errors of the kind observed for traditional deflection measurements.

Deflection measuring techniques

In general none of the standards specifies the type of deflection recording techniques, either recording a traditional cross head displacement, CMOD or setting up an LVDT with yoke arrangement placed at the neutral axis. It is necessary to prevent extraneous deflections from the specimen arising at support settlement or due to lifting of beam at the ends. In recent practice, the use of clip gauge to record CMOD is a good method of quantifying the deflection, as it is measured from the specimen and free from errors. Among all standards, the tests carried out in JCI standard claim to be independent of the type of deflection measuring technique.

CONCLUSIONS FROM THE REVIEW

It can be summarized from the review that, factors like stiffness of the testing machine, accuracy of deflection measurement, and the rate of loading determine the efficacy of the toughness measurement. In general, the various standards have similar test procedures but differ significantly in toughness measurements. The accuracy of any toughness measurement depends upon the true deflection obtained from either un-notched flexural specimen or notched specimens. Also the limit state of serviceability criteria has to be considered for the maximum deflection measured and this maximum limit depends upon the type of application the structure is subjected for use.

REFERENCES

ACI Committee 544 (1988). "Measurement of properties of fiber reinforced concrete". ACI Mater. J. 85:583-593.

ASTM (1992). Standard Test Method for Flexural Toughness and First-Crack Strength of Fiber-Reinforced Concrete (using Beam with Third-Point Loading), ASTM C 1018-92, ASTM Annual Book of Standards, Vol. 04.22, ASTM Philadelphia, USA, pp. 510-516.

Barr B, Gettu R, Al-Oraimi SKA, Bryana LS (1996). "Toughness Measurement - The Need to Think Again". Cem. Concrete Compos. 18:281-297.

EFNARC (1993). Specification for Sprayed Concrete, Final Draft Published by the European Federation of National Associations of Specialist Contractors and Material Suppliers to the Construction Industry (EFNARC), Hampshire, UK, p. 35.

Gopalaratnam VS, Gettu R (1995). "On the Characterization of Flexural Toughness in Fiber Reinforced Concretes". Cem. Concrete Compos. 17:239-254.

JCI (1984). Method of Tests for Flexural Strength and Flexural Toughness of Fiber Reinforced Concrete. JCI Standard SF-4, Japan Concrete Institute Standards for Test methods of Fiber Reinforced Concrete, Tokyo, Japan, pp. 45-51.

RILEM Draft Recommendation (1985). 50FMC, "Determination of fracture energy of mortars and concrete by means of three-point bend tests on notched beams". Mater. Struct. 18:285-290.

Taylor M, Lydon FD, Barr BIG (1997). "Toughness Measurements on Steel Fibre-reinforced High Strength Concrete". Cem. Concrete Compos. 19:329-340.

10

Studies on influence of water-cement ratio on the early age shrinkage cracking of concrete systems

A. Sivakumar

Structural Engineering Division, Vellore Institute of Technology (VIT) University, India. E-mail: sivakumara@vit.ac.in.

Concrete is sensitive to environmental changes, especially during the first few days after casting. Volumetric changes due to drying, temperature, plastic and autogenous shrinkage are often experienced. These changes are critical during early ages when the concrete is most vulnerable to cracking. Reinforcement and joints are used to control shrinkage and leads to cracking. Bad cracking leaves the reinforcement exposed to air and moisture, which may cause it to rust and weaken concrete. Plastic shrinkage is the result of a very rapid loss of moisture from freshly laid concrete within a few hours of placement, while the concrete is still plastic and before it gains any significant strength. This loss is generally caused by various material parameters such as cement content, aggregate content, water-cement ratio and the admixtures. In addition to these environmental factors such as air and concrete temperature, relative humidity (RH) and wind velocity also cater to shrinkage. When the moisture from the surface of freshly placed concrete evaporates faster than the moisture which is replaced by bleed water, the surface concrete shrinks. Plastic shrinkage cracking of concrete is a widespread problem in concrete construction, particularly in thin applications such as highway pavements, slabs cast on grade, surface repairs, overlays, patching, and shotcrete tunnel linings. If premature surface cracks occur, they may accelerate the ingress of aggressive agents, salts and moisture and reduce long-term durability. The current research is motivated to study the influence of various material parameters on the degree of plastic shrinkage cracking in concrete and to quantify more reliable crack estimation.

Key words: Plastic shrinkage, concrete, cracks.

INTRODUCTION

Plastic shrinkage occurs at early age and when bleeding rate is less than evaporation rate (Senthilkumar and Natesan, 2005). It is a type of shrinkage which occurs soon after the concrete is placed in the forms while the concrete is still in the plastic state (Chengqing et al., 2000). A negative capillary pressure will develop within the pore structure of hydrating concrete as water evaporates from the capillary pores (Sivakumar and Manu Santhanam, 2007). This negative pressure results in the development of an overall compressive force within the dehydrating surface, ultimately causing the top layer of fresh concrete to shrink (Anna et al., 1995). Once the fresh concrete's tensile strength is exceeded, shallow cracks of varying depth and length develop throughout the surface (Mane et al., 2003). Plastic shrinkage cracks are typically observed in thin concrete elements with a high surface area to volume ratio (Zhen et al., 2004). It is of major importance because it impairs the structural integrity. These cracks are going to cause further growth due to structural overloading.

Plastic shrinkage cracking does not initially affect the structural capacity but it may lead to accelerated corrosion of embedded reinforcing steel, compromising the structural capacity of the product at a later age. Plastic shrinkage cracks are the short irregular cracks that form on the surfaces of fresh concrete. The present study aimed at studying the degree of early age cracking in cementitious systems and the methodology adopted

Table 1. Concrete mixture proportions used in the study.

Cement (kg/m^3)	Fine aggregate (kg/m^3)	Coarse aggregate (10 mm) (kg/m^3)	Water (kg/m^3)	Superplasticizer (kg/m^3)
350	810	1218	168	8

Table 2. Mix proportions of the plain concrete.

Mix ID	Material proportions			
	Cement	Fine aggregate	Coarse aggregate	Water-cement ratio
M1	1	2.051	3.093	0.3
M2	1	2.051	3.093	0.4
M3	1	2.051	3.093	0.5

Figure 1. Shrinkage crack of Mix 1.

includes slabs with stress concentrators, a bolt and nut arrangement was provided at the ends. Image analysis was done using PCI Geomatics software and used to report the cracked area in a more appropriate method than manual crack measurement to study cracking at different scales. With the use of image analysis, it becomes easier and reliable for monitoring the crack development. A commercially available image analysis software (PCI Geomatica) was used in the present study for measuring the crack parameters such as crack width and crack length.

EXPERIMENTAL METHODOLOGY

Ordinary Portland cement conforming to IS 12269 of 53 grade was used for producing concrete admixtures and the specific gravity was found to be 3.37. River sand with a specific gravity of 2.69 with fineness modulus of 2.55 was used as fine aggregate. Coarse aggregates of 10 mm size and of specific gravity 2.75 was used. A naphthalene based Super plasticizer was used to obtain desired workability in the range of 50 to 75 mm. The concrete mixture proportions used in the study are provided in Tables 1 and 2.

Experimental setup

The slab mould shown in Figure 1 of dimension 500 × 250 × 75 mm was fabricated as per ASTM-C1579 using a wooden board. The slabs were provided with a stress riser of 55 mm height at the center and two base restraints of 35 mm height at 35 mm from both ends, along the transverse direction. An additional of a bolt and nut arrangement was provided at the ends to restrict the longitudinal movement of the concrete slab from the edges and to provide additional restraint, increasing the potential of cracking at the notch. The slabs were checked visually for any signs of cracking at approximately 60-min intervals. For a concrete mix with fine aggregate/coarse aggregate ratio of unity concrete mix was chosen and listed in Table 2. This rare type of concrete mixture has been considered because it produced more plastic shrinkage cracks. The degree of cracking in various concrete systems is shown in Figures 2, 3 and 4.

RESULTS AND DICUSSION

Crack observations

The time of occurrence of first crack was noted for all slabs during the experiments and the crack measurements are provided in Table 3. In the case of plain concrete, approximately 180 min after mixing with water, a fine hairline crack running throughout the width of the slab was observed above the central stress riser. This fine crack, which is caused due to settlement was found to widen upon further drying. In the case of plain concrete, a single crack over the central stress riser was found to run almost straight throughout the width of the specimen. The experiments were conducted under realistic on site conditions where temperature and humidity parameters fluctuate. Plastic shrinkage cracks occurred during the first few hours after the casting of

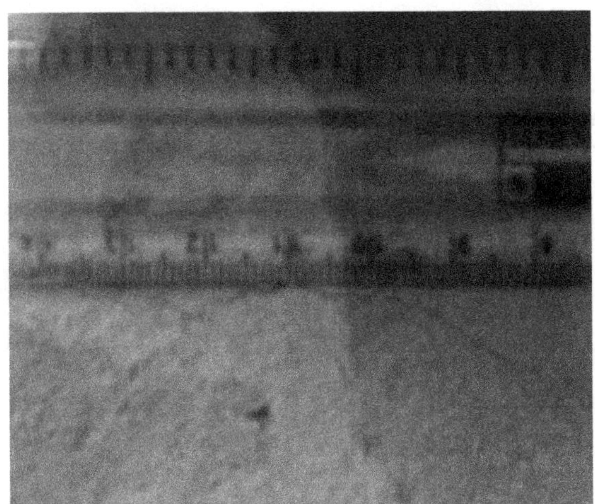

Figure 2. Shrinkage crack of Mix 1.

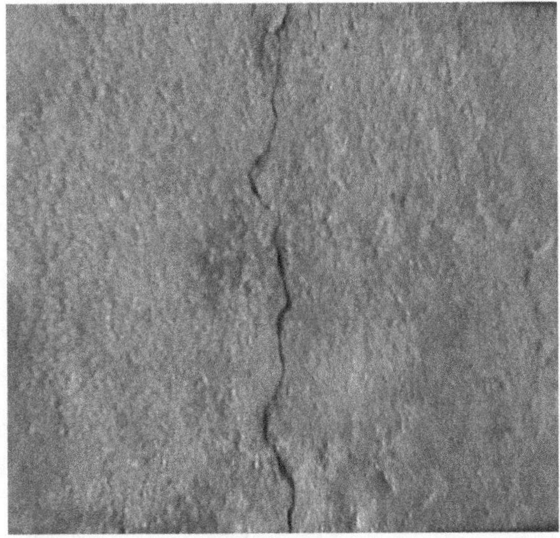

Figure 3. Shrinkage crack of Mix 2.

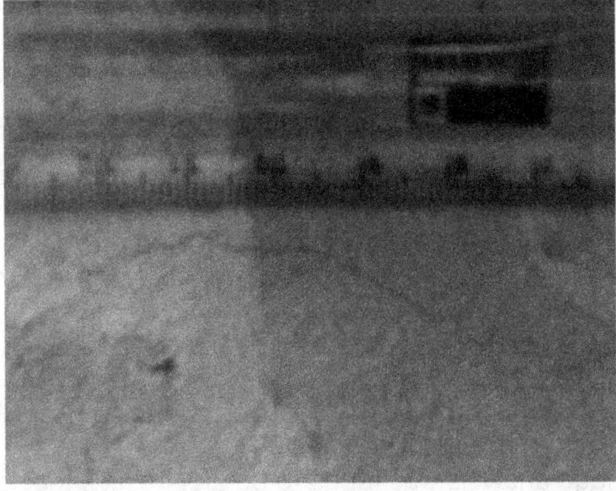

Figure 4. Shrinkage crack of Mix 3.

concrete was completed, and the concrete is still in plastic state. The width and the length of cracks were measured after 14 h. In the initial period, the specimen has sufficient moisture stored in the capillary pores, which is being used for hydration process. Thereby stresses may initiate and propagate cracks. Early age cracks are severe for low water cement ratio systems. It can be observed from experimental results that at low water cement ratio, the cracks occurs at 4 h and the crack at later stage are stabilized. The crack after a day was observed to be slightly lower than higher water cement systems. The crack length increased with the increase in water-cement ratio which is evident from Figure 5. The increase in crack length for high water content in the concrete systems could be due to continuous propagation of cracks. It can be observed from Figure 6 that, in high water cement ratio systems cracking width of concrete was observed to be less than at low water cement ratio systems. This could be possible due to enough water available for evaporation and provides better residual strength before complete drying. After 8 h the cracking propagated and later stabilized. However the cracking width increased at lower water cement ratio systems. This has resulted in higher crack width and corresponding for a medium water cement ratio concrete systems. The cracking in concrete specimen is a function of water cement ratio and the subsequent bleeding in the concrete. So it can be realized that at low water cement ratio the bleeding is less compared to high. Hence cracking is severe in low water cement ratio systems than high water cement ratio due to inadequate strength in concrete. When concrete has high water cement ratio systems the drying of concrete takes longer time and this can an advantage to gain sufficient tensile strength in plastic state. The crack width increases from 0.3 water cement ratio and reaches a maximum value at 0.4 water cement ratio and starts decreasing with an increase in water content which can be observed from Figure 5. This could be due to the reason that concrete attains early age cracking whereas in 0.5 the concrete has lesser crack width comparatively as it has sufficient water which does not lead to the formation of early age cracks. The cracks are highly tortuous for high water cement ratio systems compared to low water cement ratio systems since cracks occur at delayed time and has attained sufficient strength, this results in for a longer path taken by the cracks which can be observed from Table 3. The total crack area is the product of total crack length and mean crack width and the experimental results exhibited an increase in crack area with an increase in water cement ratio (Figure 7). It can be concluded that early age cracking is phenomenal for low water cementitious systems due to less bleed water available for compensating evaporation. Whereas in the case of high water cementitious systems enough water is available on the top surface and also results in the reduction of crack width as well as crack area. Crack length depends on the tortuosity of initiation and

Table 3. Crack measurements on the concrete slabs.

Specimen ID	Total crack length (mm)	Mean crack width (mm)	Total crack area (mm^2)	Standard deviation	Coefficient of variance (%)
M1	209.15	0.23	48.10	0.35	1.52
M2	240.66	0.41	98.67	0.52	1.26
M3	330.58	0.12	39.67	0.27	2.25

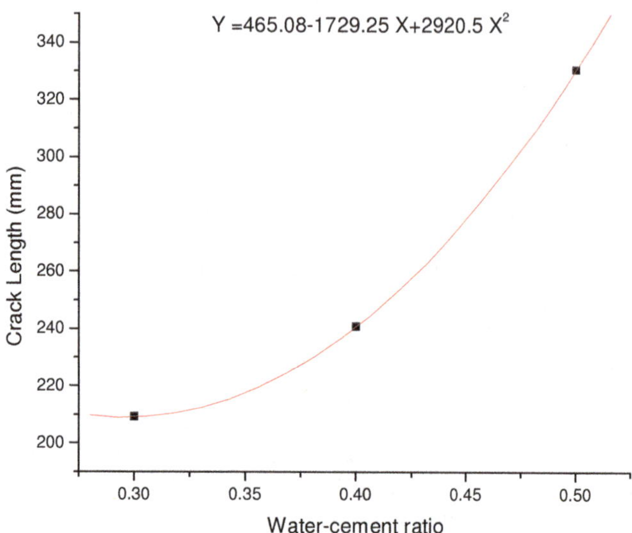

Figure 5. Water cement ratio verses Crack length.

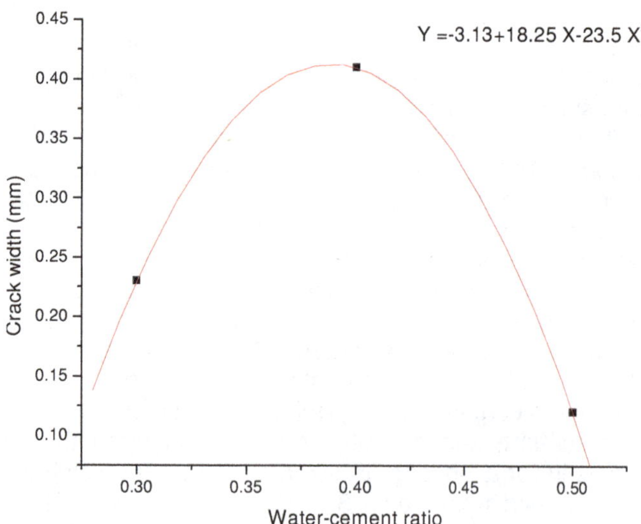

Figure 6. Water cement ratio verses width.

propagation and this certainly high for low water cementitious systems than for high water cementitious systems. This results in continuous propagation of cracks without any discontinuity in concrete systems with high water content.

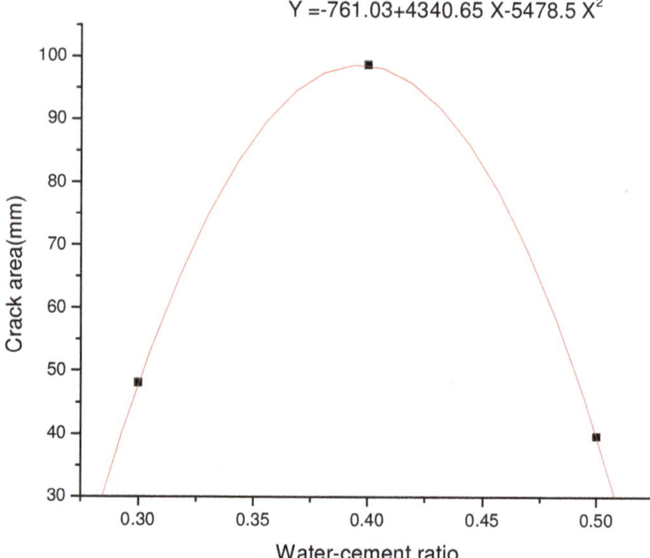

Figure 7. Water cement ratio verses Crack area.

Conclusions

The following observations are derived from the present study:

1. The image analysis measurement technique was capable of assessing the shrinkage characteristics of any concrete system.
2. Plastic shrinkage cracking is greatly influenced by the cement/total aggregate ratio. As the water cement ratio increases the total crack length increases whereas the mean crack width is not increased.
3. The maximum width of crack is moderate for any type of concrete mix with different water/cement ratios subjected to realistic site conditions.
4. The increase in fineness of the material will often reduce the amount of bleed water that will reach the concrete's surface.
5. The present experimental method is appropriate for drying that takes place on the site conditions and subsequent propensity for the formation of plastic shrinkage cracking in concrete.
6. Image analysis was found to be an effective tool for the bias free estimation of plastic shrinkage cracking.

REFERENCES

Anna K, Markku L, Pekka S (1995). "Experimental study on the basic phenomena of shrinkage and cracking of fresh mortar." J. Cem. Concrete Res. 25:1747-1754.

Chengqing QI, Jason W, Jan O (2000). "Characterization of Plastic Shrinkage Cracking in Fiber Reinforced Concrete Using Image Analysis and A Modified Weibull Function." J. Cem. Concrete Res. pp. 135-139.

Mane SA, Desai TK, Kingsbury D, Mobasher B (2003). "Modelling of Restrained Shrinkage Cracking in Concrete Materials". J. Compos. pp. 230-236.

Senthilkumar SRR, Natesan SC (2005). "Prediction of restrained Plastic Shrinkage Cracking in plain cement concrete". Mag. Concrete Res. 57:579–587.

Sivakumar A, Manu Santhanam (2007). "A quantitative study on the Plastic Shrinkage Cracking in high strength hybrid fibre reinforced" J. Cem. Concrete Compos. 29:575–581.

Zhen He, Xiangming Z, Zongjin LI (2004). "New Experimental Method for Studying Early-Age Cracking of Cement Based Materials." J. Cem. Concrete Res. 101:50-56.

Characteristic studies on the mechanical properties of quarry dust addition in conventional concrete

A. Sivakumar* and Prakash M.

Department of civil engineering, Vellore Institute of Technology (VIT) University, India.

Currently India has taken a major initiative on developing the infrastructures such as express highways, power projects and industrial structures etc., to meet the requirements of globalization, in the construction of buildings and other structures. Concrete plays the key role and a large quantum of concrete is being utilized in every construction practices. River sand, which is one of the constituents used in the production of conventional concrete, has become very expensive and also becoming scarce due to depletion of river bed. Quarry dust is a waste obtained during quarrying process. It has very recently gained good attention to be used as an effective filler material instead of fine aggregate. In the present study, the hardened and durable properties of concrete using quarry dust were investigated. Also, the use of quarry dust as the fine aggregate decreases the cost of concrete production in terms of the complete replacement for natural river sand. This paper reports the experimental study which investigated the influence of 100% replacement of sand with quarry dust. Initially cement mortar cube was studied with various proportions of quarry dust (CM 1:3, CM 1:2, and CM 1:1). The experimental results showed that the addition of quarry dust for a fine to coarse aggregate ratio of 0.6 was found to enhance the compressive properties as well as elastic modulus.

Key words: Concrete, quarry dust, fillers, elastic modulus, compressive strength.

INTRODUCTION

In the recent past good attempts have been made for the successful utilization of various industrial by products (such as flyash, silica fume, rice husk ash, foundry waste) to save environmental pollution. In addition to this, an alternative source for the potential replacement of natural aggregates in concrete has gained good attention. As a result reasonable studies have been conducted to find the suitability of granite quarry dust in conventional concrete.

However, recycled concrete aggregate, fly ash, blast furnace slag, as well as several types of manufactured aggregates have been studied by many researchers, Zain et al. (2000); Neville (2002); Gambhir (1995). Quarry dust, a by-product from the crushing process during quarrying activities is one of those materials that have recently gained attentions to be used as concreting aggregates, especially as fine aggregates. Quarry dust have been used for different activities in the construction industry, such as road construction, and manufacture of building materials, such as lightweight aggregates, bricks, tiles and autoclave blocks.

Researches have been conducted in different parts of the world, to study the effects of incorporation of quarry dust into concrete. Galetakis and Raka (2004) studied the influence of varying replacement proportion of sand with quarry dust (20, 30 and 40%) on the properties of concrete in both fresh and hardened state (Nevillie, 2002).

Saifuddin et al. (2001) investigated the influence of partial replacement of sand with quarry dust and cement with mineral admixtures on the compressive strength of concrete (Gambhir, 1995), whereas Celik and Marar investigated the influence of partial replacement of fine aggregate with crushed stone dust at varying percent-tages in the properties of fresh and hardened concrete (Safiuddin et al., 2001; Celik and Marar, 1996). The present study is intended to study the effects of quarry dust addition in conventional concrete and to assess the rate of compressive strength development for different quarry dust to coarse aggregate ratio, (Goble (1999); De Larrard and Belloc, 1997).

*Corresponding author. E-mail: sivakumara@vit.ac.in.

Table 1. Physical properties of cement.

S/N	Properties	Results obtained
1.	Specific gravity	3.37
2.	Initial Setting Time (minutes)	33.00
3.	Final Setting Time (minutes)	480.00
4.	Standard Consistency (%)	32%

METHODOLOGY

Cement

In the present study an ordinary Portland cement (OPC 53 grade) was used. The physical properties of the cement tested according to Indian standards procedure confirms to the requirements of IS 12269 and the physical properties are given in Table 1.

Fine aggregate

The river sand conforming to zone II as per IS-383-1987 was used for making reference concrete and its specific gravity was found to be 2.3. The loose and compacted bulk density values of sand were 1455 and 1726 kg/m^3 respectively.

Quarry dust

The basic tests on quarry dust were conducted as per IS-383-1987 and its specific gravity was around 1.95. Wet sieving of quarry dust through a 90 micron sieve was found to be 78% and the corresponding bulking value of quarry dust was 34.13%.

Coarse aggregate

Crushed granite coarse aggregate conforming to IS 383-1987 of size 12 mm and down having a specific gravity of 2.6 was used. The loose and compacted bulk density values of coarse aggregate were 1483 and 1680 kg/m^3, respectively, for different grade of concrete.

Chemical admixture

Workability of concrete mixtures was tested using slump cone test to obtain a target slump of 75 -90 mm. In the present study, a naphthalene based super plasticizer was used at an optimum dosage, not exceeding 1.0% by weight of cement.

Water

Water is an important ingredient of concrete as it initiates the chemical reaction with cement, and the mix water was completely free from chlorides and sulfates. Ordinary potable water was used throughout the investigation as well as for curing concrete specimens.

Concrete mixture proportions

Concrete consists of a two phase material, namely mortar and aggregate phases. Hence in the present study the two phases were studied separately. Three types of mortar mixtures were taken in the ratio of 1:3, 1:2 and 1:1. The cement mortar cubes containing 100% replacement of sand with quarry dust were compared with conventional mortar cube prepared with natural river sand. Concretes, namely M1, M2 and M3, were prepared with the w/c ratio of 0.32. For all concrete mixtures, the fine to coarse aggregate ratio varied at 0.6, 0.7 and 0.8. The performance characteristics of all quarry dust concretes were compared with reference concrete containing without quarry dust. Three different binder contents were chosen at 300, 350 and 400 kg/m^3. The dosages of the super-plasticizers were kept constant by monitoring the desired flowability in trial mixtures. The details of the mixture proportions were presented in Table 2.

Mixing details

A rotating pan mixer (capacity 0.05 m^3) was used for mixing the constituent material. Fresh concrete was then casted in a 100 mm cube mould, 150 × 300 mm and 100 × 200 mm cylindrical mould. Immediately after demoulding, the specimens were kept in the curing tank at 25 ± 3°C. The compressive strength of the concrete specimens was determined in accordance to IS: 516-1959 (Table 3). The experimental test results presented are the average of five specimens.

RESULTS AND DISCUSSION

Compressive strength on mortar cube

The 7.05 × 7.05 × 7.05 mm mortar cubes were tested to determine the compressive strength properties of different specimens. The compressive strength was obtained as per IS: 516-1959. The 3, 7 and 28 days compressive strength of mortar cube were shown in Figure 1. The 28 days compressive strength of 100% replacement of sand with quarry dust of mortar cube (CM 1:1) was 11.8% higher than the controlled cement mortar cube. The 3, 7 and 28 days compressive strength of cement mortar had shown a decreasing trend compared to the reference concrete.

Effect of quarry dust on compressive strength of concrete

The 100 mm size concrete cubes were used as test specimens to determine the compressive strength. The test results of the cubes are compiled in Figures 2 to 7. Compressive strength was obtained as per IS: 516-1959. The compressive strength of concrete cube (100% replacement of sand with quarry dust) and natural sand were compared. The compressive strength of concrete cube containing 100% replacement of sand with quarry dust the value ranges between 21.3 to 33.63 MPa. For a binder content of 300 kg/m^3. The maximum compressive strength of concrete cube containing 100% replacement of sand with quarry dust was found to be 33.63 MPa at 56 days for a fine to coarse aggregate ratio of 0.7. The reference compressive strength of concrete cube ranged between 18.70 to 42.10 MPa for different fine to coarse

Table 2. Concrete mixture proportions.

Mix ID	Cement (kg/m^3)	F/C (kg/m^3)	Fine Aggregate (kg/m^3)	Coarse aggregate (kg/m^3)	Water (kg/m^3)	Super plasticizer dosage (kg/m^3)	w/c
		0.6	787.50	1312.50	420.00		
Mix 1	300	0.7	864.71	1235.29	395.29	2.40	0.32
		0.8	933.30	1166.67	373.33		
		0.6	768.75	1281.25	410.00		
Mix 2	350	0.7	844.12	1205.88	385.88	2.80	0.32
		0.8	911.00	1138.89	364.44		
		0.6	750.00	1250.00	400.00		
Mix 3	400	0.7	823.50	1176.47	376.47	3.20	0.32
		0.8	889.00	1111.11	355.55		

Table 3. Compressive strength results of various concrete mixtures.

Mix ID	Cement (kg/m^3)	F/C (kg/m^3)	Compressive strength of concrete (MPa)					
			without quarry dust			with quarry dust		
			7 days	28 days	56 days	7 days	28 days	56 days
		0.6	23.05	40.20	42.10	21.3	28.2	29.9
Mix 1	300	0.7	18.70	25.63	29.10	23.53	25.43	33.63
		0.8	33.36	35.60	37.20	22.33	25.33	26.10
		0.6	36.56	38.50	39.60	32.90	36.40	36.46
Mix 2	350	0.7	34.99	36.33	38.80	23.76	29.27	30.40
		0.8	31.50	36.70	38.10	32.76	34.50	38.10
		0.6	24.90	26.90	31.10	13.80	21.13	37.66
Mix 3	400	0.7	21.40	22.50	24.50	18.80	27.30	29.60
		0.8	31.10	24.50	36.00	22.90	23.93	31.00

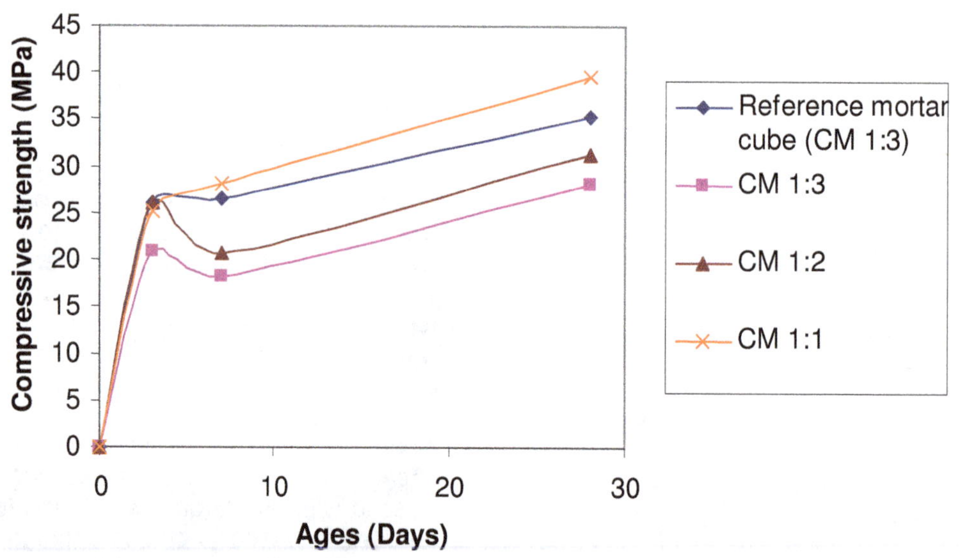

Figure 1. Compressive strength of mortar cube.

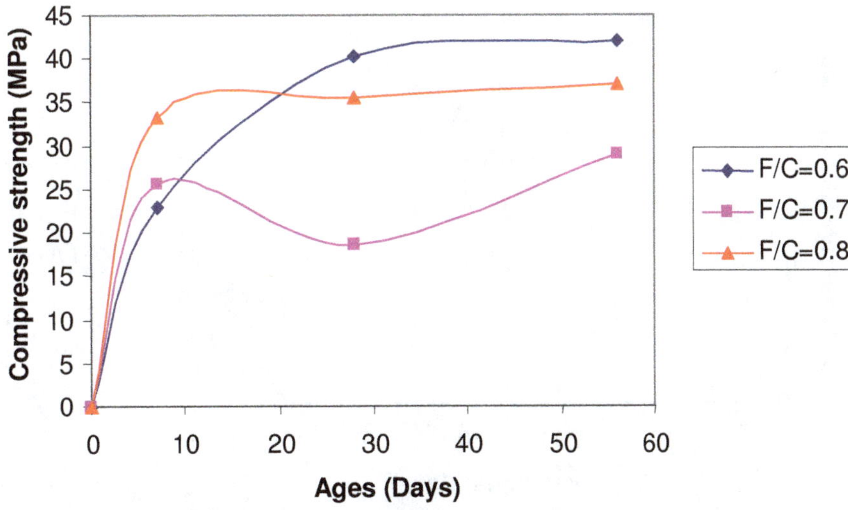

Figure 2. Reference compressive strength of concrete at 300 Kg/m^3

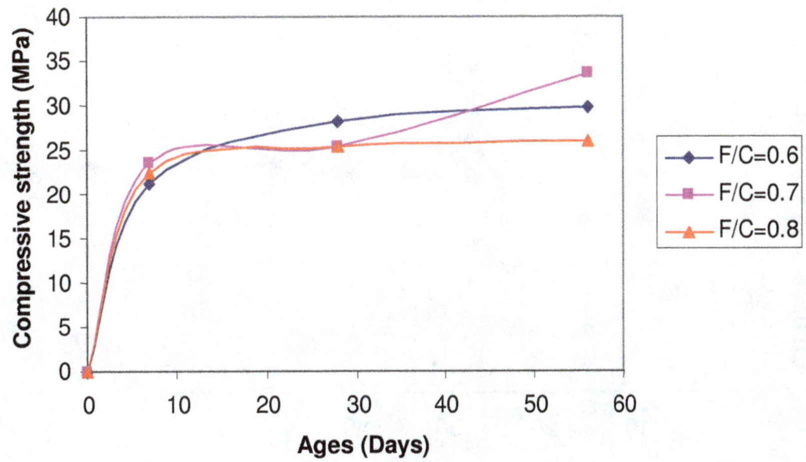

Figure 3. Effect of quarry fines on compressive strength of concrete at 300 Kg/m^3.

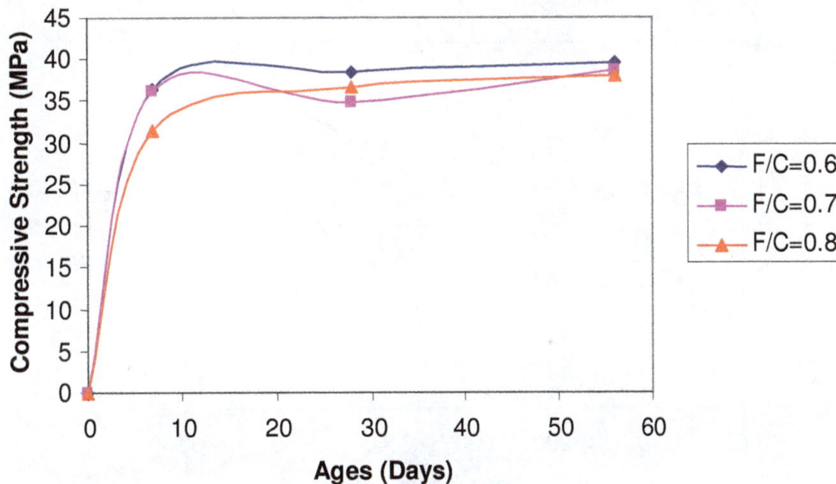

Figure 4. Reference compressive strength of concrete at 350 kg/m^3.

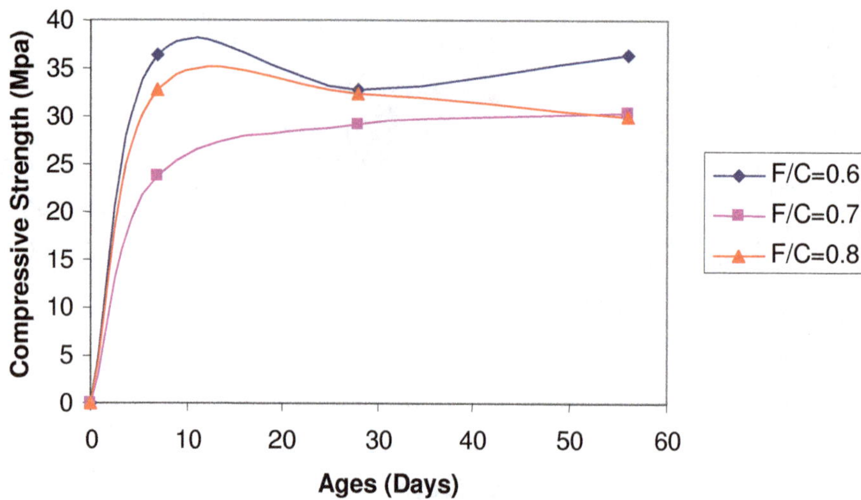

Figure 5. Effect of quarry fines on compressive strength of concrete at 350 kg/m^3

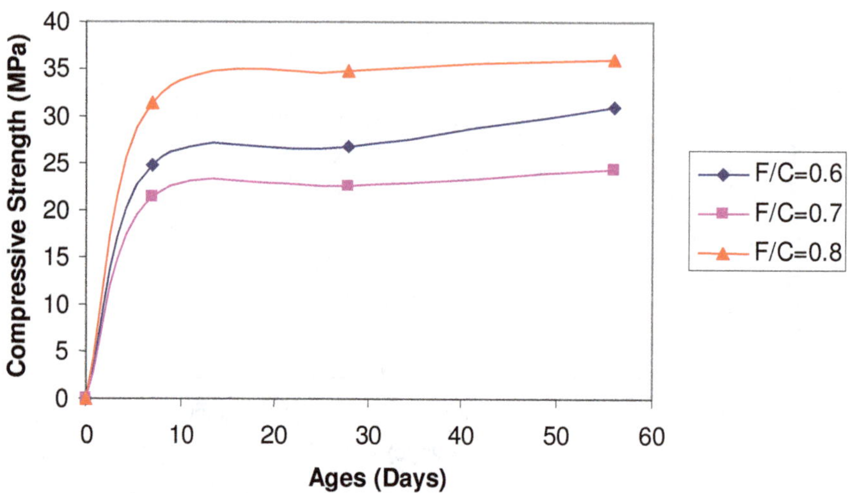

Figure 6. Reference compressive strength of concrete at 400 kg/m^3.

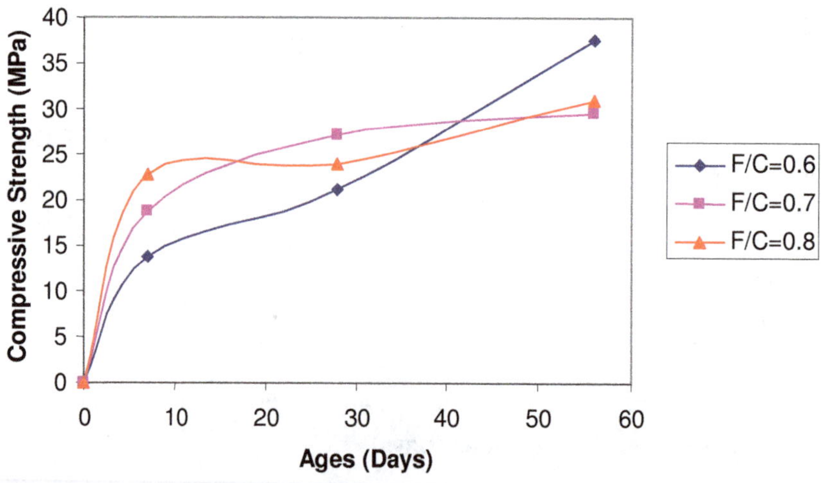

Figure 7. Effect of quarry fines on compressive strength of concrete at 400 kg/m^3.

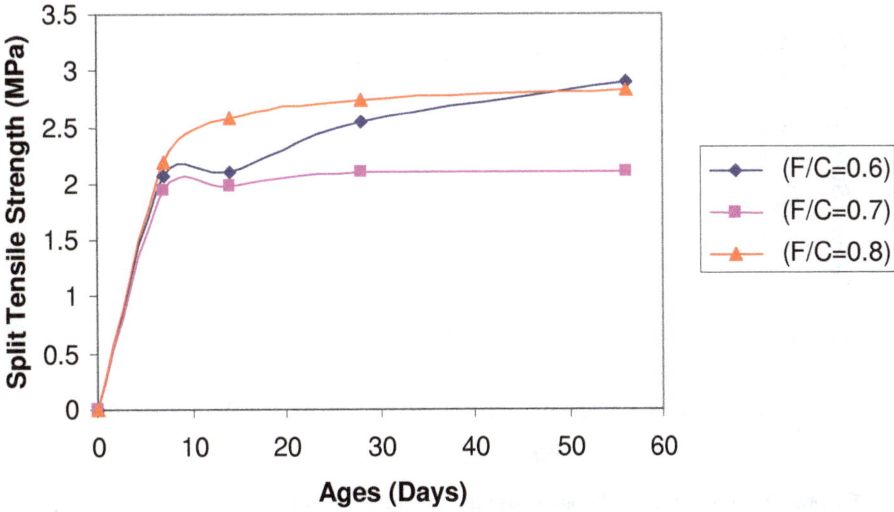

Figure 8. Reference split tensile strength concrete at 300 kg/m^3.

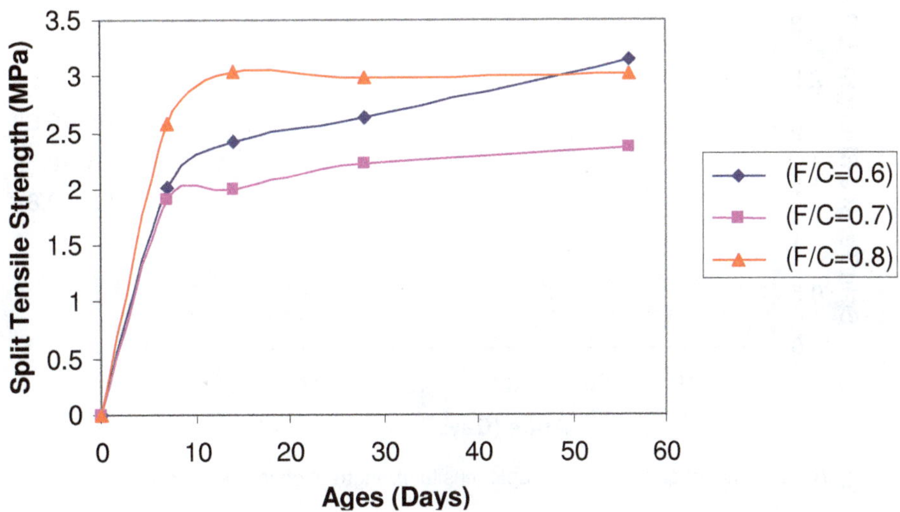

Figure 9. Effect of quarry fines on split tensile strength of concrete at 300 kg/m^3.

aggregate ratio. Whereas, for a binder content of 350 kG/m^3 the compressive strength of concrete (100% replacement of sand with quarry dust) ranged between 23.76 to 36.46 MPa. The maximum compressive strength was observed to be 36.46 MPa at 56 days for an F/C ratio of 0.6. The reference compressive strength of concrete specimens ranged between 31.50 to 39.60MPa. The maximum compressive strength of reference concrete was 39.6 MPa at 56 days for F/C of 0.6. It was also noted that at a binder content of 400 kg/m^3 the 56 days compressive strength (100% quarry dust) was 10.7% higher than all other concrete mixtures. From the experimental test results it can be inferred that with an increase in quarry dust percentage the strength was found to be increasing upto an optimal value of the fine to coarse aggregate ratio of 0.6.

Effect of quarry dust on split tensile strength

The split tensile properties of concrete specimens with and without quarry dust for different binder content and F/C ratio were shown in Figures 8 to 13. It was observed from the experimental test results that for a binder content of 300 kg/m^3, the split tensile values varied from 1.91 to 3.15 MPa and a maximum split tensile strength of 3.15 MPa was observed at 56 days for an F/C ratio of 0.6. Similarly for a binder content of 350 kg/m^3, the maximum split tensile strength varied from 2.48 to 3.33 MPa (Table 4). In general, maximum split tensile strength of 3.33 MPa was obtained for quarry dust concrete with an F/C ratio of 0.6. The addition of quarry dust has shown significant increase compared to plain cement concrete with natural sand. This can be safely concluded that

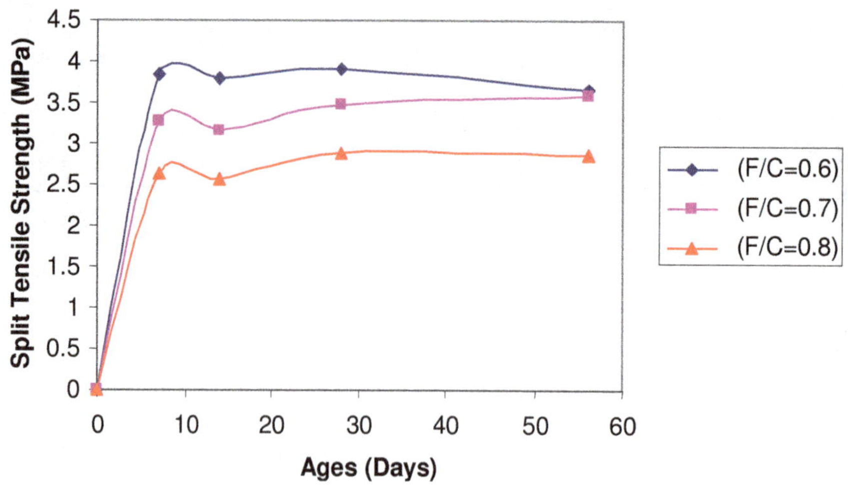

Figure 10. Reference split tensile strength concrete at 350 kg/m^3.

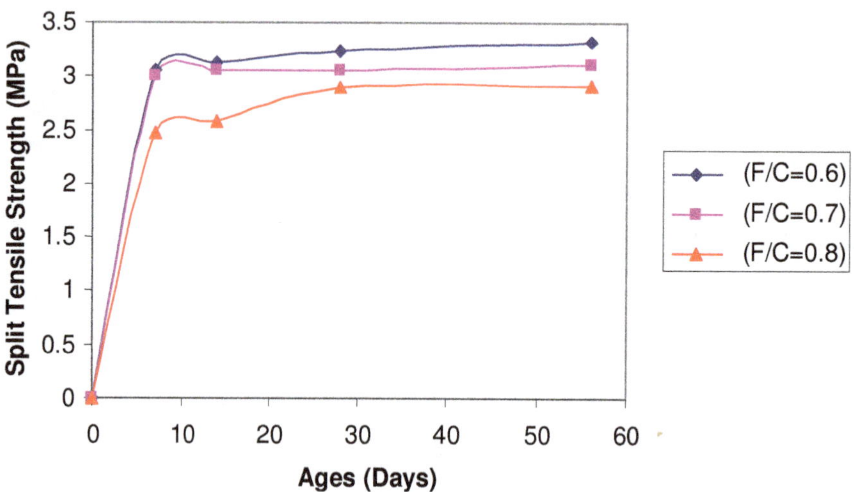

Figure 11. Effect of quarry fines on split tensile strength of concrete at 350 kg/m^3.

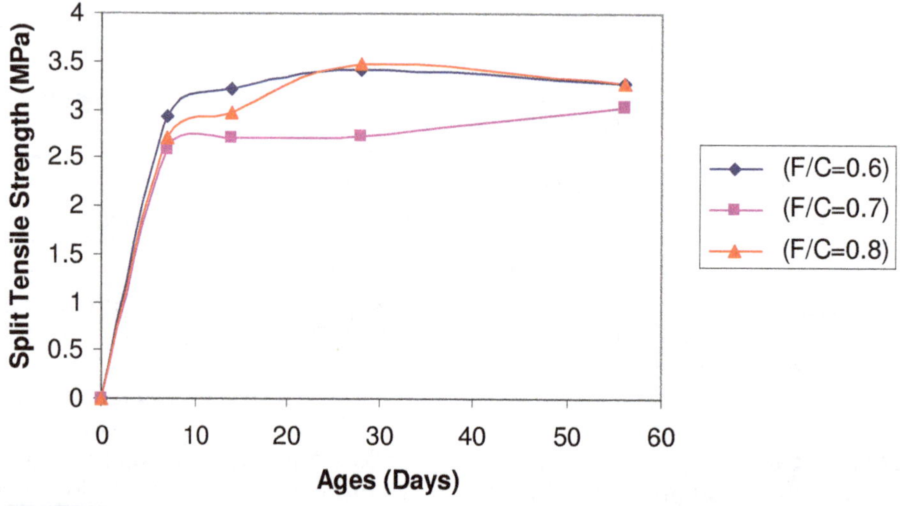

Figure 12. Reference split tensile strength concrete at 400 kg/m^3.

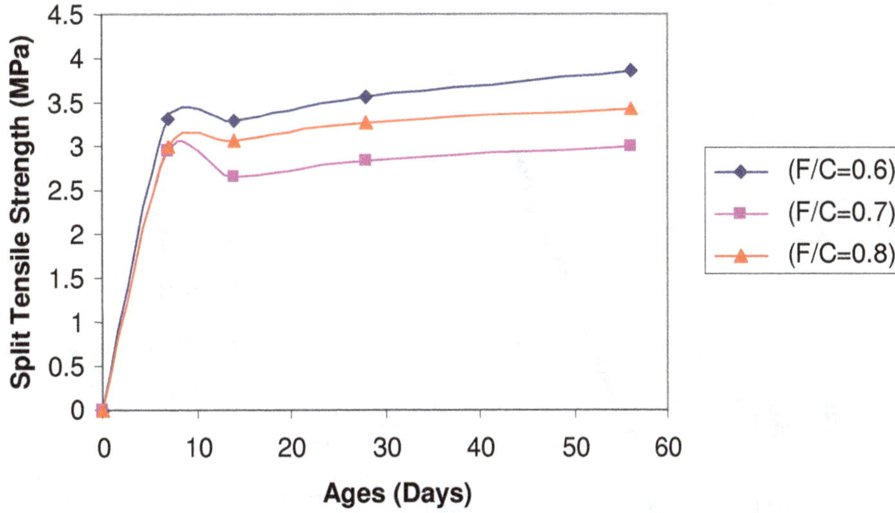

Figure 13. Effect of quarry fines on split tensile strength of concrete at 400 kg/m^3.

Table 4. Split tensile strength results of various concrete specimens.

Mix ID	Cement (kg/m^3)	F/C (kg/m^3)	Compressive strength of concrete (MPa)							
			without quarry dust				with quarry dust			
			7 days	14 days	28 days	56 days	7 days	14 days	28 days	56 days
		0.6	2.07	2.10	2.55	2.90	2.02	2.42	2.64	3.15
Mix 1	300	0.7	1.94	1.97	2.10	2.10	1.91	1.99	2.23	2.37
		0.8	2.2	2.58	2.73	2.83	2.58	3.64	2.99	3.03
		0.6	3.85	3.79	3.92	3.66	3.06	3.14	3.23	3.33
Mix 2	350	0.7	3.28	3.15	3.47	3.66	3.00	3.06	3.06	3.12
		0.8	2.64	2.58	2.90	2.87	2.48	2.58	2.90	2.91
		0.6	2.93	3.21	3.41	3.28	3.33	3.29	3.57	3.87
Mix 3	400	0.7	2.58	2.71	2.74	3.02	2.94	2.66	2.85	3.01
		0.8	2.71	2.96	3.47	3.28	2.99	3.07	3.26	3.42

quarry dust concrete can lead to significant improvement in microstructure due to different size fractions. Also, the filler effects of quarry dust can lead to significant increase (18.6%) in the split tensile strength compared to reference concrete.

Effect of quarry dust on modulus of elasticity

The modulus of elasticity of concrete with and without quarry dust for different binder content and F/C ratio are presented in this section. The results of elasticity modulus of concrete are shown in Figures 14 to 31. At 300 kg/m^3, the maximum elastic modulus of 29.31 GPa was obtained for an F/C ratio of 0.6. The elasticity modulus of concrete for 100% replacement of sand with quarry dust varied from 14.27 to 29.31 GPa. For the reference concrete, the maximum modulus of elasticity of concrete is 33.03 GPa at 56 days for an F/C ratio of 0.7 (Table 5). At 350 kg/m^3, the maximum modulus of elasticity of concrete (100% replacement of sand with quarry dust) ranged from 14.27 to 28.87 GPa. It can be noted that the effects of quarry dust on elastic modulus was observed at a binder content of 400 Kg/m^3 and had showed 15% higher than other concrete specimens. The effects of quarry dust on the elastic modulus property were found to be consistent with conventional concrete containing natural sand.

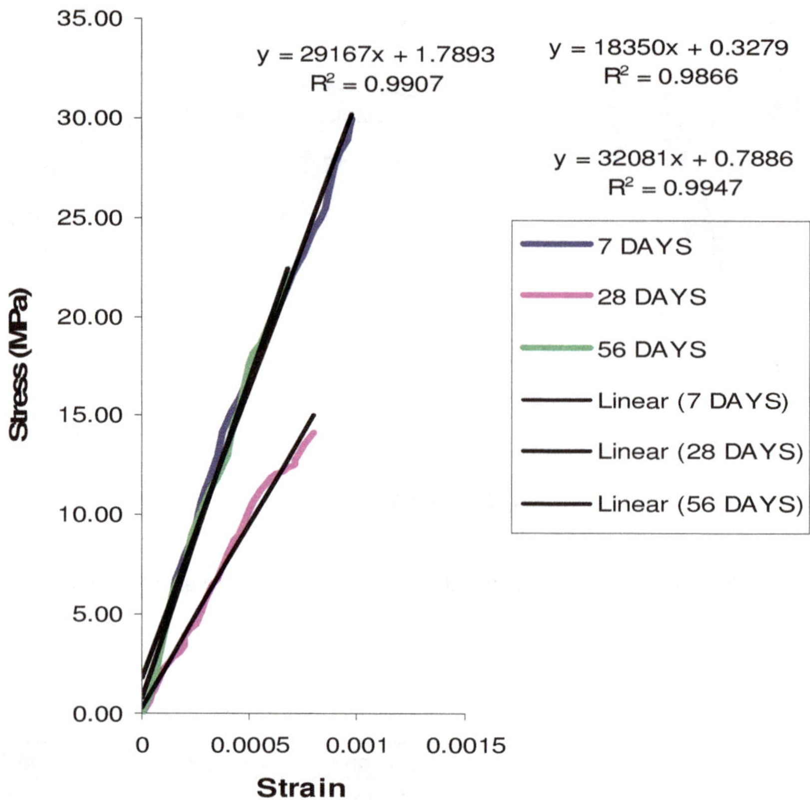

Figure 14. Reference modulus of elasticity 300 kg/m^3 at F/C=0.6.

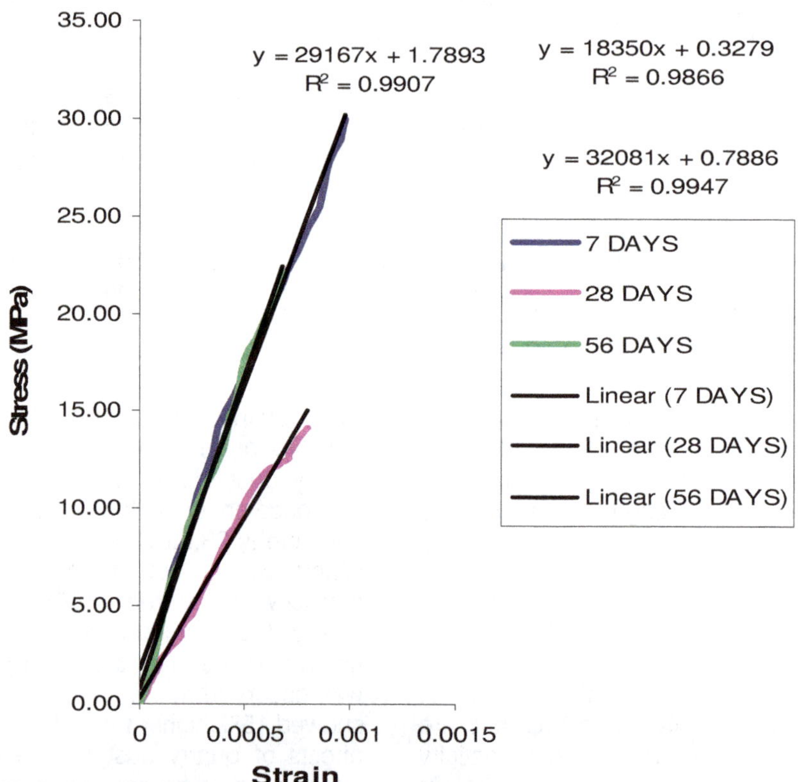

Figure 15. Effect of quarry fines on modulus of elasticity 300 kg/m^3 at F/C=0.6.

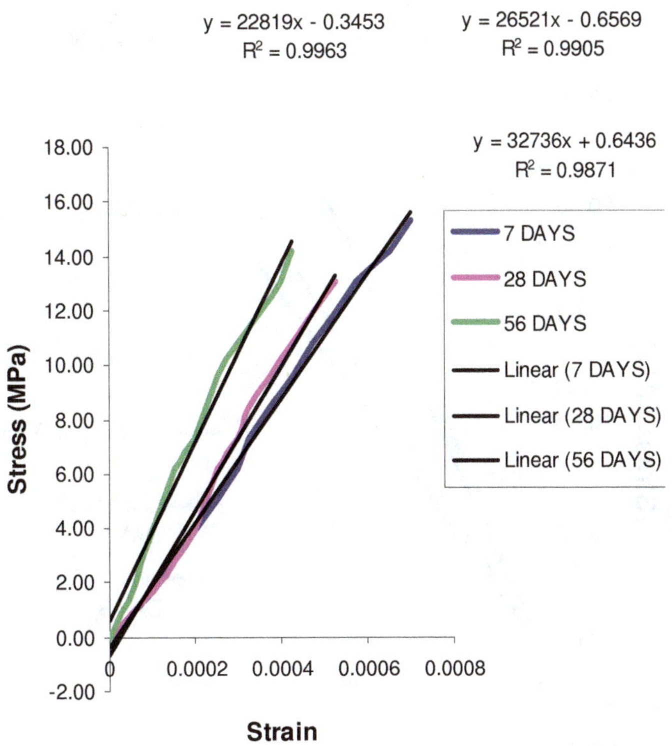

Figure 16. Reference modulus of elasticity 300 kg/m^3 at F/C=0.7.

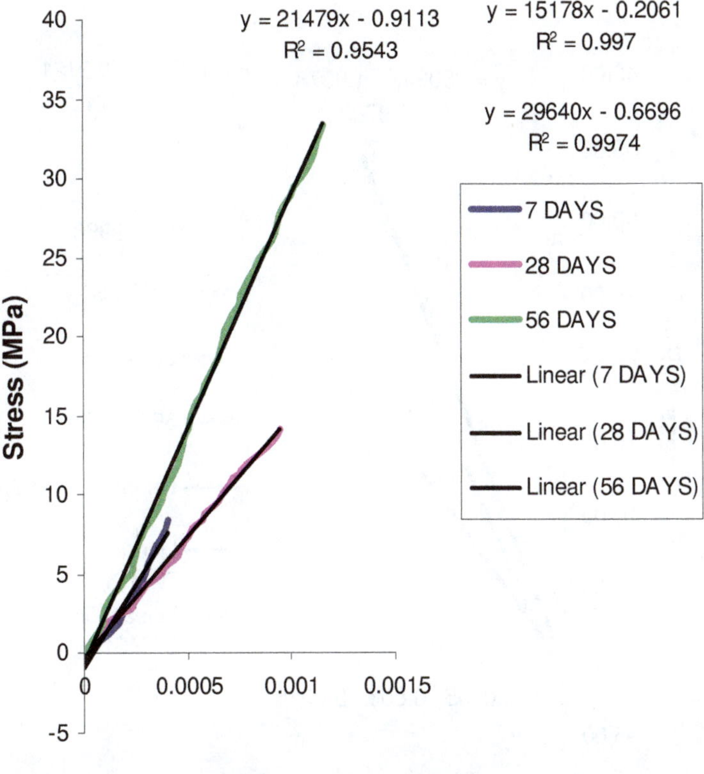

Figure 17. Effect of quarry fines on modulus of elasticity 300 kg/m^3 at F/C=0.7.

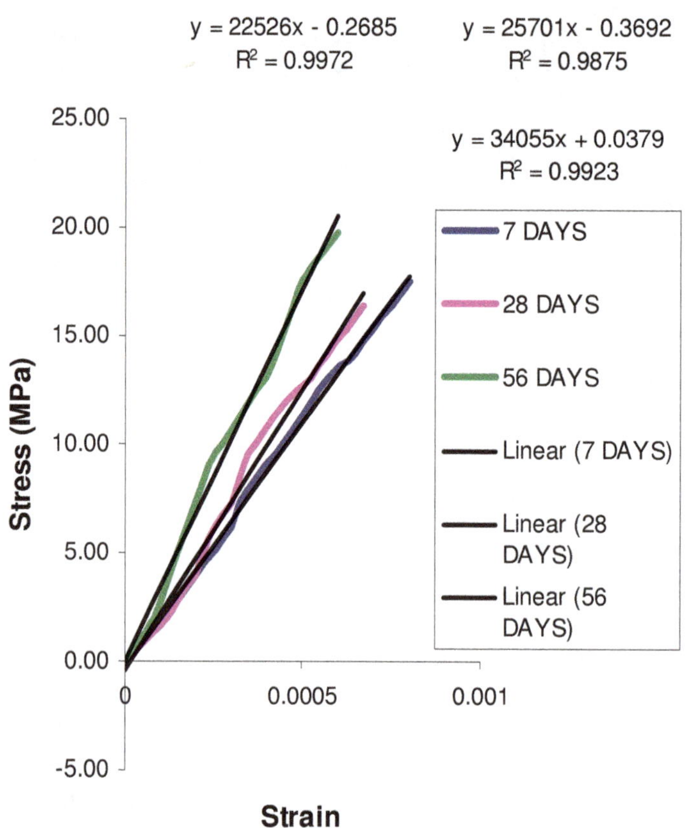

Figure 18. Reference modulus of elasticity 300 kg/m^3 at F/C=0.8.

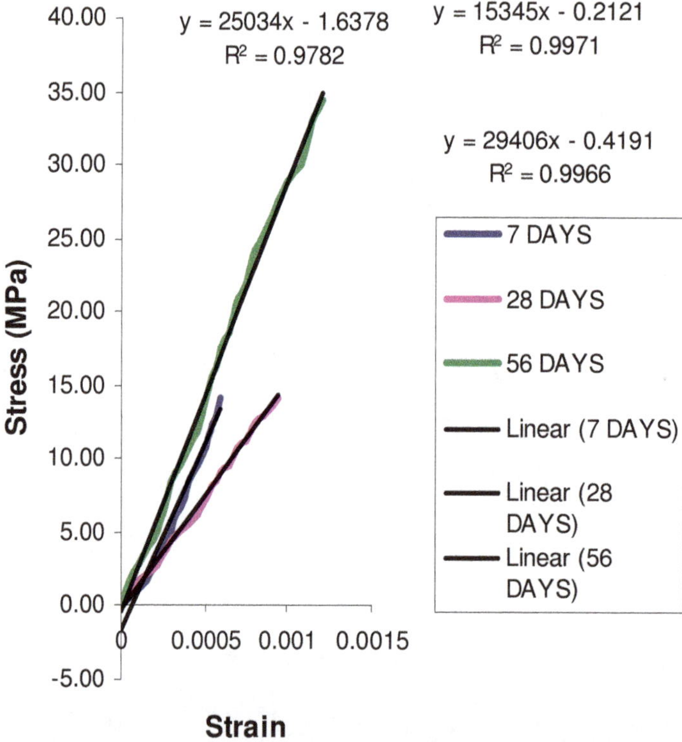

Figure 19. Effect of quarry fines on modulus of elasticity 300 kg/m^3 at F/C=0.8.

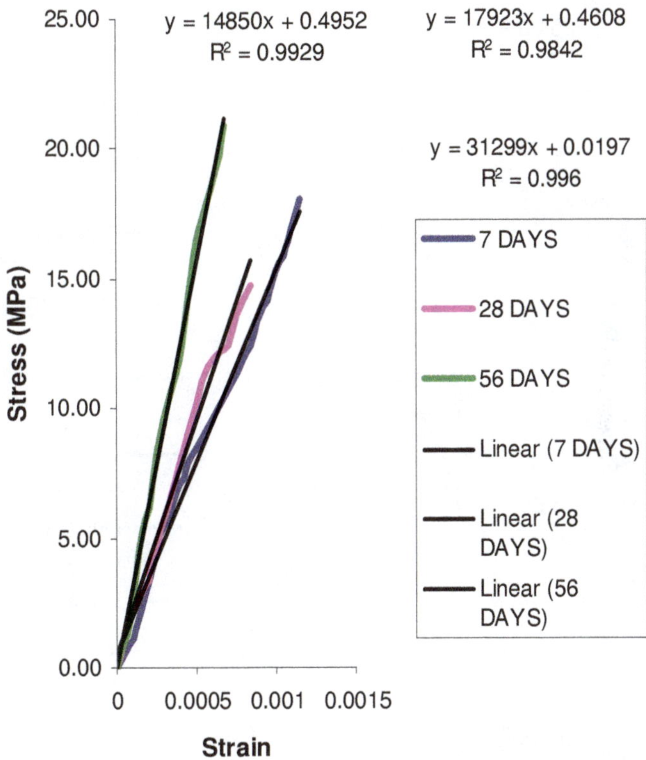

Figure 20. Reference modulus of elasticity 350 kg/m^3 at F/C=0.6.

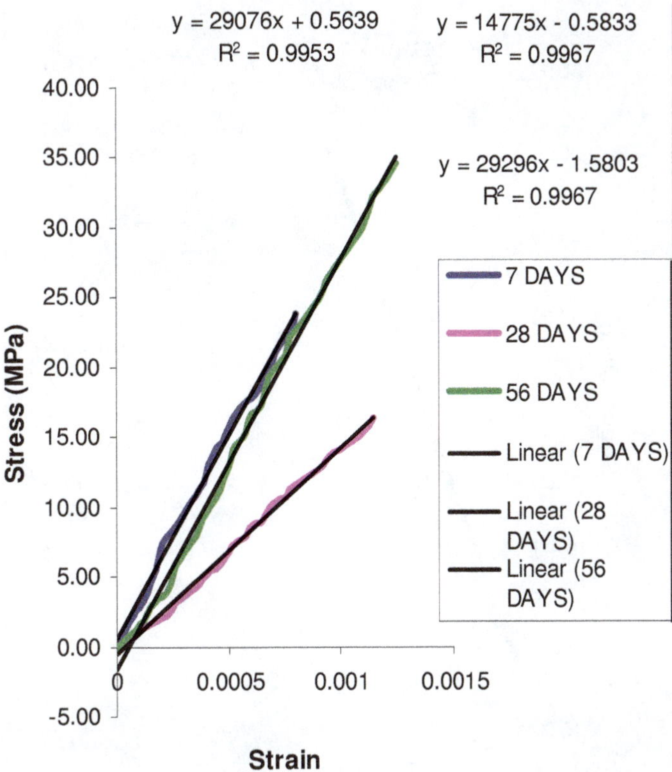

Figure 21. Effect of quarry fines on modulus of elasticity 350 kg/m^3 at F/C=0.6.

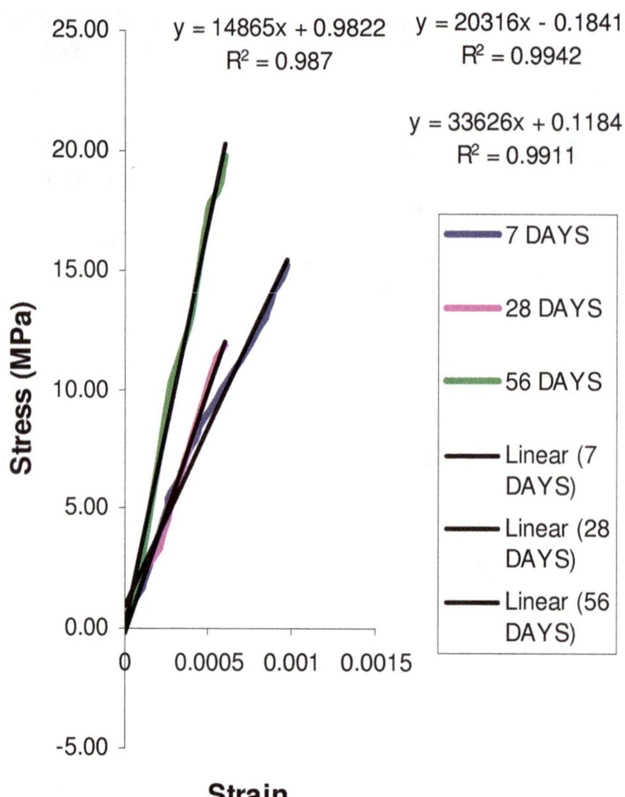

Figure 22. Reference modulus of elasticity 350 kg/m^3 at F/C=0.7.

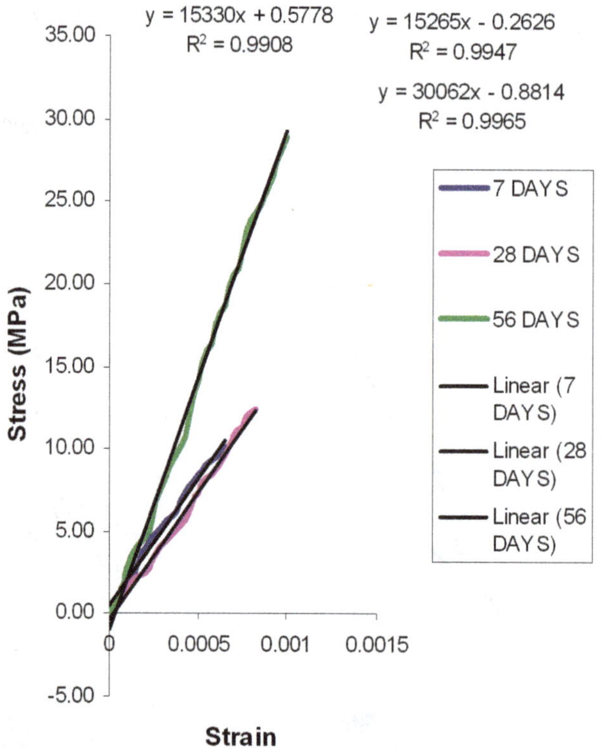

Figure 23. Effect of quarry fines on modulus of elasticity 350 kg/m^3 at F/C=0.7.

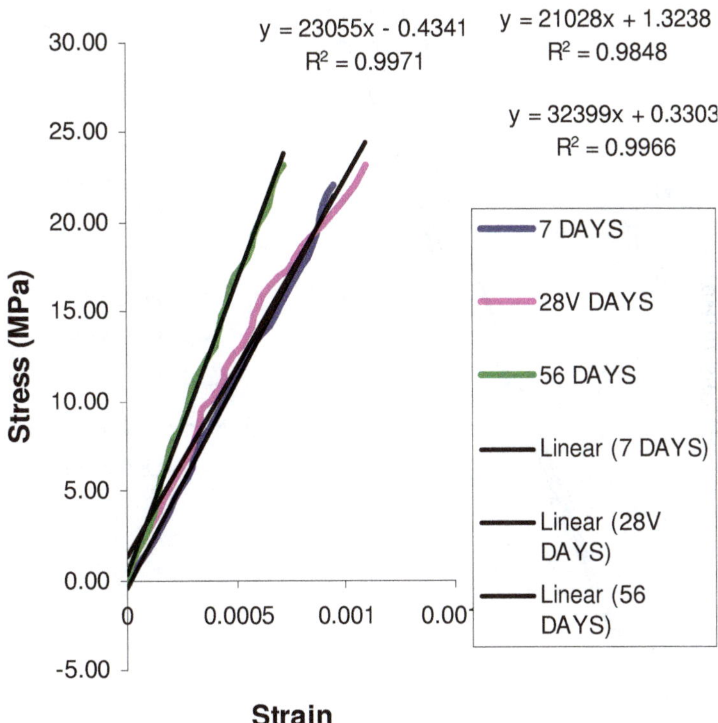

Figure 24. Reference modulus of elasticity 350 kg/m^3 at F/C=0.8.

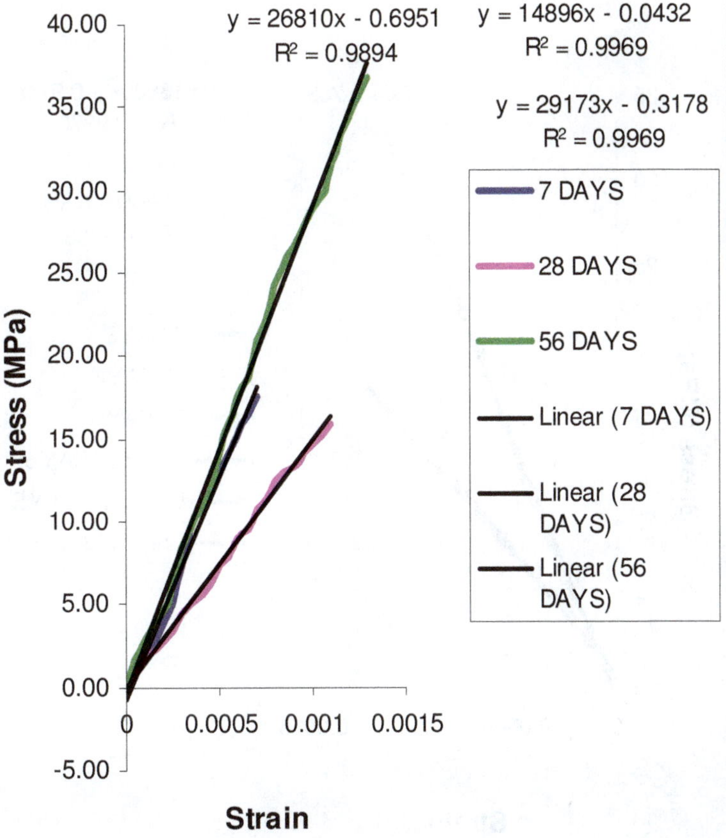

Figure 25. Effect of quarry fines on modulus of elasticity 350 kg/m^3 at F/C=0.8.

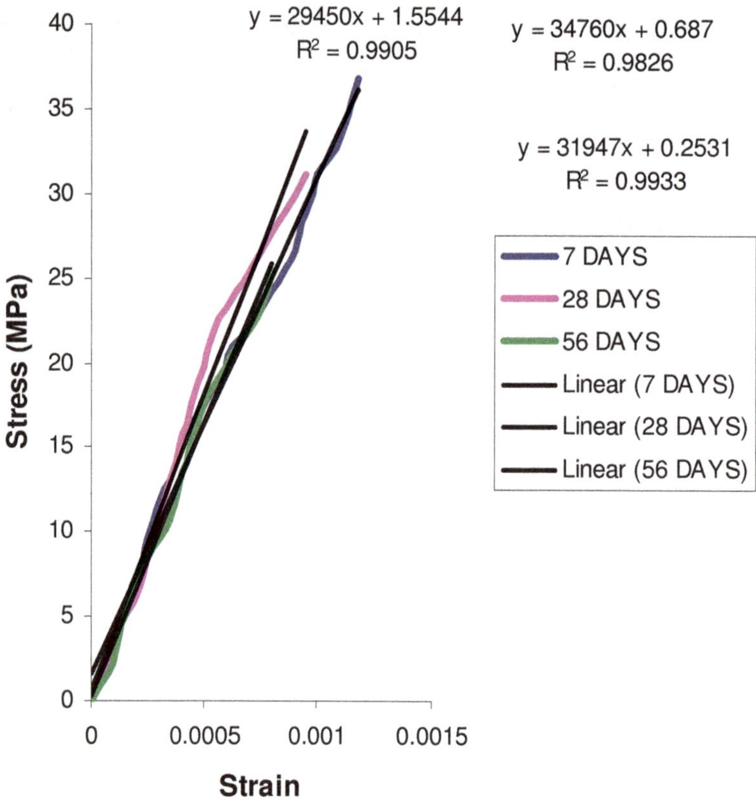

Figure 26. Reference modulus of elasticity 400 kg/m^3 at F/C=0.6.

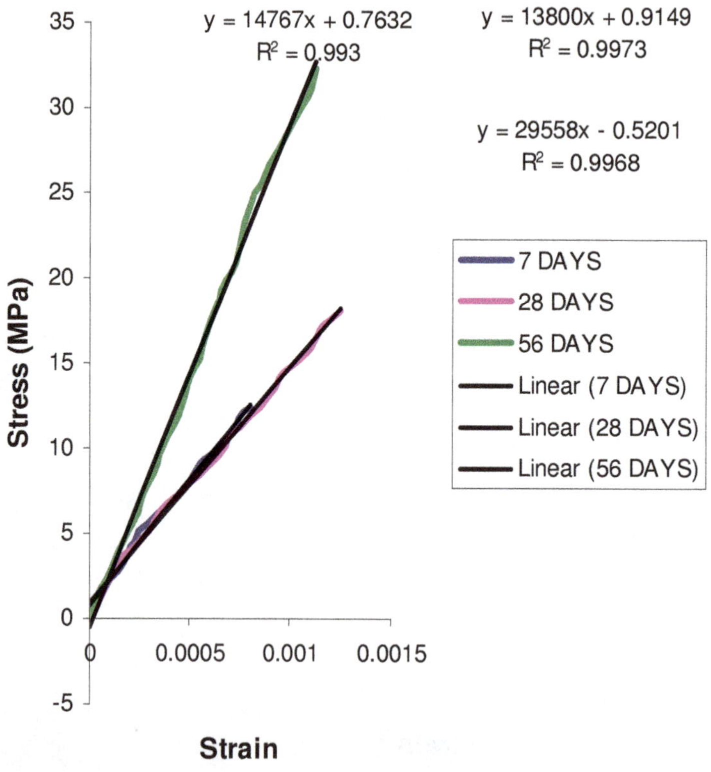

Figure 27. Effect of quarry fines on modulus of elasticity 400 kg/m^3 at F/C=0.6.

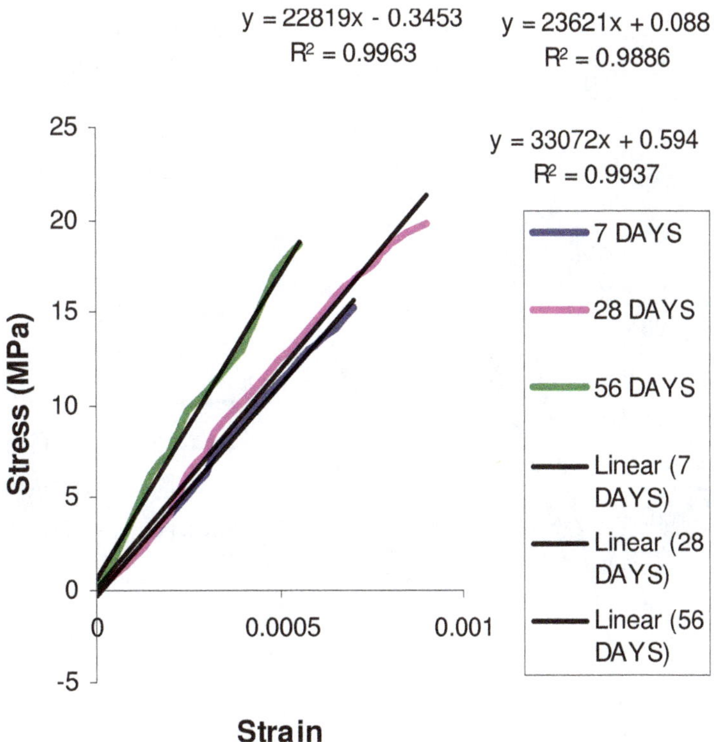

Figure 28. Reference modulus of elasticity 400 kg/m^3 at F/C=0.7.

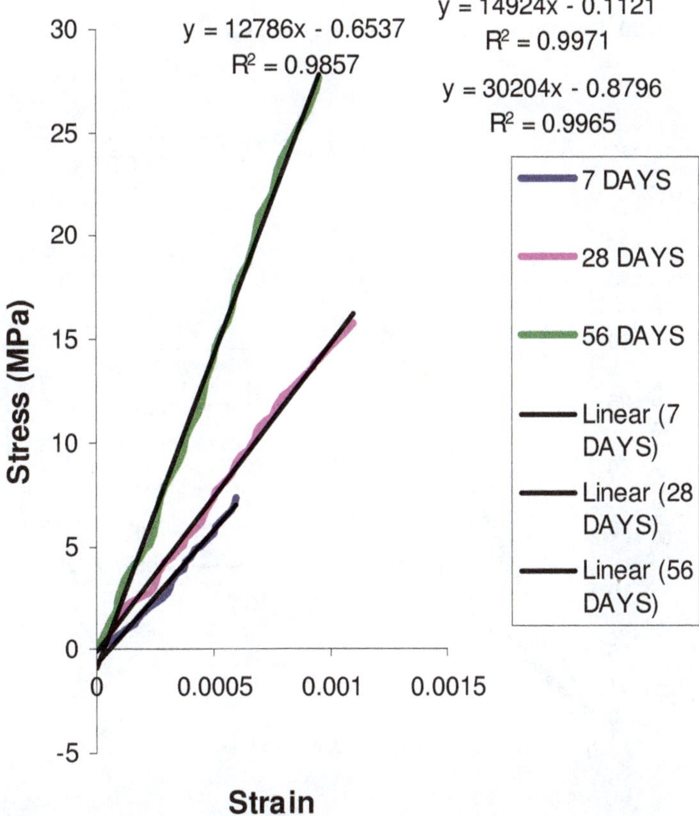

Figure 29. Effect of quarry fines on modulus of elasticity 400 kg/m^3 at F/C=0.7.

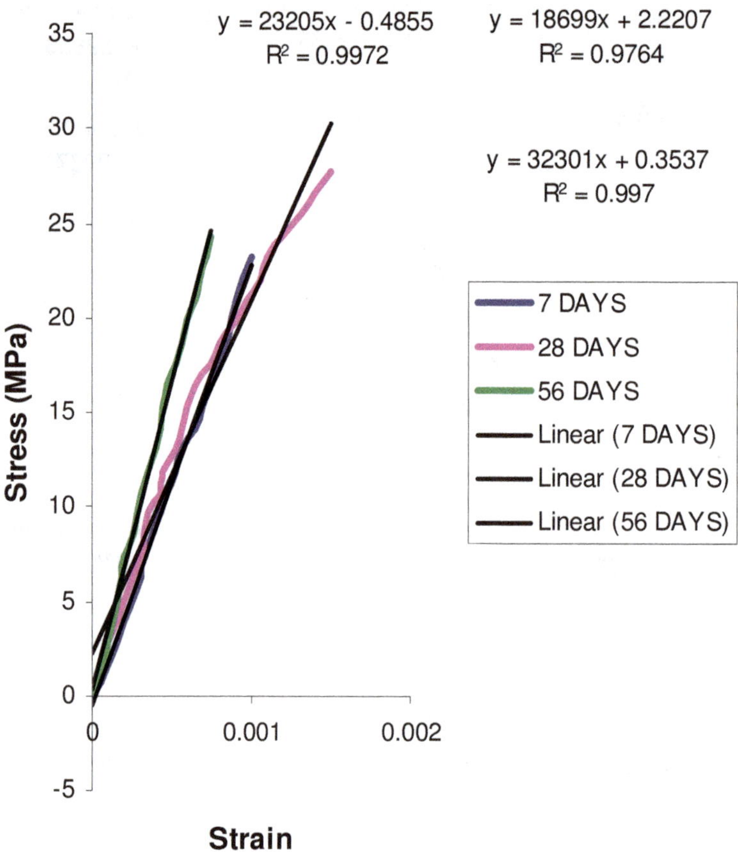

Figure 30. Reference modulus of elasticity 400 kg/m^3 at F/C=0.8.

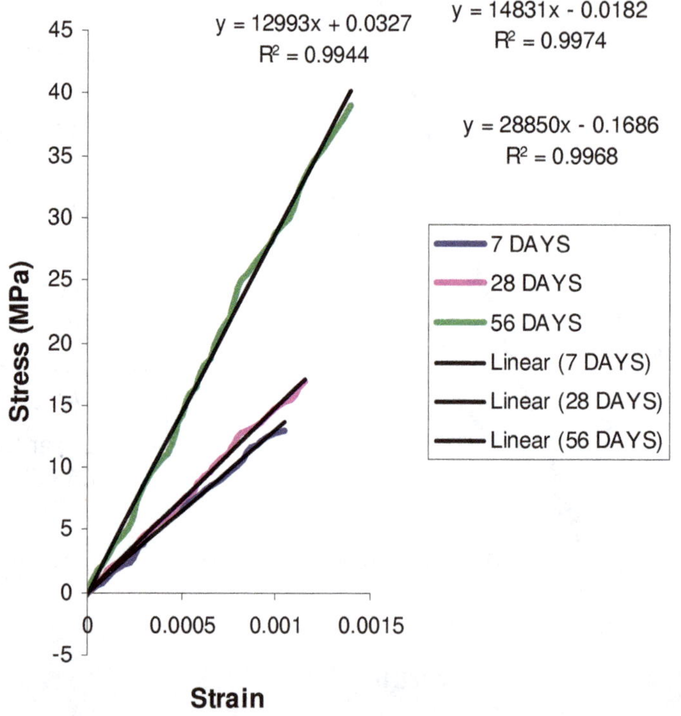

Figure 31. Effect of quarry fines on modulus of elasticity 400 kg/m^3 at F/C=0.8.

Table 5. Modulus of elasticity of concrete for various concrete specimens.

Mix ID	Cement (kg/m^3)	F/C (kg/m^3)	Modulus of elasticity of concrete (GPa)					
			without quarry dust			with quarry dust		
			7 days	28 days	56 days	7 days	28 days	56 days
		0.6	30.77	17.69	32.71	30.49	21.0	28.53
Mix 1	300	0.7	21.84	24.80	33.33	21.23	14.90	29.05
		0.8	21.94	24.32	33.03	23.59	14.89	28.78
		0.6	15.09	27.74	31.45	29.31	14.27	27.67
Mix 2	350	0.7	15.68	19.81	33.03	15.67	15.09	28.87
		0.8	23.24	21.10	32.02	25.07	14.40	28.31
		0.6	31.30	32.78	31.84	15.56	14.49	28.68
Mix 3	400	0.7	21.84	22.02	33.97	12.27	14.41	29.21
		0.8	23.21	18.49	31.85	12.40	14.76	27.90

Conclusion

The analysis of experimental data showed that the addition of the quarry dust improved the strength properties of concrete which was on par with that of conventional concrete. From above testing results, it is inferred that the quarry dust may be used as an effective replacement material for natural river sand. The increase of cement content in the mortar phase shows an increase in the strength. The fine quarry dust tends to increase the amount of superplasticizers needed for the quarry mixes in order to achieve the rheological properties. The 28 days compressive strength of 100% replacement of sand with quarry dust of mortar cube (CM 1:1) is 11.8% higher than the controlled cement mortar cube. The 56 days maximum Compressive strength for 100% replacement of sand with quarry dust of 400 kg/m^3 at F/C=0.6 was 17.45% higher than the reference concrete. At 56 days the maximum split tensile strength 100% replacement of sand with quarry dust of 400 kg/m^3 at F/C=0.6 was 15.25% higher than the reference concrete. The maximum modulus of elasticity at 100% replacement of sand with quarry dust of 300 kg/m^3 at F/C=0.6 is 10.24% higher than the reference concrete. When the quarry dust has high fineness, its usage in the normal concrete is limited because it increases the water demand.

REFERENCES

Celik T, Marar K (1996). Effects of crushed stone dust on some properties of concrete, Cement Concrete Res., 26(7):1121-1130.

De Larrard F, Belloc A (1997). The influence of aggregate on the compressive strength of normal and high-strength concrete. ACI Mater J., 94(5):417–426.

Galetakis M, Raka S (2004). Utilization of limestone dust for artificial stone production: an experimental approach. Miner. Eng., 17:355–357.

Gambhir ML (1995). Concrete Technology-Second Edition, Tata McGraw-Hill Publishing Company Limited, New Delhi.

Goble CF, Cohen MD (1999). Influence of aggregate surface area on mechanical properties of mortar. ACI Mater. J., 96(6):657–662.

IS: 516-1959. Indian Standard Methods of Test for Strength of concrete. Bureau of Indian Standards, New Delhi.

Nevillie AM (2002). Properties of Concrete –Fourth and Final Edition, Pearson Education Limited, Essex.

Safiuddin M, Zain MFM, Mahmud MF, Naidu RS (2001). Effect of quarry dust and mineral admixtures on the strength and elasticity of concrete, Proceedings of the Conference on Construction Technology, Kota Kinabalu, Sabah, Malaysia, pp. 68-80.

12

Study of coir reinforced laterite blocks for buildings

Aguwa J. I.

Department of Civil Engineering, Federal University of Technology, Minna, Nigeria.

Some index property tests which include natural moisture content, sieve analysis, specific gravity, compaction, Atterberg limits were carried out on the laterite sample for the purpose of identification and classification. Twenty blocks of size 150 mm × 150 mm × 150 mm of coir reinforced laterite were moulded at coir content of 0, 0.063, 0.125, 0.188 and 0.25% by mass of the laterite respectively, and they were cured in the laboratory under atmospheric conditions. These blocks were subjected to compressive strength test at the curing ages of 7, 14, 21 and 28 days respectively. Five blocks were tested for each mix at each curing age, making a total of one hundred blocks. It was found that the 28 days compressive strength of the coir reinforced laterite blocks were 2.11, 2.18, 2.18 and 2.26 N/mm^2 for 0.063, 0.125, 0.188 and 0.25% coir content respectively. The determined strengths of coir reinforced laterite blocks are higher than those for ordinary laterite blocks.

Key words: Buildings, coir, compressive strength, laterite blocks.

INTRODUCTION

The need for locally manufactured building materials has been emphasized in many countries of the world. There is imbalance between the expensive conventional building materials coupled with depletion of traditional building materials. To address this situation, attention has been focused on low-cost alternative building materials (Agbede and Manasseh, 2008).

Often one of the problems encountered in the study of laterite is the basic definition of what is laterite. Many people usually define laterite as a type of red soil used in road construction especially in the tropics. Although laterite physically has element of red colour, the above definition is not a clear and true one in that some soils such as red sandy-clay soil can easily be mistaken as laterite. Laterite may be defined as that class of pedogenics in which the cementing materials are the sesquioxides and constitute not less than 50% of its constituents when the sample is chemically analyzed.

Sesquioxides are those chemical substances with empirical formula M_2O_3 where M=Potassium, K, Rubidium, Rb or Cesium, Cs. At ordinary temperatures and pressure below 100 mm mercury, Potassium peroxide combines with oxygen to give the sesquioxides, K_2O_3.

Study of laterite shows that laterite contains hydrated aluminum and iron oxides and the presence of iron can be noticed by the characteristic colour produced by iron in the soils. The aluminum is generally in the form $Al_2O_3.nH_2O$, which is called Bauxite, an ore of aluminum. The ore appears to be developed when intense and prolonged weathering removes the silicon from the clay minerals and leaves a residue of hydrous aluminum oxides.

Before carrying out chemical analysis for complete definition of laterite, a soil may be suspected to be laterite by observing some of the physical properties. Usually, laterite is reddish brown in colour and gravelly in texture. The reddish colour becomes predominant when wet while the brown colour becomes distinct when dry. Some of the particles stick to the palm of the hand when wet and they can easily be dusted off when dried. Also some laboratory index property tests are used to classify laterite. Such tests are the Atterberg limit test, grain size analysis, compaction test etc. The results of these tests are compared with already determined standard results of the index properties.

Soil stabilization and improvement can be achieved by altering the soil properties to conform to the desired characteristics. The objectives of soil stabilization include increase in strength, reducing compressibility, improving stability, decreasing heave due to frost or swelling and increasing or decreasing permeability. Soil stabilization has been extensively used in the construction of roads, airfields, earth dams and embankment. Stabilization includes compaction, pre-consolidation and protection of the surface from erosion and moisture infiltration. Investigation of soil and its groundwater condition will indicate whether soil improvement or stabilization is needed and the technique to be employed. Chemical additives such as Portland cement, hydrated lime, gypsum, alkalis, sodium chloride, calcium chloride, aluminium compounds and industrial waste products have the potentials for soil stabilization. Magafu (2010) reported that lime has proved to permanently increase strengths (compressive, tensile, flexural and shear) of soils, reduce to minimum volumetric expansion of soils and create excellent freeze-thaw resistance (durability). He further added that however, due to cheapness and availability in most developing countries, the most widely used stabilizers are the Portland cement, lime, bitumen and agricultural waste are used to a lesser extent.

Coconuts are agricultural products and edible in at least all the tropical countries including Nigeria. The husks and shells are thrown away as wastes and they pose a lot of environmental waste problems. Within the husks are the strands like fibres called coir, which are removed from the husks by beating, decorticating or defibring (Dutch Plantin). These coconut fibres are strong, tough and extremely resistant to fungal and bacteria decomposition. Fibre length varies from 0.3 to 250 mm but the average length ranges from 100 to 200 mm. Coir cross sections are highly elliptical and non uniform with average diameter of 0.25 mm. It has high degree of crystallinity with spiral angle of micro fibre ranging between 30 and 40° and this imparts greater extensibility compared to other natural fibres. Coir has a high lignin content and thus a low cellulose content, as a result of which it is resilient, strong and highly durable (TIS). In spite of low cellulose content, coir are very closely arranged and this accounts for its better durability compared to other natural fibres.

Chandra et al. (2008) reinforced three types of soil; clay, silt and silty sand with polypropylene fibre of 0.3 mm diameter and found that their uniaxial compressive strengths increased appreciably. This result is synonymous to that of Fatani et al. (1991) who reinforced soils with metallic fibres. Fibres play significant role when soil is subjected to tension forces.

A major advantage of the use of laterite instead of sand in moulding building blocks is the low cost, due to little or no quantity of cement is required to produce blocks with adequate compressive strengths (Aguwa, 2010).

Laterite bricks were made by the Nigerian Building and Road Research Institute (NBRRI) and used for the construction of a bungalow, Madedor (1992). From the study, NBRRI proposed the following specifications as requirements for laterite bricks: bulk density of 1810 kg/m^3, water absorption of 12.5%, compressive strength of 1.65 N/mm^2 and durability of 6.9% with maximum cement content fixed at 5%.

The aim of this study is to investigate the use of coconut fibres in reinforcing laterite blocks in order to improve strength for buildings. Some of the major objectives of this work are; to determine the physical properties of laterite and coir, to mould laterite blocks reinforced with coir at varying percentage of coir content and to determine the compressive strengths of these blocks at 7, 14, 21 and 28 days. The use of laterite blocks in buildings is very common especially in rural areas. With the abundance of laterite and coconut fibres in this country, coir reinforced laterite can be used for various construction works such as load bearing and non-load bearing walls for resisting static loading. This will reduce the cost of construction materials compared to sandcrete blocks which are not economical and also the environmental problem of coir deposition can be reduced by converting these wastes to treasure.

MATERIALS AND METHODS

Laterite

The laterite sample used was collected at a depth of 1.5 m to 2.5 m from an existing borrow pit behind works department of Federal University of Technology Gidan Kwano campus Minna, Nigeria (Latitude 9° 37'N and Longitude 6° 33'E), using the trial pit method of disturbed sampling.

Coir

The coconut fibre was obtained from coconut farm at Garatu Bida road Minna, Nigeria. They were sun dried properly, removed carefully and cut into smaller pieces, not exceeding 50mm in length to allow for proper mixing with the laterite.

Water

Tap water was used for the mixing and it was properly examined to ensure that it was clean, free from particles and good for drinking as specified in BS 3148 (1980).

Soil index properties

Index property tests on the laterite soil for the purposes of

Plate 1. Coir reinforced laterite blocks under curing.

characterization and classification, which include natural moisture content, particle size distribution, Atterberg limits, specific gravity, linear shrinkage and compaction tests were carried out in accordance with BS 1377 (1990).

Preparation of specimens

The respective quantities of laterite, coir as well as water required for the mix were proportioned and batched by mass. In quality controlled mix such as this, measurement by mass was adopted, (Vazirani and Chandola, 1997). Manual mixing and moulding were used and proper mixing was achieved by turning the mixture from one side to the other for six times (Neville, 2000). Five different mixes were prepared for coir reinforced laterite using coir contents of 0, 0.063, 0.125, 0.188 and 0.25% by mass of laterite respectively. One hundred blocks of coir reinforced laterite mixes of size 150 mm × 150 mm × 150 mm were moulded by filling the mould in three layers and each layer was given 30 blows of compaction using a standard rammer of weight 2.5 kg falling from a height of 30 cm. To ensure even distribution of blows, approximately 150 mm square sheet of plywood was placed on the mixture in the mould and compaction was done on it. The freshly moulded blocks were carefully extruded on a clean and flat surface and were cured under laboratory conditions. The coir reinforced laterite blocks under curing are shown in Plate 1.

Compressive strength test

An electrically operated Seidner compression machine was used for the compressive strength test on the coir reinforced laterite blocks in accordance with BS 1881 Part 116 (1983). The blocks were subjected to compressive strength test at 7, 14, 21 and 28 days age of curing in five replications for each mix and the average compressive strength was calculated. In crushing test, care was taken to ensure that the blocks were properly positioned and aligned with the axis of the thrust of the compression machine to ensure uniform loading on the blocks (Neville, 2000).

RESULTS AND DISCUSSION

Identification of laterite soil

The index properties of the laterite used for the study are summarized in Table 1 while Figure 1 shows the particle size distribution. The laterite was well graded in accordance with BS 882 (1983) and classified to be A-7-6 according to AASHTO (1986) classification based on the geotechnical properties. The laterite is reddish-brown in colour with specific gravity of 2.28 and linear shrinkage of 10. The percentage of the soil passing BS sieve No 200 is 34.32 and the plasticity index is 20.5. An investigation into the geotechnical and engineering properties of sample as well as study of soil maps of Nigeria after Akintola (1982) showed that the sample collected belong to the group of ferruginous tropical soils derived from acid igneous and metamorphic rocks.

Figure 2 shows the compaction characteristics for the laterite used in the study and a normal compaction curve for most laterite is depicted, indicating the maximum dry density (MDD) of 1430 kg/m^3 and the optimum moisture content (OMC) of 29.8%.

Compressive strength

The relationship between compressive strength and percentage of coir content is shown in Figure 3. It was observed that the compressive strength of coir reinforced

Table 1. Properties of the Natural Laterite used for the study.

Characteristic	Laterite
Natural moisture content (%)	17.25
Percentage passing BS No 200 sieve (%)	34.32
Liquid limit (%)	50
Plastic limit (%)	18.33
Plasticity Index (%)	31.67
Linear shrinkage	10
AASHTO classification	A-7-6
Maximum dry density (kg/m^3)	1430
Optimum moisture content (%)	29.8
Specific gravity	2.28
Condition of sample	Air-dried
Colour	Reddish-Brown

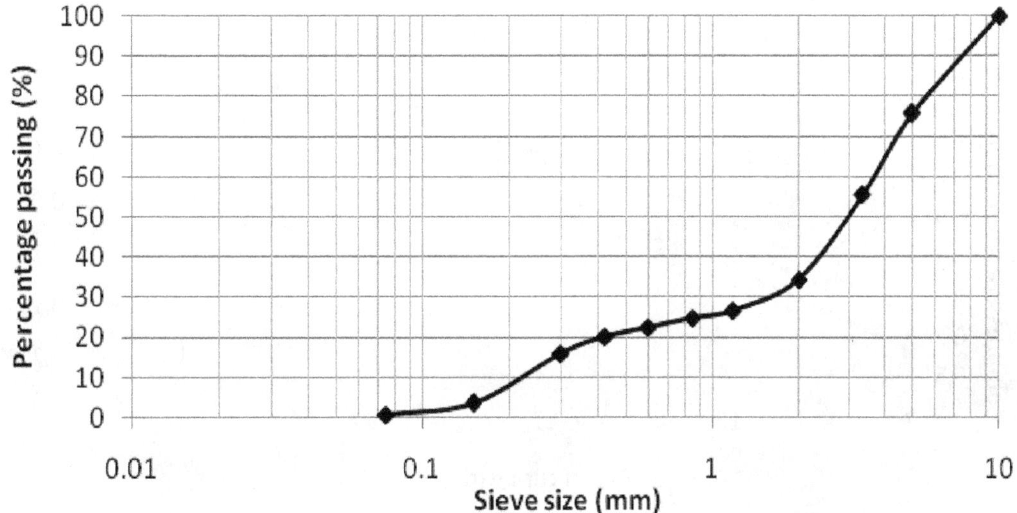

Figure 1. Particle size distribution curve for the Laterite.

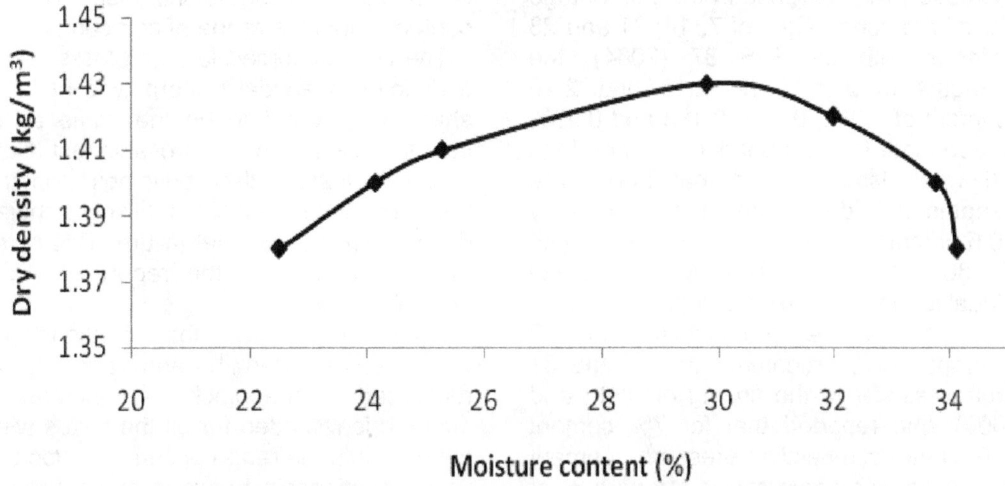

Figure 2. Compaction characteristics for the Laterite.

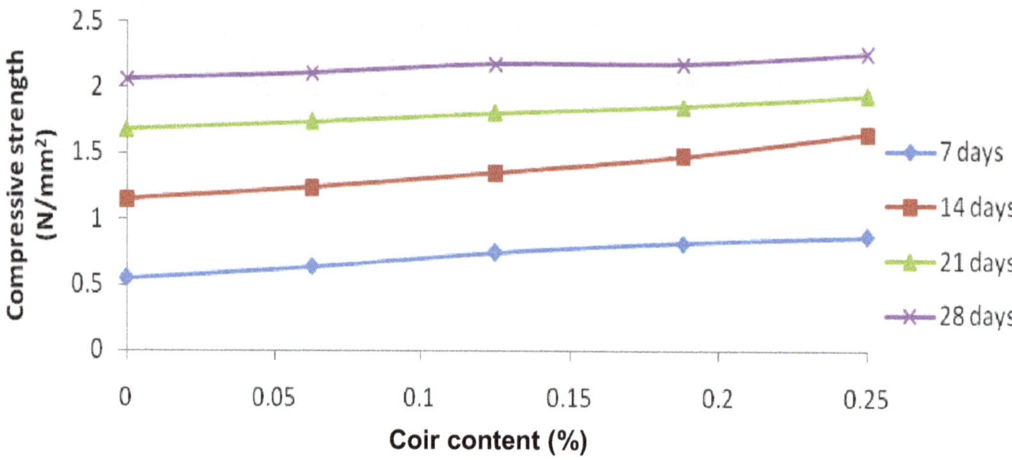

Figure 3. Compressive Strength-Coir content relation for coir reinforced laterite blocks.

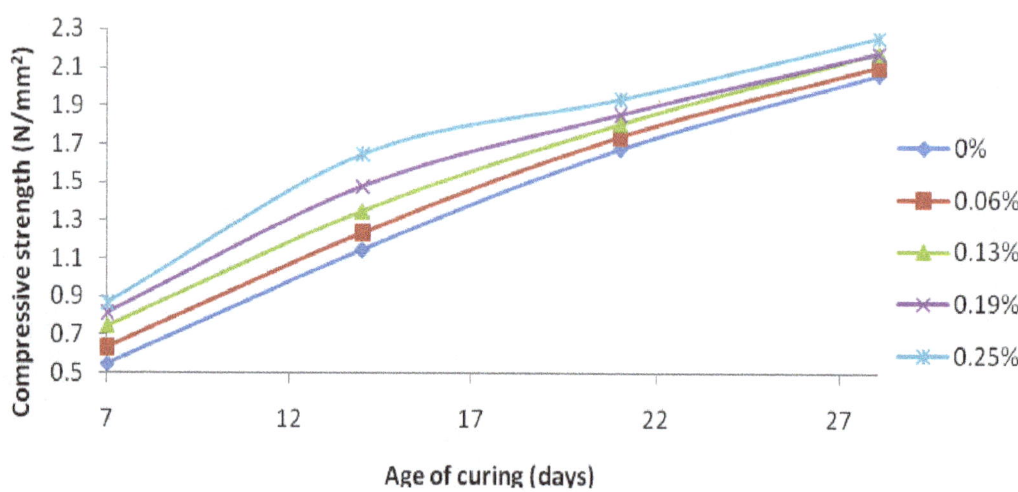

Figure 4. Compressive Strength-Age of curing relation for coir reinforced laterite blocks.

laterite blocks increased with increase in the percentage content of coir for all the curing ages of 7, 14, 21 and 28 days. In accordance with the NIS 87 (2004), the compressive strengths of 2.11, 2.18, 2.18 and 2.26 N/mm^2 for coir content of 0.063, 0.125, 0.188 and 0.25% respectively, are adequate for load bearing and non-load bearing walls. It was also observed that the failure compressive strengths at 28 days curing age did not vary by more than ±0.5 N/mm^2 and this is in agreement with clause 9 of BS 1881-127 (1990). The lowest crushing strength of individual load bearing blocks shall not be less than 2.5 N/mm^2 for machine compaction and 2.0 N/mm^{2}for hand compaction as recommended by NIS 87 (2004). This result is similar to the finding of Alutu and Oghenejobo (2006) who reported that for 7% cement content and13.76 N/mm^{2}compactive pressure, cement stabilized laterite blocks with compressive strength of at least 2.0 N/mm^2 at 28 days could be produced. The compressive strengths increased with increase in coir content within the range of coir content studied.

The coir reinforced laterite blocks consistently showed a definite pyramidal pattern type of failure. The conical shape was found to be the same as that of concrete cubes subjected to compression test (Neville, 2000). This is an indication that buildings constructed with coir reinforced laterite blocks will likely withstand considerable deformation before total failure. This pyramidal pattern of failure conforms to the report by Aguwa and Tsado (2011).

Figure 4 shows the relationship between the compressive strength and age of curing for coir reinforced laterite blocks. An increase in compressive strength is recorded for all the mixes with longer days of curing within the range of curing period tested.

The relationship between mass and age of curing for coir reinforced laterite blocks is shown in Figure 5. The

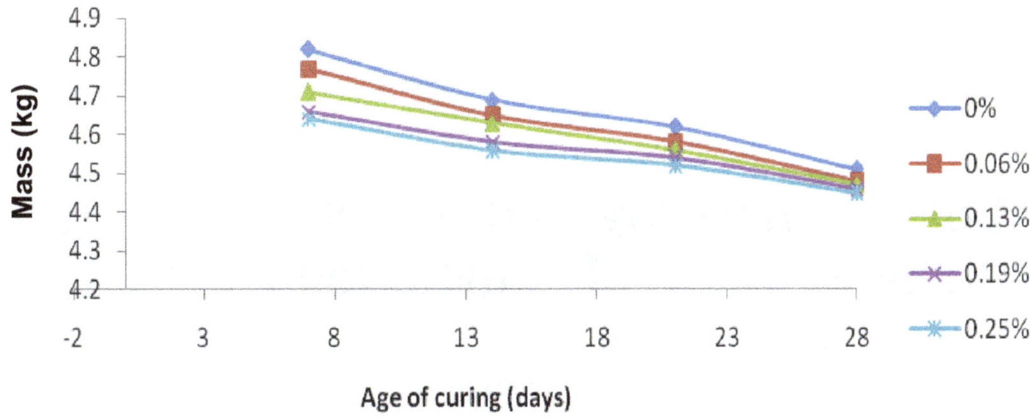

Figure 5. Mass Age of curing relation for coir reinforced laterite blocks.

decrease in mass with longer days of curing could be attributed to continued water absorption by the dry coir. Also gradual shrinkage as a result of water absorption by the coir could lead to loss in weight and increase in bonding.

Conclusion

Coir has the potential to increase the compressive strength of laterite blocks by ten percent (10%) at 28 days curing duration and reduction in mass by two percent (2%). Increase in strength and reduction in mass are two essential qualities of a structural material. Also the use of coir in reinforcing laterite blocks will minimize the environmental problem of waste deposition in addition to reduction in the cost of building blocks.

REFERENCES

AASHTO (1986). Standard Specifications for Transportation Materials and Method of Testing and Sampling, American Association of State Highway and Transportation Officials, Washington D.C., U.S.A.

Agbede IO, Manasseh J (2008). Use of Cement-Sand Admixture in Laterite Bricks Production for Low Cost Housing, Leonardo Electron. J. Pract. Technol. 12:163-174. ISSN 1583-1078.

Aguwa JI (2010). Comparative Study of Compressive Strengths of Laterite-Cement and Sandcrete Blocks as Building Materials. NSE Technical Transactions - A Technical Publication of the Nigerian Society of Engineers, 45(4):23–34.

Aguwa JI, Tsado TY (2011). Effect of Mixing Water Content on the Compressive Strength of Laterite-Cement Blocks as Walling Units in Buildings. Environ. Technol. Sci. J. (ETSJ) 4(1):60–67.

Akintola FA (1982). Geology and Geomorphology. In Nigeria in Maps, edited by K.M Barbours, Hodder and Stoughton, London

Alutu OE, Oghenejobo AE (2006). Strength, Durability and Cost Effectiveness of Cement-Stabilized Laterite Hollow Blocks. J. Eng. Geol. Hydrogeol. 39(1):65-72.

BS 3148 (1980). "Tests for Water for Making Concrete." British Standard Institution, 2 Park Street, London, WIA 2BS.

BS 882 (1983). Aggregates from Natural Sources for Concrete, British Standards Institution, 2 Park Street, London.

BS 1881 Part 116 (1983). Methods for Determining Compressive Strengths of Concrete Cubes, British Standards Institution, 2 Park Street, London.

BS 1881 Part 127 (1990). Method of Verifying the Performance of Concrete Cube Compression Machine Using the Comparative Cube Test, British Standards Institution, 389 Chiswick High road London, W4 4AL.

BS 1377 (1990). Methods of Testing Soils for Civil Engineering Purposes. British Standards Institution, 2 Park Street, London.

Chandra S, Viladkar MN, Nagrale PP (2008). Mechanistic Approach for Fibre -Reinforced Flexible Pavements. J. Transp. Eng. 134(1):15-23.

Dutch Plantin http://www.dutchplantin.com/en/company/produ Accessed on 15th December, 2012 by 6.00 AM.

Fatani MN, Bauer GE, Al-Joulani N (1991). Reinforcing Soil with Aligned and Randomly Oriented Metallic Fibres. Geotech. Test. J. ASTM, 14(1):78-87.

Madedor AO (1992). The Impact of Building Materials Research on Low Cost Housing Development in Nigeria, Engineering Focus April–June, Publication of the Nigerian Society of Engineers, 4(2):37-41.

Magafu FF (2010). Utilization of Local Available Materials to Stabilize Native Soils (Earth Roads) in Tanzania-Case Study Ngana, doi: 10.4236/eng.2010.27068 Published Online July 2010 (http://www.SciRP.org/journals/eng), Accessed on 20th October, 2012 by 6.30AM, Engineering, 2010, 2, 516-519.

Neville AM (2000). Properties of Concrete, 4th Edition, 39 Parker Street, London, Pitman Publishing Ltd.

NIS 87 (2004). Standards for Sandcrete Blocks. Standard Organization of Nigeria, Lagos, Nigeria.

Transport Information Service (TIS) (2012). (Cargo Loss Prevention Information from German, Marine Insurers), (http://www.tis-gdv.de/tis_e/ware/fasern/kokosf).

Vazirani VN, Chandola SP (1997). Concise Handbook of Civil Engineering. Chand & Company Ltd, 7361 Ramnagar, New Delhi-110055, p.1135

13

Effect of soil moisture in the analysis of undrained shear strength of compacted clayey soil

Rohit Ghosh

Department of Construction Engineering, Jadavpur University, Kolkata, India. E-mail: rohitghosh29@gmail.com.

An experimental study was undertaken to study the principles of soil compaction and establish a relation between the undrained shear strength of compacted clay and its moisture content. Compaction of soil is an important prerequisite for the construction of man-made structures like bridges, roads, dams, embankments etc. In the present study fine grained saturated clay finer than 2×10^{-6} m were used for all purposes. An important property of cohesive soils is that compaction increases their shear strength and compressibility. The shear strength of clay is the maximum shear stress it can sustain. It is helpful as the common soil failures are due to shear failures (Iannacchione et al., 1994). Undrained condition of clayey soil occurs when pore water pressure of soil changes due to external loading. If the soil is sheared without changing the water content its strength remains the same. During the study the optimum moisture content was analyzed from the compaction curve. Finally, the shear strength curve was drawn for different compaction efforts which clearly showed an exponential decrease in the shear strength of clayey soil with gradual increase in the water content. Factors other than the change in water content and compaction energy were not considered.

Key words: Undrained shear strength, optimum moisture content, dry density, compaction curve.

INTRODUCTION

Soil compaction is a process of mechanical densification of soil by pressing the soil particles close to each other and removing the air between them. It is of utmost importance in the broad science of Geotechnical engineering playing a significant role in all types of Geotechnical investigations. Compacted soils are widely used in the construction of geotechnical and geo-environmental structures and the durability and stability of these structures are directly related to the achievement of proper soil compaction. Compaction is not appropriate for granular soils. The principal soil properties affected by compaction include settlement, shearing resistance, water movement and volume change. The constitutive equations for volume change, shear strength and flow for unsaturated soil have been generally accepted in Geotechnical engineering (Freedlund and Rahardjo, 1993a). Hence, undrained shear strength analysis of compacted soils is of relevance in dealing with these structures. The shear strength of an unsaturated clayey soil and soil-water characteristic curve depend on the soil structure or the aggregation which in turn depends on the initial water content and the method of compaction. Cohesion appears to be largely due to the intermolecular bond between the adsorbed water surrounding each grain, especially in fine grained soils. Therefore, the value of the cohesion will thus vary with the soil water content, grain size of soils and its compaction. As moisture content increases cohesion decreases because of greater separation of clay particles. The bearing capacity of all types of soils and clayey soils in particular, by and large, depend on their shear strength. The drained and undrained shear strengths of clayey soil are different due to varying soil structures. Shear strength of clayey soil is its tendency of resisting shear movement along the soil surface. Most geotechnical failures involve shear failures of the soil. The shear strength parameters of soil are known as cohesion, c and angle of friction, Ø and are defined by the Mohr-Coulomb failure envelope:

$$\tau = \sigma \ \tan(\phi) + c$$

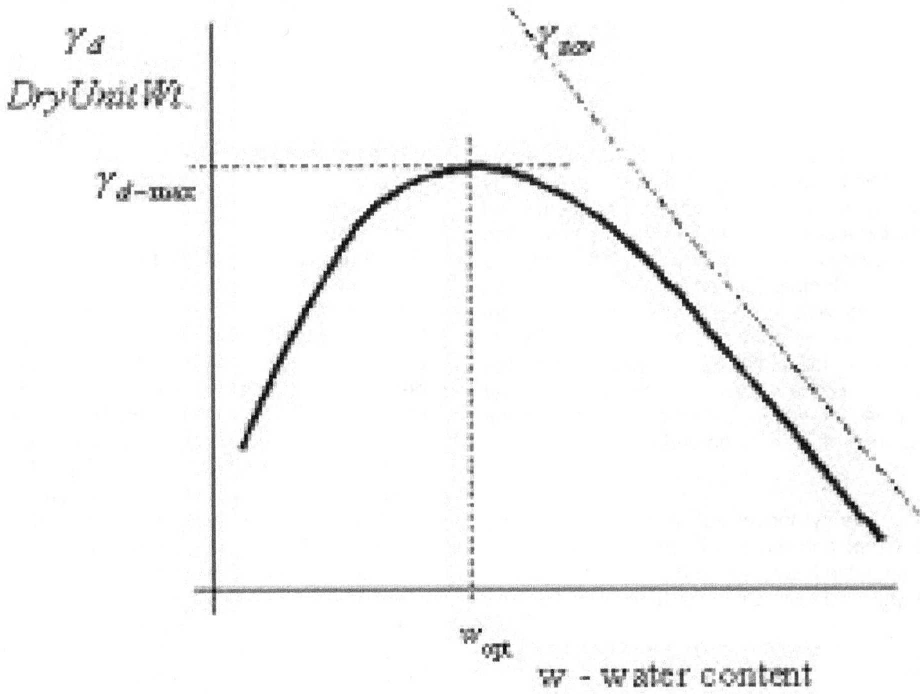

Figure 1. Relation between moisture content and dry unit weight of soil.

Where, **σ** is the normal stress.

BACKGROUND INFORMATION

The moisture content of a soil has a major impact on how well the soil will compact. When a soil is completely dry it will not compact to it greatest possible density because of friction between the soil particles. As the moisture content increases, the water lubricates the soil, allowing it to move more easily into a compact state and the density increases. Eventually the soil is compacted to its greatest possible dry density (the maximum dry density) and the moisture content at which this happens is referred to as the 'Optimum Moisture Content'. If the soil is wetted further, the extra water replaces some of the solid soil particles and the dry density reduces as there is less material present. Soil compacted at moisture content greater than the optimum moisture content has exactly the opposite characteristics to the one compacted below it. For a particular compaction effort, the dry density of soil increases with the moisture content of the soil up to the optimum moisture content beyond which it decreases. When the compaction effort increases, the optimum moisture content decreases. The relationship between dry density and water content is usually represented by a graph (Figure 1).

The zero-air-void unit weight is represented by the following equation and the zero-air-void line is shown beside.

$$\gamma(z.a.v) = \frac{G_s \gamma_w}{1 + wG_s}$$

γ(zav)

W

It is clear from the given equation that the zero-air-void density is inversely proportional to the moisture content w of the soil. For a given soil and moisture content the best possible compaction is represented by the zero-air-voids curve. The actual compaction curve will always be below. As more water is added and the moisture content exceeds the optimum value the void spaces get filled with water and further compaction is not possible. The specific gravity of water and dry density of soil are assumed constant for a particular moisture content.

CLASSIFICATION OF CLAYEY SOILS

According to their moisture content, clayey soils can be classified into 4 categories namely solid, semisolid, plastic and liquid. As the soil moves from being solid to liquid, the cohesion value (c) decreases. Clay particles are less than 0.00015 inch (0.004 mm) in diameter and are much smaller than ordinary sand. Clayey soils are virtually cohesive soils with high cohesive strength and are fine grained in nature.

PROCEDURE OF COMPACTION

For compacting the clayey soil, STANDARD PROCTOR COMPACTION TESTS were used. These tests are prescribed in ASTM D698. Designed by Proctor (1933), this laboratory test is performed to determine a relationship between the dry density and the moisture content of a soil sample for a particular compactive effort. The compactive effort is the amount of mechanical energy that is applied to the soil mass. In this test the soil is compacted using a 5.5 lb hammer falling a distance of one foot into a solid filled mould in three layers. For every value of moisture content the soil is compacted with 25 blows. Compaction effort designed in this laboratory test is comparable with that obtained in the field. Dry density achieved by mixing soil with different water contents were obtained to determine the maximum dry density and the corresponding optimum moisture content. For a particular compactive effort the dry density depends on the moisture content of the soil. The following apparatus are used during the compaction process:

(i) Moulds – They should be cylindrical with a diameter of 101.6 mm and height of 116.4 mm and therefore a volume of 944 cc. They are fitted with a detachable base plate and a removable collar.
(ii) A metal Rammer – It should have a circular face of 50 mm and weighing 2.49 kg.
(iii) Balances – They should be accurate to .01 g.
(Iv) Sieves – 75 mm, 19 mm and 4.75 mm sieves are required.
(v) Mixing Tools –Mixing pan, spoon,spatula etc.
(vi) Metal tray – It should have dimensions 600 × 800 × 80 mm.
(vii) Cans
(viii) Oven
(ix) Sample Extruder

After compaction the collar and base plate are detached carefully and the weight of the compacted soil in the mould is measured. The soil sample is extruded and cut at the middle from where a chunk is broken and put into the oven in a can for determining the water content in the soil.

Computations: The bulk density ρ in kg/m^3 of each compacted specimen is calculated as

$$\rho = (M_2 - M_1)/V,$$

Where M_1 is the mass of the mould and base in kg, M_2 is the mass of the mould, base and soil in kg and V is the volume of the mould. The dry density of the soil sample is given as:

$$\rho_d = \rho/(1 + w)$$

The amount of water to be added with air-dried soil at the commencement of the test is about 8% to 10% below the plastic limit of the soil. The water should be mixed thoroughly and adequately with the soil. The moisture content is calculated as:

$$\mathbf{w = (w_4 - w_3) / (w_4 - w_2)}$$

Where w_4 is the weight of the can + wet soil, w_3 is the weight of the can + dry soil and w_2 is the weight of the empty can.

Effect of compaction energy: As the compaction energy increases the maximum dry unit weight of compaction increases and the optimum moisture content decreases to some extent and maximum dry density increases.

Preparing the sample: The clay was mixed with sufficient amount of water to reach saturation point. The degree of saturation of compacted soil was found to be 100% through back calculation from the measured value of bulk unit weight, water content and specific gravity as given below:

$$S_r = (w.G/e) \times 100\%$$

It may kindly be noted that Khing et al. (1994) following the procedure similar to that achieved 100% saturation in the clay bed (Figure 2).

Determination of undrained shear strength

The undrained shear strength of clayey soil is determined by the Laboratory and Field Vane Shear Tests as per IS: 2720. This test is prescribed in ASTM D4648/ D4648 M-10. It provides a rapid determination of shear strength on undisturbed or remolded or reconstituted soils. This test method covers the miniature vane test in very soft to stiff saturated fine-grained clayey soils ($\phi = 0$). Knowledge of the nature of the soil in which each vane test is to be made is necessary for assessment of the applicability and interpretation of the test results. It is suitable for the determination of undrained shear strength of cohesive soils. It consists of a torque head adjustable in height by means of a lead screw rotated by a drive wheel to enable the vane to be lowered into the specimen. The vane diameter, vane size, rod diameter and vane height are in accordance with the IS codes. The test apparatus consists of a four-blade stainless steel vane which is lowered into the mould containing the compacted soil. Generally, the height of the vane is two times its diameter. It is driven by an external torque supplied by an electric motor. The vane should be inserted into the soil to a depth at least two times the height of the vane. The vane rotates at a slow speed of 0.1º per second. It determines the torsional force required to cause a cylindrical surface to be sheared by the vane; this force is then converted to a unit shearing resistance of the cylindrical surface. It is of basic importance that the friction of the vane rod and instrument be accounted for; otherwise, the friction would be improperly recorded as soil strength. The torque measured at the failure gives the undrained shear strength of the soil at that moisture level. This test is performed on clay compacted with 5, 10 and 15 blows in three layers .The torque is measured using a spring which controls the degree of rotation of the metallic dial. Essentially, the torque is dependent on the degree of rotation of the spring and varies with it. The shear strength of the soil is a function of the torque generated at failure. It is given by:

$$\mathbf{T = \tau \times K}$$

Here, T is the torque in N-m (Lambe and Whitman, 1979), τ is the undrained shear strength at failure in Pa and K is the vane blade constant in m^3.

From experiments, it is found that K= $\{11\ D^2H(1+D/3H)\} / (2 \times 10^6)\}$ in SI system, where D and H are the diameter and height of the vane respectively.

Among the basic drawbacks of this method is the assumption of full mobilization of the strength along a cylindrical failure surface as reported by Ladd et al. (1977). However, several studies have shown that mobilization of strength can be:

(i) Triangular: Shear strength mobilization is the maximum at the periphery and it decreases linearly to zero at the center.
(ii) Uniform: Shear strength mobilization is constant from the periphery to the soil center.
(iii) Parabolic: Shear strength mobilization is the maximum at the periphery and decreases parabolically to zero at the center.

Bjerrum (1974) showed that as the plasticity of soils increases, shear strength values obtained from Vane Shear Test may give results that are unsafe for foundation design. For this reason he suggested the correction:

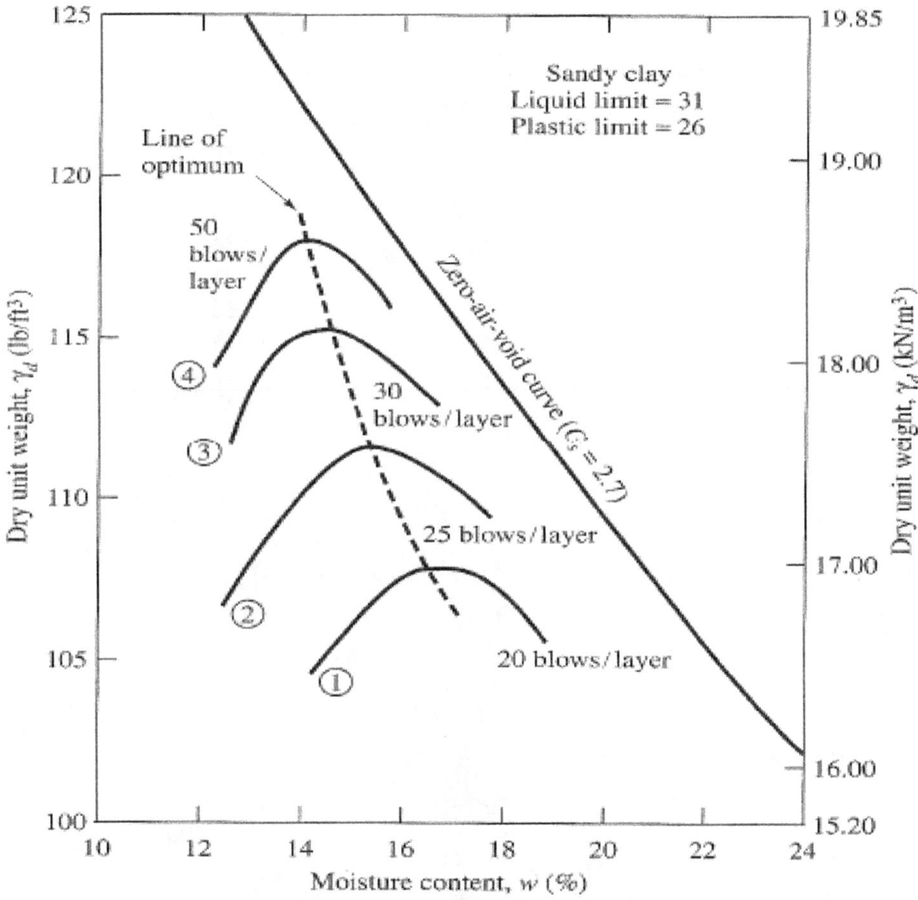

Figure 2. Effect of compaction energy on moisture content and dry unit weight.

$C_{u\,(design)} = \mu .\ C_{u\,(vane\ shear)}$
μ = correction factor = 1.7 – 0.54 log (PI)

Recently, Morris and Williams (1994) gave the correlation of μ as:

$\mu = 1.18\ e^{-.08(PI)} + 0.57$
$\mu = 7.01\ e^{-.08(LL)} + 0.57$, LL = Liquid Limit (%)

Empirical relation between the effective overburden pressure and undrained shear strength

The first such relation was proposed by Skempton (1957) and is given as: $(C_{u\,(VST)}) / \sigma_0' = 0.11 + 0.0037(PI)$, where PI is the Plasticity Index of the soil.

Ladd et al. (1977) proposed : $(C_u / \sigma_0')_{overconsolidated} / (C_u / \sigma_0')_{normally\ consolidated} = (OCR)^{0.8}$, OCR is the Overconsolidation ratio and is given as $OCR = \sigma_c' / \sigma_0'$, σ_c' is the preconsolidation pressure.

RESULTS

Proctor compaction test

Estimation of optimum moisture content

The optimum moisture content was found to be around 16% from the compaction curve. The maximum dry density recorded was 17.53 g/cc (Table 1 and Figure 3).

Atterberg's Test

The results of the Atterberg Limit Test are given below:

Liquid Limit: 84%
Plastic Limit: 46%

Procedure of vane shear tests

For the laboratory miniature vane shear test the compaction is done so as to account for the overburden pressure in the field. The following were noted:

For compaction of sample in mould:

(i) Weight of hammer: 17.65 N
(ii) Height of drop: 210 mm
(iii) Number of layers: 3
(iv) Diameter of vane: 1.2 cm

Table 1. Proctor compaction test results.

Weight of mould + soil (kg)	Weight of empty mould (kg)	Weight of wet soil (kg)	Density of soil (g/cc)	Moisture content (%)	Dry density (g/cc)
6.364	4.625	1.739	17.39	9	15.955
6.488	4.625	1.863	18.63	12	16.63393
6.613	4.625	1.988	19.88	14	17.4386
6.659	4.625	2.034	20.34	16	17.53448
6.626	4.625	2.001	20.01	18	16.95763
6.611	4.625	1.986	19.86	20	16.55
6.578	4.625	1.953	19.53	22	16.0082
6.522	4.625	1.897	18.97	24	15.05556

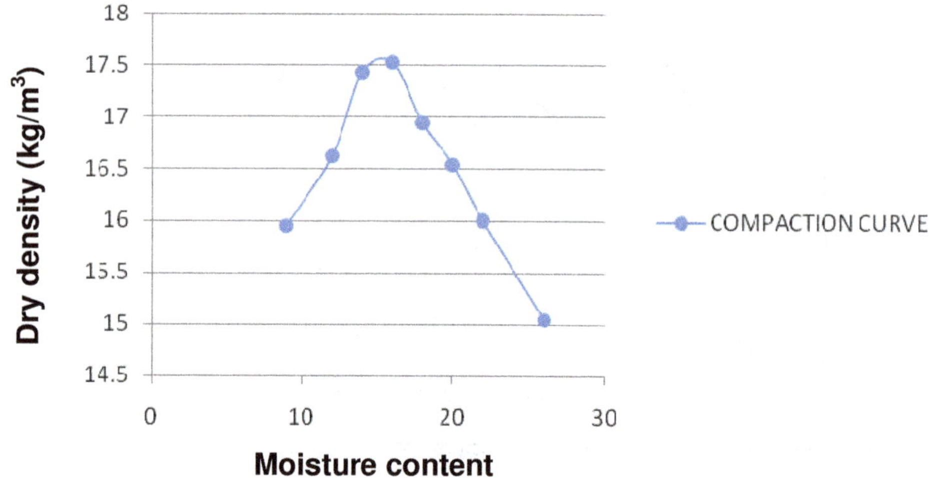

Figure 3. Compaction curve.

Table 2. Atterberg's limits.

Depth (m)	Moisture content (%)	Liquid limit	Plastic limit
1.0	20.44	76	32
2.0	28.18	72	34
3.0	34.52	72	28
4.0	37.86	68	30
5.0	39.12	68	28
6.0	42.52	70	27
7.0	45.40	68	30
8.0	49.22	68	29

(v) Height of vane- 2.4 cm
(vi) Mould diameter- 3.8 cm
(vii) Mould height – 10 cm
(viii) Volume of mould – 113.46 cc

Calculation of compaction energy:

(a) For 5 blows: E = (5 * 3 * 0.01765 * 0.210) / 113.46 * 10^{-6} = 490 kN-m / m^3
(b) For 10 blows: E = (10 * 3 * 0.01765 * 0.210) / 113.46 * 10^{-6} = 980 kN-m / m^3
(c) For 15 blows: E = (15 * 3 * 0.01765 * 0.210) / 113.46 * 10^{-6} = 1470 kN-m / m^3

For the field test the undrained shear strength is calculated for both the undisturbed and remolded state. First, the vane is pushed into the soil. The torque is applied at the top of the torque rod to rotate the vane at a uniform speed. The soil is classified as A-7-6 according to the AASHTO system. The Atterberg's Limits are recorded as shown in Tables 2 to 4.

The sensitivity of the clay is the ratio of the undrained shear strength in the undisturbed state to the remolded state and is found to be 3.04. The plot is drawn with the undrained shear strength (kN/m^2) along Y-axis and moisture content along X-axis (Figure 4).

The shear strength curve is plotted with the moisture content (%) along X-axis and the undrained shear

Table 3. Undisturbed field shear strength.

Depth (m)	Moisture content (%)	Undrained shear strength (kN/m^2)
1.0	20.44	48.914
2.0	28.18	44.784
3.0	34.52	57.932
4.0	37.86	32.119
5.0	39.12	21.659
6.0	42.52	24.406
7.0	45.40	18.847
8.0	49.22	18.754

Table 4. Remolded field shear strength.

Depth (m)	Moisture content (%)	Undrained shear strength (kN/m^2)
1.0	20.44	26.876
2.0	28.18	21.225
3.0	34.52	22.542
4.0	37.86	14.870
5.0	39.12	5.114
6.0	42.52	5.790
7.0	45.40	5.215
8.0	49.22	5.208

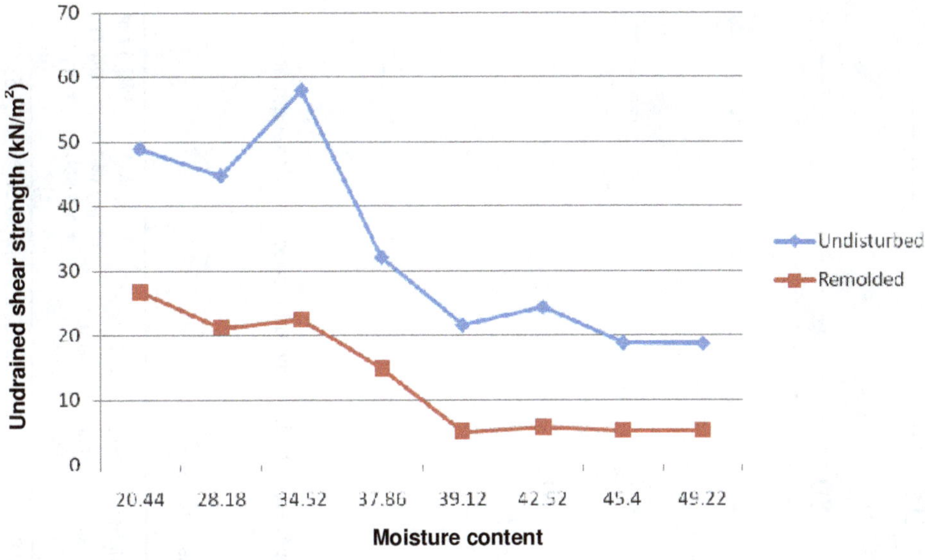

Figure 4. Relation between undrained shear strength and moisture content.

strength (kN/m^2) along Y-axis.

The blue line of the Figure 5 is for 5 blows, red line for 10 blows and green line for 15 blows. The three curves clearly indicate an exponential decrease in the shear strength of compacted clayey soil with gradual increase in moisture content (Tables 5 to 7).

DISCUSSION

The undrained shear strength of soil is a function of its moisture content and mineralogical properties. It is seen that the Laboratory Vane Shear Test shows steeper decrease in the undrained shear strength than the Field

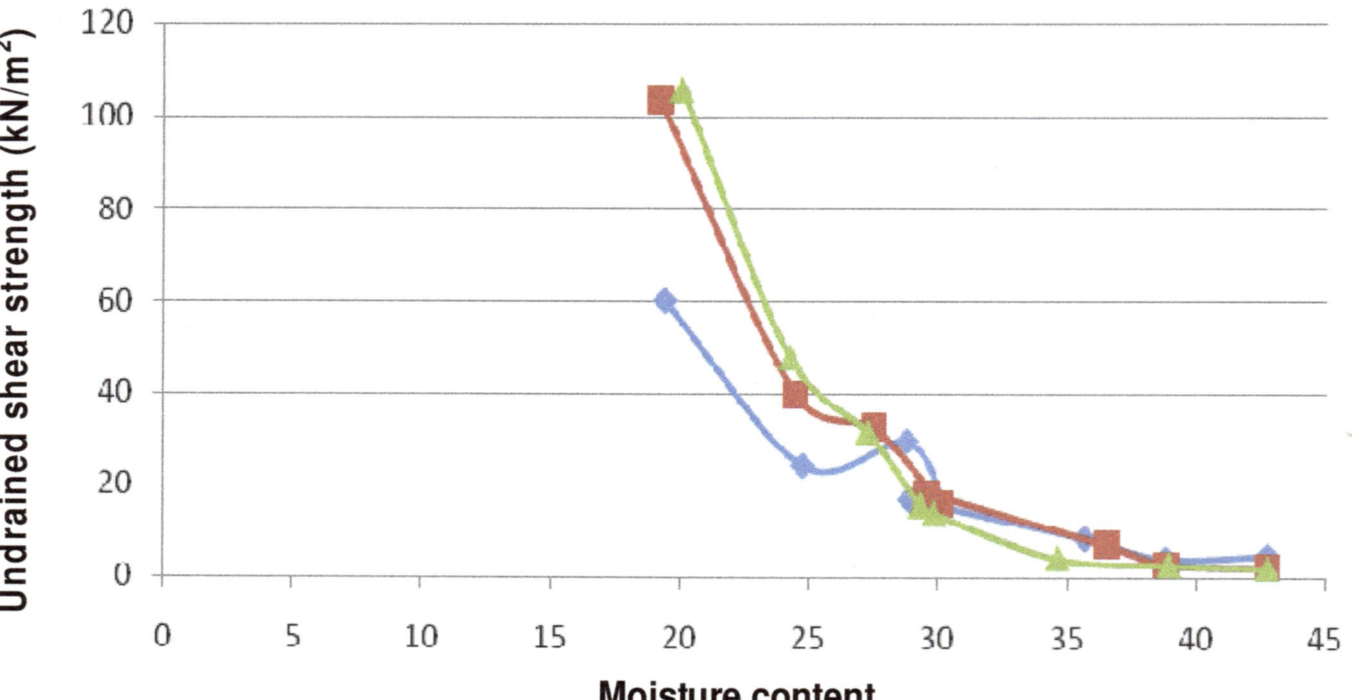

Figure 5. Relation between undrained shear strength and moisture content.

Table 5. Laboratory Vane Shear result for 5 blows.

Weight of empty mould (g)	Empty mould + wet soil (g)	Weight of wet soil	Bulk density (g/cc)	Weight of empty can (g)	Weight of empty can + wet soil (g)	Weight of empty can + dry soil (g)	Dry soil weight (g)	Weight of water (g)	Moisture content (%)	Spring type	Initial angle (°)	Final angle (°)	Rotation (°)	Torque (kg-cm)	Shear strength (kN/m²)
1544	1733	189	1.66583	35.671	56.517	53.131	17.46	3.386	19.3929	F	170	289	119	3.8125	60.1974
1544	1752	208	1.83329	30.537	54.398	49.665	19.128	4.733	24.7438	E	135	227	92	1.53846	24.2915
1544	1764	220	1.93906	30.551	63.079	55.808	25.257	7.271	28.7881	F	160	209	49	1.875	29.6053
1544	1758	214	1.88617	30.474	56.171	50.22	19.746	5.951	30.1377	F	169	182	13	0.96154	15.1822
1544	1763	219	1.93024	30.479	52.841	47.83	17.351	5.011	28.8802	F	345	362	17	1.075	16.9737
1544	1753	209	1.84211	40.113	73.939	65.044	24.931	8.895	35.6785	E	132	147	15	0.55128	8.70445
1544	1747	203	1.78922	40.259	59.398	54.046	13.787	5.352	38.8192	E	138	144	6	0.27273	4.30622
1544	1746	202	1.78041	31.234	57.037	49.308	18.074	7.729	42.7631	E	130	137	7	0.31818	5.02392

Table 6. Laboratory Vane Shear result for 5 blows.

Weight of empty mould (g)	Empty mould + wet soil (g)	Weight of wet soil	Bulk density (g/cc)	Weight of empty can (g)	Weight of empty can + wet soil (g)	Weight of empty can + dry soil (g)	Dry soil weight (g)	Weight of water (g)	Moisture content (%)	Spring type	Initial angle (°)	Final angle (°)	Rotation (°)	Torque (kg-cm)	Shear strength (kN/m²)
1544	1751	207	1.82448	35.44	55.907	52.61	17.17	3.297	19.2021	H	161	243	82	6.55556	103.509
1544	1770	226	1.99194	34.293	60.219	55.123	20.83	5.096	24.4647	E	138	304	166	2.5	39.4737
1544	1764	220	1.93906	33.175	59.264	53.64	20.465	5.624	27.4811	H	345	357	12	2.0625	32.5658
1544	1757	213	1.87736	31.941	61.567	54.715	22.774	6.852	30.0869	F	168	182	14	1	15.7895
1544	1761	217	1.91262	40.138	64.404	58.867	18.729	5.537	29.5638	F	343	363	20	1.15	18.1579
1544	1750	206	1.81566	36.169	51.027	47.053	10.884	3.974	36.5123	E	136	146	10	0.45455	7.17703
1544	1737	193	1.70108	30.539	52.725	46.521	15.982	6.204	38.8187	E	139	143	4	0.18182	2.87081
1544	1747	203	1.78922	34.294	56.926	50.15	15.856	6.776	42.7346	E	140	143	3	0.13636	2.15311

Table 7. Laboratory Vane Shear result for 15 blows.

Weight of empty mould (g)	Empty mould + wet soil (g)	Weight of wet soil	Bulk density (g/cc)	Weight of empty can (g)	Weight of empty can + wet soil (g)	Weight of empty can + dry soil (g)	Dry soil weight (g)	Weight of water (g)	Water content (%)	Spring type	Initial angle (°)	Final angle (°)	Rotation (°)	Torque (kg-cm)	Shear strength (kN/m²)
1544	1760	216	1.9038	31.681	57.319	53.044	21.363	4.275	20.0112	H	165	250	85	6.72222	106.14
1544	1773	229	2.01838	31.23	66.81	59.872	28.642	6.938	24.2232	F	168	243	75	3.025	47.7632
1544	1765	221	1.94787	31.682	59.174	53.29	21.608	5.884	27.2307	H	348	359	11	2	31.5789
1544	1760	216	1.9038	33.175	60.953	54.663	21.488	6.29	29.2722	F	169	183	14	1	15.7895
1544	1757	213	1.87736	35.44	67.961	60.496	25.056	7.465	29.7933	F	349	360	11	0.88462	13.9676
1544	1716	172	1.51599	32.293	54.655	48.906	16.613	5.749	34.6054	E	139	145	6	0.27273	4.30622
1544	1744	200	1.76278	32.301	70.849	60.053	27.752	10.796	38.9017	E	138	142	4	0.18182	2.87081
1544	1747	203	1.78922	34.294	56.926	50.15	15.856	6.776	42.7346	E	140	143	3	0.13636	2.15311

Test. This could be partly attributed to the fact that due to the absence of any overburden pressure the sample in the lab exhibits the maximum shear strength at a moisture content which is close to the Optimum Moisture Content. It will be slightly more than the one obtained from the Proctor Compaction Test because of different compaction energy. It is found that the Lab Shear Test values are significantly higher than the Field Shear Test values for the same moisture content and hence correction as suggested by Bjerrum (1973) has to be applied for checking the safety against shear failure during foundation design. It is also noted that the corrected values for the shear strength obtained in lab and field are about 80% lower than the actual values (Mohd et al., 1997).

REFERENCES

ASTM D698a (1933). Standard Test Methods for Laboratory Compaction Characteristics of Soil Using Standard Effort (12400 lb/ft^3-600 kN-m/m^3). Annual Book of ASTM Standards, Volume 04.08.

Bjerrum L (1974). Problems of soil mechanics and construction on soft clays. Norwegian Geotechn. J. Vol. 110.

Freedlund DG, Rahardjo H (1993). Soil Mechanics for unsaturated soils. John Wiley and Sons Inc New York, NY.

Iannacchione AT, Vallejo LE (1994). Shear strength evaluation of Clay-Rock Mixtures.

Khing KH (1994). The bearing-capacity of a strip foundation on geogrid-reinforced sand.

Ladd CC, Foot R, Ishihara K, Poulos HG, Schlosser F (1977). Stress deformation and Stress Characteristics. 33(3): 21-26.

Mohd J, Mohd A, Taha R (1997). Prediction and Determination of Undrained Shear Strength of Soft Clay at Bukit Raja, Pertanika. J. Sci. Technol. 5(1):111-126.

Vanapalli SK (1994). "Simple Test Procedures and their Interpretation in evaluating the Shear strength of Unsaturated Soils". University of Saskatchewan.

14

Durability based suitability of bagasse-cement composite for roofing sheets

Omoniyi, T. E. and Akinyemi B. A.

Department of Agricultural and Environmental Engineering, Faculty of Technology, University of Ibadan. Oyo State, Nigeria.

Accelerated and natural weathering of bagasse reinforced cement composite filled with rice ash pozollan used as roofing sheets were studied. In this paper, the durability of natural fibers such as sugarcane bagasse used as roofing sheets has been reported by conducting an experimental investigation. This investigation includes determination of mechanical strength properties such as compressive, tensile, modulus of rupture and flexural properties of the roof once every 3 months for a period of 8 years under alternate wetting and drying conditions and was exposed to ultraviolet light for the same period. The 8 years study showed no significant difference in the strength and sorption properties for the treated bagasse at 2% $CaCl_2$ and the 20% replacement of cement with rice husk ash. This confirms that treated bagasse cement composite is suitable for both external and internal construction purposes.

Key words: Bagasse, durability, weathering, pozollan and roofing sheets.

INTRODUCTION

Natural fibers are prospective reinforcing materials and their use until now has been more traditional than technical. They have long served many useful purposes, but the application of materials technology for the utilization of natural fibers as the reinforcement in concrete has only taken place in comparatively recent years. The distinctive properties of natural fiber reinforced concretes are of improved tensile and bending strength, greater ductility, and greater resistance to cracking and hence improved impact strength and toughness. Besides its ability to sustain loads, natural fiber reinforced concrete is also required to be durable.

Durability of vegetable fiber reinforced concrete is related to the ability to resist both external (temperature and humidity variations, sulfate or chloride attack etc) and internal damage (compatibility between fibers and cement matrix, volumetric changes etc). The degradation of natural fibers immersed in Portland cement is due to the high alkaline environment which dissolves the lignin and hemi-cellulose phases thus weakening the fiber structure (Silva and Rodrigues, 2007). Gram and Skarendahl (1978) was the first author to study the durability of sisal and coir fiber reinforced concrete. The fiber degradation was evaluated by exposing them to alkaline solutions and then measuring the variations in tensile strength. This author reported a deleterious effect of Ca^{2+} elements on fiber degradation. He also stated that fibers were able to preserve their flexibility and strength in areas with carbonated concrete with a pH of 9 or less. Toledo Filho et al. (2000) also investigated the durability of sisal and coconut fibers when immersed in alkaline solutions. Sisal and coconut fibers conditioned in a sodium hydroxide solution retained respectively 72, 7 and 60.9% of their initial strength after 420 days. As for the immersion of the fibers in a calcium hydroxide solution, it was noticed that original strength was completely lost after 300 days. According to those authors the explanation for the higher attack by $Ca(OH)_2$ can be related to a crystallization of lime in the fibers pores.

*Corresponding author. E-mail: temidayoomoniyi@gmail.com or bantonbows@gmail.com.

Ramakrishna and Sundararajan (2005) also reported degradation of natural fiber when exposed to alkaline medium. Other authors studied date palm reinforced concrete reporting low durability performance which is related to fiber degradation when immersed in alkaline solutions (Kriker et al., 2008). Ghavami (2005) reported the case of a bamboo reinforced concrete beam with 15 years old and without deterioration signs. Lima et al. (2008) studied the variations of tensile strength and modulus of elasticity of bamboo fiber reinforced concrete expose to wetting and drying cycles, reporting insignificant changes, thus confirming its durability.

Ismail (2007) worked on the compressive and tensile strength of natural fiber-reinforced cement based composites and the results showed that the tensile strength of composite increases, this increase in strength is about 53% while the compressive strength decreases as the fiber volume fraction is increased. It has been observed that composites with roselle particle reinforcement showed more tensile strength which was followed by short fiber and long fiber reinforced composites and compressive strength of urea-formaldehyde resin matrix has been found to increase when reinforced with fiber. It was found that with particle reinforcement, compressive strength increases to a much more extent than short and long fiber reinforcement (Singha and Thakur, 2008).

Compressive properties results for alkaline pre-treatment done on banana fibers shows that fiber treatment is favourable for epoxy matrix composites but not favourable in the case of polyester composite matrix. The flexural strength of the banana fiber composite was found to be higher than the banana fiber alone and the higher the flexural strength and modulus of elasticity observed in the banana fiber composite, the more fiber interaction takes place (Lina-Herrera et al., 2006).

Fiber length has profound impact on the properties of composites. Besides holding the fibers together, the matrix has the important function of transferring applied load to the fibers. The efficiency of a fiber reinforced composite depends on the fiber-matrix interface and the ability to transfer stress from the matrix to the fiber (Karnani et al., 1997).

Extensive research has been conducted on the use of some of the natural fibers for cement particle board (CPB) production. The research, however, is limited to sorption and strength characteristics of boards at early age of about 28 days. This work examines the effects of production variables on the hydration of the fibers mixed with Portland cement. The work also compares and contrasts the functions and performance of these fibers. If these materials are suitable, the work can lead to production of a wide range of CPB with different durability properties and with diverse levels of long-term performance under internal and external exposure conditions. The work may provide economic and environmental benefits to local communities in West Africa.

MATERIALS AND METHODS

Experimental study

Mix and specimen casting

Materials used in the production of roofing sheets were bagasse, ordinary Portland cement, aggregates (sharp and soft sand), potable water, calcium chloride and pozzolans. Bagasse was obtained from Bodija in Ibadan, Oyo State as shown in Figure 1. The raw bagasse was received at about 30% moisture content. It was sun-dried for two weeks, manually depithed and further sun-dried for two weeks to a moisture content range of 7 to 10%. Part of the sun-dried bagasse was manually shredded to generate flakes while the rest was hammer-milled to produce bagasse particles as shown in Figure 2. Bagasse flakes were divided into three groups: short flakes (2.4 to 20 mm long), medium flakes (21 to 30 mm long) and long flakes (31 to 76 mm). The hammer-milled particles were passed through sieves of sizes 2.4 mm, 850 and 600 μ m. Particles that passed through 2.4 mm but were retained on the 850 μ m sieve were categorized as coarse particles while those that passed through 850 μ m and were retained on the 600 μ m sieve were classified as fine particles.

Bagasse moisture content was determined by the oven drying method using a representative small sample. The oven drying moisture content of the bagasse was determined at temperature of 103 ± 2°C in accordance with the ASTM D 1037 (1991) using three replicates.

Calcium chloride ($CaCl_2$) was obtained from the chemical laboratories at Ibadan as a cement accelerator. $CaCl_2$ was used in preference to other accelerators because it is cheaper and more effective (Li et al., 1997).

Ordinary Portland cement was procured from the local market in Ibadan, Oyo State. The cement meets the specifications of the British Standards for ordinary Portland cement BS 12 (1991). It was stored in air-tight containers and was used up soon after delivery to prevent strength deterioration.

Water source was from the University of Ibadan supply. The quantity of water required for the manufacture of the roofing sheets was determined using the relationship applied by researchers such as Fuwape (1995) and Sudin and Swamy (2000).

$$R_q = 0.35C + (0.30 - M)W \quad (1)$$

Where
R_q = water required (litres)
C = weight of cement (kg)
M = percentage moisture content of Bagasse on dry basis
W = oven-dry weight of the Bagasse (kg)

Rice husk used as pozzolan was obtained from rice processors in Erio Ekiti, Ekiti State, Nigeria while ashing of rice husk was done by both open air burning which involved the setting ablaze of the rice husk inside an open metal container and burning was supported by the use of kerosene while gently stirring the contents, this was later transferred to a furnace at 760°C. Sharp sand was obtained from the river flowing through the Nnamdi Azikiwe Hall while soft sand was obtained from a construction site at the Faculty of Technology both at the University of Ibadan. The sand was washed and dried to reduce the soluble matter and fine particle contents.

Two water/cement ratios (0.4 and 0.5) and three sand/cement ratios (1.0, 2.0 and 3.0) were investigated. The factorial

Figure 1. Unprocessed Bagasse

Figure 2. Processed Bagasse.

combination of the sand and water ratios gave a total of six treatments and three replications of each treatment were produced making a total of 18 samples. Bagasse contents were varied from 1 to 4% by mass of cement to determine the influence of bagasse mass fractions on the properties of roofing sheets. The factorial combination gave a total of 15 treatments with three replications.

Roofing sheet production processes involved blending together of cement, sand, water, $CaCl_2$ and bagasse (flakes and particles). These constituent materials were batched in proportions determined in experimental design depicted by the flow chart of Figure 3.

Measured quantities of cement and sand were dry-mixed until a high level of uniformity was achieved. The water was slowly added while mixing was continued for about 10 min until uniform consistency and colour were achieved. Bagasse was added at the end of mixing cycle to minimize damage. The roofing sheets were produced by vibration. The sand/cement/bagasse slurry was then placed on a polythene sheet spread on the surface of the moulding table and vibrated for about 40 s to ensure adequate consolidation and compaction with removal of void spaces (Dahunsi, 2000). The mixture was then transferred onto a corrugated plastic mould, covered with wet cloth, placed in a cool dry place for 24 h and then de-moulded. The specimens were subsequently cured in water for 28 days.

Testing

Compression tests

Figure 4 shows compression test in progress. An increasing compressive load at a rate of 1 mm/min was applied until failure occurred. The tests were conducted at 3, 7, and 28 days after curing.

Flexural test

The flexural test was carried out in accordance to BS 5669 (1993). The testing was carried out for 3, 7, 14 and 28 days after curing. Three replicates of each were tested. Flexural strength was determined in accordance with ASTM C 78 - 91 (1991). Three specimens (250 mm × 75 mm × 25 mm) were tested for each mixture, and the average strength reported.

Breaking strength test

Breaking Load/Strength: This test method consisted of supporting the sheet specimens on the ends of three cylindrical rods, arranged in an equilateral triangle form and applying load until the sheet specimen failed. The sheet strength was the load necessary to cause such roofing sheet failure. The roofing sheet strength recorded was based on average values of 3 replicates of the same composition mixes.

Impact strength test

The roofing sheets were tested for impact strength according to ASTM D 1037 (1991). The test was carried out at Forest Research Institute of Nigeria (FRIN) in Ibadan. Each test piece was cut into 300 mm × 300 mm × 6 mm and evenly supported in rebated square frame without fastenings.

Porosity test

Porosity was determined on 75 mm × 75 mm × 25 mm specimens using the method of vacuum saturation. The specimens were dried in the oven at 105 ± 3°C until no change in measured weight was observed. The specimens were kept dry in a vacuum chamber for 3 h before water was introduced to the chamber under vacuum. The vacuum was maintained for 6 more hours after which time the specimens were left in water for 18 h. The saturated surface dried weight was then determined. For the fiber reinforced specimens, the water absorbed by the fiber was accounted for in the vacuum saturated weight so as to obtain the effective porosity. Porosity of a plain (control) mortar was also determined for comparison.

Durability tests

Durability tests conducted included long term exposure to natural weathering, accelerated and grave-yard tests.

Long term exposure test: A building of area 1.83 × 1.83 m^2 (6ft × 6ft) was erected behind the Department of Agricultural and Environmental Engineering, University of Ibadan on the 5th of February 2007 to investigate the effects of natural weathering on the roofing sheets (Figure 13). The roofing sheets in Table 2 were installed by the use of specially designed nibs incorporated during the production (Table 3). Other samples were stored outdoors on rooftops. The properties of these roofing sheets were determined

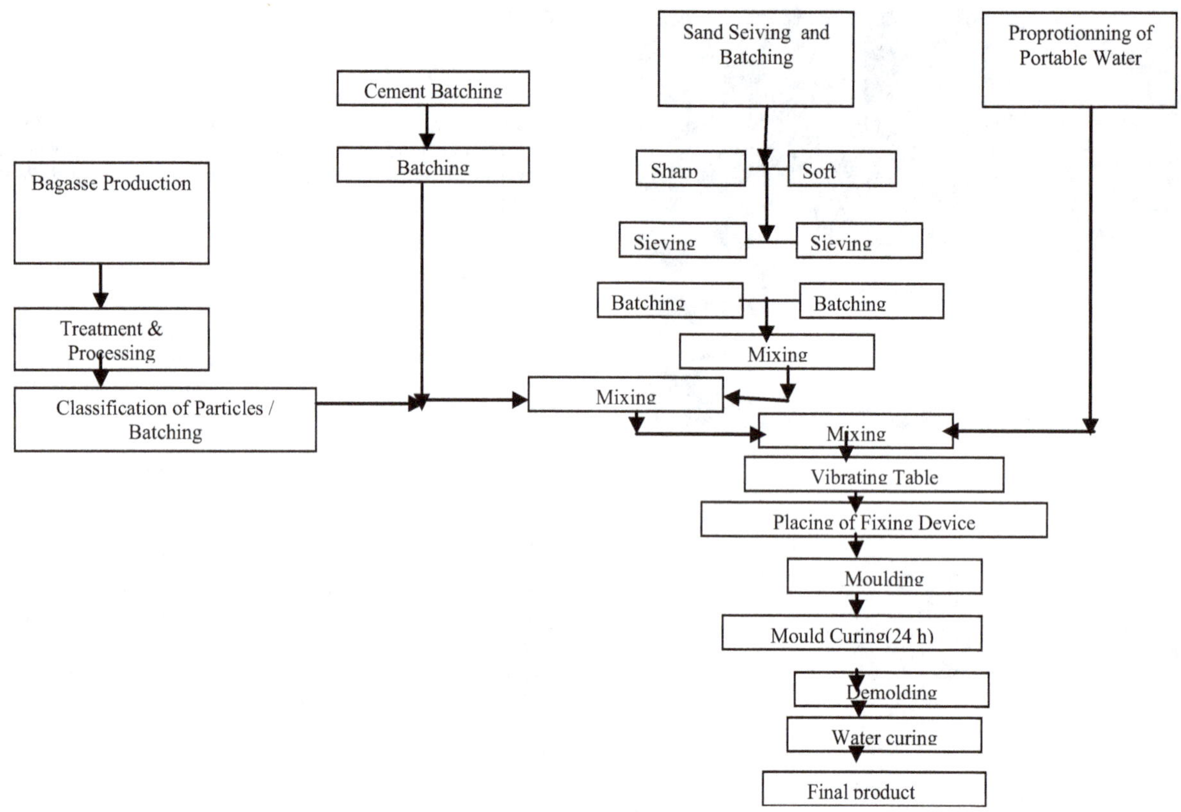

Figure 3. Flowchart of production process using randomly oriented flakes.

Figure 4. Compression test.

progressively over the duration of the experiment.

Hot water immersion test: The hot water immersion test method investigates the long term chemical interaction between the constituents in the composite. Wet conditions and elevated temperature were used to accelerate the deterioration. Specimens were saturated in water maintained at 60°C. The test procedure was in accordance with ASTM C 1185-91 with the specimen in hot water for 56 ± 2 days. Strength properties were evaluated at the end of the test duration and the result was compared to the control specimen.

Accelerated ageing test (Soak and Dry Cycles): Other specimens for each formulation were cured in the same way for 28 days of age and then they were submitted to the accelerated aging test (soak-dry cycles). This test consists of submerging the specimens into water for 18 h and after they were put into an oven at 60°C of temperature during the 6 h to complete 24 h. The aging test composed of 50 cycles and it was based on the methodology of the European Standards EN-494-98 section 7.3.5. After ageing, the samples were conditioned and tested for static bending, dimensional stability (thickness swelling and water absorption) and compression shear according to ASTM D – 1037 (1991).

Exposure cycling test for exterior use: This test was carried out to estimate the weathering qualities of roofing sheets under severe exposure conditions (.IS: 2380-1963). Each specimen was subjected to six complete cycles of accelerated ageing. Each cycle consisted of the following:

(i) Immersion in water at 49°C for 1 h.
(ii) Spraying with steam and water vapour at 93°C for 3 h.
(iii) Storing at -12°C for 20 h.
(iii) Heating in dry air at 99°C for 3 h.
(iv) Spraying again with steam and water vapour at 93°C for 3 h.
(v) Heating again in dry air at 99°C for 18 h.

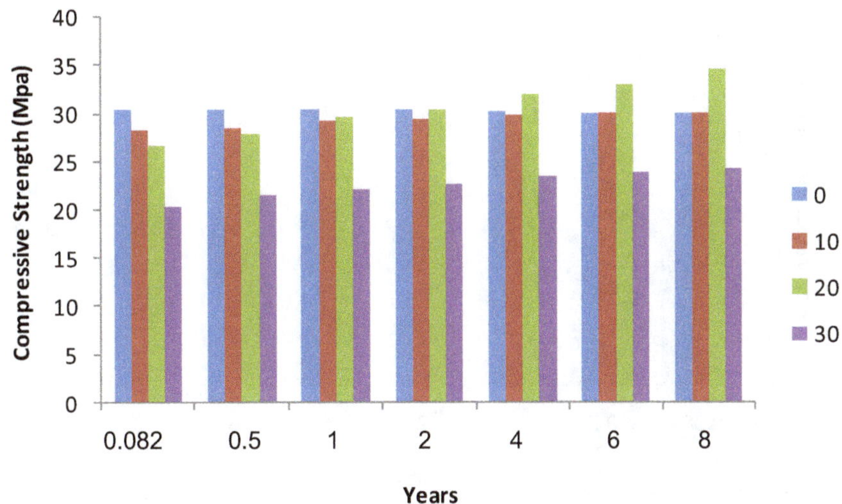

Figure 5. Eight years compressive strength of composites (0, 10, 20 and 30 are rice husk ash replacement levels).

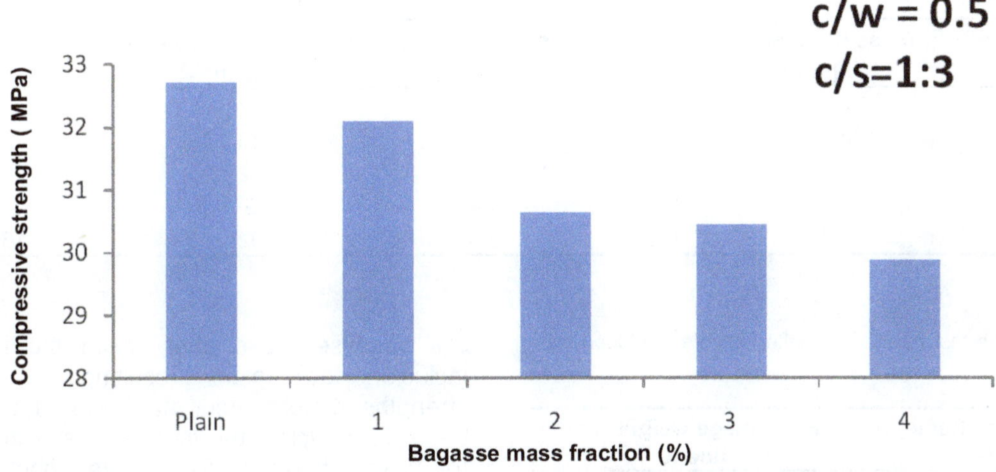

Figure 6. Comparison of 28-day compressive strength values of sheets with different bagasse contents.

After the completion of the six cycles of exposure, the material for test was further conditioned at a temperature of 27°C and relative humidity of 65% for at least 48 h before being subjected to tests such as water absorption, static bending, compression and flexural tests. Frequent inspections of the material were made during the ageing cycles for any signs of delamination or other disintegration.

RESULTS AND DISCUSSION

The compressive strength

Compressive strength decreased with the increase in the fiber contents. Highest compressive strength of 32.71 MPa was obtained for the plain concrete while the least value of 29.89 MPa was obtained at 4% bagasse contents as shown in Figures 5 and 6. The reduction in compressive strength due to increase in bagasse content could be attributed to the fact that the elastic modulus of bagasse is lower than that of the cement matrix. The study was conducted every six months for the 8 years study and it showed no significant difference in the compressive strength of samples treated with 2% $CaCl_2$ and the 20% replacement of cement with rice husk ash.

The flexural strength

Flexural strength was found to increase with fiber contents until an optimal fiber mass was reached and then the flexural strength decreased. The reason could be that, after curing and conditioning, there was an increased fiber-to-matrix bond, and the failure was

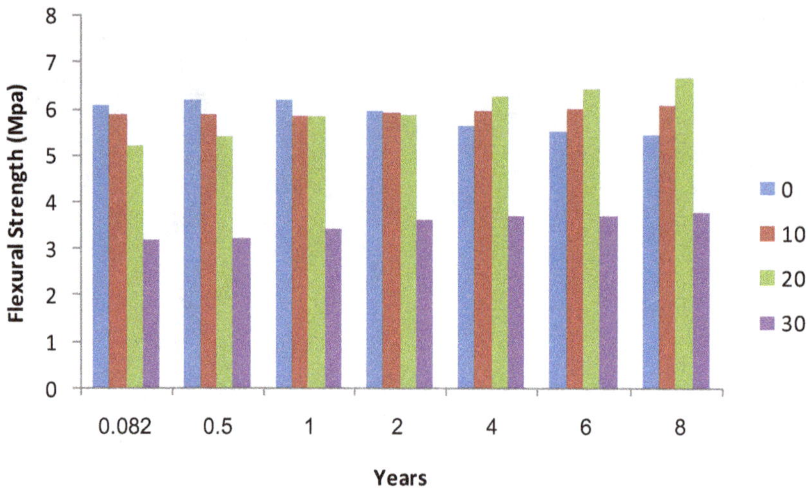

Figure 7. Eight years flexural strength.

Table 1. Impact strength of composites (%).

Percentage of fibers	Mass dropped (kg)	Height of failure (m)	Impact energy (mgh) Joules	Impact toughness KJ/m^2	Percentage improvement-over plain
0	4.5	0.03	1.32	1.47	0
1	4.5	0.05	2.21	2.46	67.42
2	4.5	0.08	3.53	3.92	167.42
3	4.5	0.10	4.41	4.9	234.1
4	4.5	0.08	3.53	3.92	167.42

Table 2. Graveyard test results for roofing sheets produced with pozolan.

Roofing sheets (Cement: Sand)	Percentage weight loss due to
50:50	16.1
60:40	14.8
70:30	14.2
80:20	12.9
90:10	12.9

Table 3. Porosity.

Percentage of bagasse	Porosity (%)
1	1.85
2	3.11
3	5.13
4	6.81
5	9.48

probably due to fiber fracture rather than fiber pull-out. The increase in the flexural strength was dominant up to 3% bagasse mass after which the influence of soft inclusion took place resulting in reduced flexural strengths at higher fiber mass. For samples incorporating bagasse particles, the trend was similar to that obtained for flakes, however for a given fiber mass, bagasse particles showed higher flexural strength. The reason could be that bagasse flakes because of larger surface area (larger aspect ratio) provided larger areas of preferential weakness in the matrix, resulting in a reduction in flexural strength. With increase in fiber mass beyond 3%, the flexural strength reduced and at about 4% fiber mass the flexural strength was essentially the same. At higher fiber mass, there were increased chances of fiber bailing, clumping, creating voids that led to the reduction of flexural strength. At 2% fiber contents over a period of 8 years, there was no significant difference in the flexural strength as shown in Figure 7.

Impact strength

All the bagasse reinforced composites had impact energy more than twice that of un-reinforced ones as shown in Table 1, Figures 8 and 9. At 1% bagasse content, the percentage improvement over plain sheets was 67.42%.

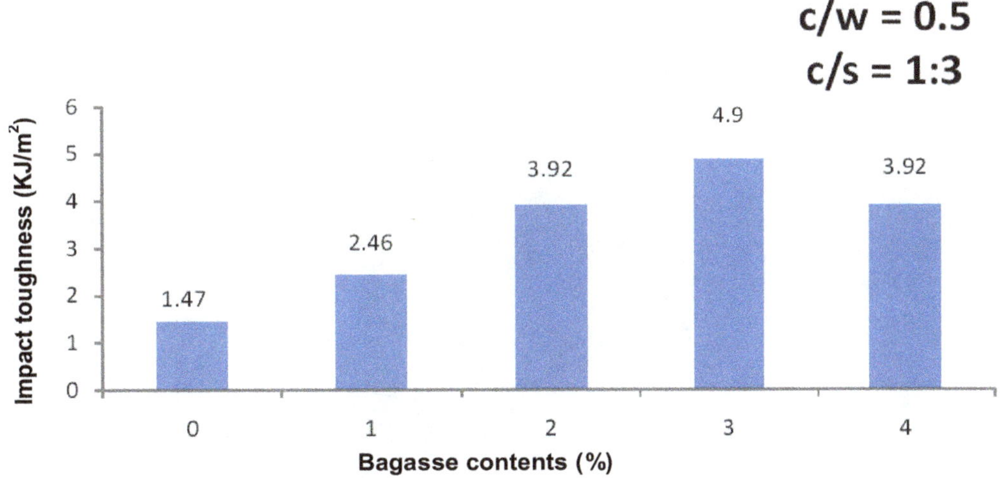

Figure 8. Impact strength test at different levels of bagasse content (%).

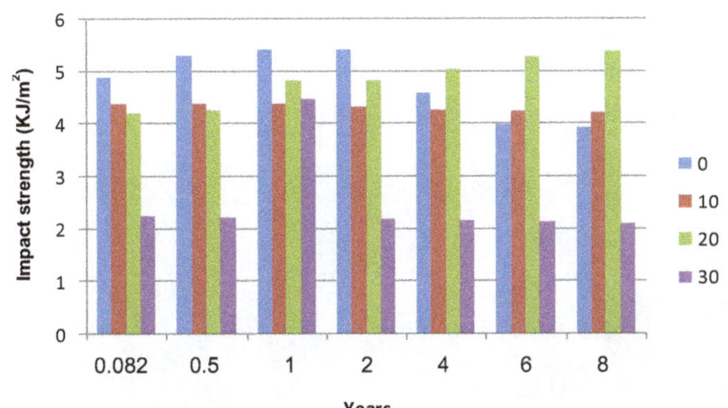

Figure 9. Eight years impact strength.

There was a progressive increase in the impact strength as the bagasse content increased until a maximum value was obtained at 3% bagasse content. The percentage improvement over plain mix at this bagasse content was 234.11%. When the bagasse contents exceeded 3% a decline in value was recorded probably due to bailing problems which contributed to lack of bonding between the matrix and the bagasse flakes.

The impact strength gives an indication of the resistance of a material to vibration or shock loading. It is also a measure of the work done in breaking a test piece (Dahunsi, 2000). The bagasse flakes in the matrix absorbed part of the shock loading on the composites, thereby reducing the load directly absorbed by the concrete. At the optimum impact energy (3% bagasse content), the reinforced mixes resisted up to four times the damage inflicted on the plain before reaching destruction. This improvement will be useful to the overall resistance of the roofing sheets to damage in storage, transportation and during installation and perhaps throughout the service life of the roofing sheet. However no marked difference was observed in the impact energy when the aspect ratios were varied by varying the lengths of the flakes (below 20 mm, 20 mm to 30 mm and above 30 mm) but at constant bagasse contents of 3%. The failure of the bagasse particles (below 20 mm) specimens took the form of complete fracture (chatter); while the long flakes (above 20 mm) specimens deformed with the flakes still holding the matrix together, The long flake is recommended to prevent complete shatter of the specimen during failure.

Accelerated ageing test results

The treated bagasse fibers offer both stiffness and strength to matrix after the initial cracking. The mechanical properties increased with increase in the fibers contents until the optimum value of 3% of bagasse fiber content. The mechanical results after 50 cycles of accelerated aging test as shown in Figure 10 showed that the formulations with higher percentage of bagasse contents were however more susceptible to the degradation after the ageing tests. This behaviour could probably be attributed to the degradation of bagasse fibers in cement matrix. Flexural toughness considerably deteriorated with ageing as shown in Figure 11. The flexural strength however increased with ageing. This was possibly due to the densification of the interfaces and petrification of bagasse flakes, thus improving the flexural strength and reducing the toughness of the roofing sheets. The flexural strength increased slightly and the toughness decreased considerably. The bagasse mass fraction also had drastic effect on the toughness reduction; the more the bagasse contents the more the toughness reduction due to accelerated wetting and drying. Values of the post cracking strength for the

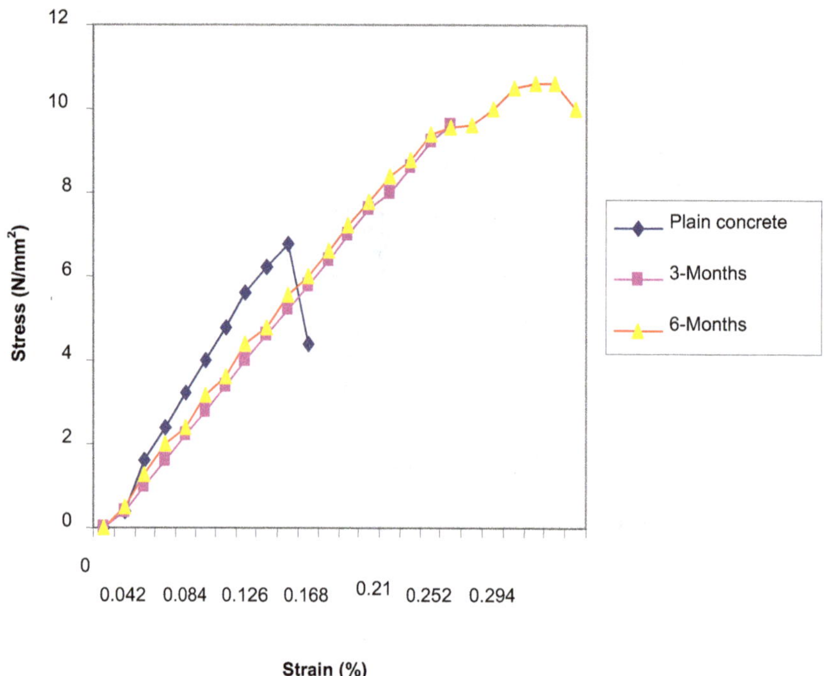

Figure 10. Effects of weathering on stress– strain behaviour of bagasse-cement roofing sheets.

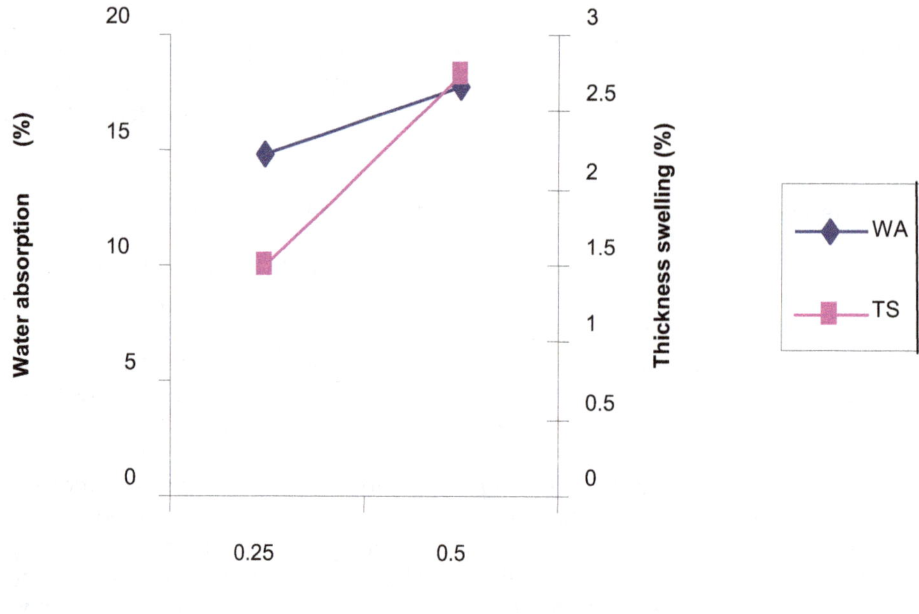

Figure 11. Influence of flake thickness on the water absorption and thickness swelling of the roofing sheets (with pozollan).

specimens aged outdoors reduced up to 30% probably due to embrittlement of the fibers in the cement matrix. This agreed with the past studies which have shown that the natural fibers are chemically decomposed in the alkaline environment of the cement matrix resulting in brittle composite which has reduced capacity to cracking. This led to the study of partial replacement of the OPC with rice husk ash as reported elsewhere in this report.

Figure 12. The roofing sheets.

Figure 13. Natural weathering test.

Performances of roofing sheets subjected to graveyard tests

The bagasse particles and flakes samples encased in concrete were observed to be safe from attack by termites. The matrix offered adequate protection for the bagasse samples. Bagasse particles and flakes that were not encased in cement matrix were completely destroyed by the termites at the end of the test period. It was observed that the performance of the samples subjected to graveyard test was independent of the treatment of the bagasse samples or flakes provided they were encased in cement matrix. However, roofing sheets produced with addition of pozzolan showed some effects of termite attack. The higher the proportion of pozzolan in the composite the more pronounced the termite attack. Figure 12 shows the result of termite attack which indicates that roofing sheets made with 50:50 cement pozzolan ratio had the highest effect of the termite attack with the mean of 16.21% while the least was obtained from the roofing sheets made cement to pozzolan ratio of 90:10. The use of pozzolan as a partial replacement for cement promotes the activities of termites when roofing sheets were exposed to termitarium the effects are however not significant.

Porosity

The increase in porosity with increasing bagasse mass can be explained by the fact that, bagasse particles in addition to being porous can absorb water. The high value of porosity with increase in bagasse content may be due to the tendency of the particles to clump together while mixing, entrapping water filled spaces, which consequently turn into voids. Increased bagasse mass enhances the potential for fibre bailing and clumping (Figure 13).

REFERENCES

American Society for Testing and Materials (ASTM) (1991). C1185 - 1991 Standard Test Methods for Sampling and Testing Non-Asbestos Fiber-Cement Flat Sheet, Roofing and Siding Shingles, and Clapboards. Accelerated. Annual book of ASTM Standards, 04.09 Wood, Philadelphia, PA pp. 1–84.

American Society for Testing and Materials (ASTM) (1991). C78 - 91 Standard Test Method for Flexural Strength of Concrete (Using Simple Beam with Third-Point Loading) beams, concrete, flexural strength testing. www.astm.org/Standards/C78.htm. pp. 1-413.

American Society for Testing and Materials (ASTM) C618 (2005). Standard Specification for Coal Fly Ash and Raw or Calcined Natural Pozzolan for Use in Concrete, fly ash, natural Pozzolan pozzolans. www.astm.org/Standards/C618.htm: pp. 1–623.

British Standards BS 12 (1991) Specification for Portland cement of internal ceramic and natural stone wall tiling and mosaics in normal conditions: BS 8297:2000 - Code of www.techstreet.com/cgi-bin/detail?product_id=10445 products.ihs.com/bs-seo/gbm42_2.htm - United Kingdom.

British Standards BS 12 (1991). Specification for Portland cement....of internal ceramic and natural stone wall tiling and mosaics in normal conditions: BS 8297:2000 - Code of ... www.techstreet.com/cgi-bin/detail?product_id=10445

British Standard**s** BS 5669 5 (1993). Particleboard. Code of practice for the selection and application of particleboards for specific purposes. BS 5881:1980* ISO 5638-1978 ... products.ihs.com/**bs**-seo/gbm42_2.htm - United Kingdom

Dahunsi BIO (2000). The Properties and Potential Application of Rattan Canes as Reinforcement Material in Concrete. A Ph.D. Thesis submitted to the Department of Agricultural Engineering, University of Ibadan pp. 1–290.

Fuwape JA (1995). The Effect of Cement-Wood Ratio on the Strength Properties of Cement-Bonded Particleboard from Spruce. J. Trop. For. Prod. 1:49–58.

Ghavami K (2005). Bamboo as reinforcement in structural concrete elements. Cem. Concr. Compos. 27:637-649.

Gram HE, Skarendahl A (1978). "A Sisal Reinforced Concrete: Study No. 1 Material," Report No. 7822, Swedish Cement and Concrete Research Institute Stockholm pp. 1-15.

Ismail MA (2007). Compressive and Tensile Strength of Natural Fiber-Reinforced Cement base Composites. Al-Rafidain Eng. 15:2.

Karnani R, Krishnan M, Narayan R (1997). Biofiber-Reinforced Polypropylene Composites. Polym. Eng. Sci. 37(2):476-483.

Kriker A, Bali A, Debicki G, Bouziane M, Chabannet M (2008). Durability of date palm fibres and their use as reinforcements in hot dry climates. Cem. Concr. Compos. 30:639-648.

Li VC, Kanda T, Lin Z (1997). The Influence of Fiber Matrix Interface Properties on Complementary Energy and Composite Damage Tolerance. Key Engineering Materials, Proceedings of 3rd Conference on Fracture and Strength of Solid, Hong Kong, pp. 215–220.

Lima HC, Willrich FL, Barbosa NP, Rosa MA, Cunha BS (2008) Durability analysis of bamboo as concrete reinforcement. Mater. Struct. 41:981–989

Lina-Herrera E, Selvum P, Uday V (2006). Banana Fiber Composites for Automotive and Transportation Applications. Mater. Sci. Eng. pp. 4391-4398.

Ramakrishna G, Sundararajan T (2005). Studies on the durability of natural fibres and the effect of corroded fibres on the strength of mortar. Cem. Concr. Compos. 27:575-582.

Silva J, Rodrigues D (2007). Compressive Strength Of Low Resistance Concrete Manufactured With Sisal Fiber. 51º Brazilian Congress of Ceramics, Salvador, Brazil.

Singha AS, Vijay KT (2008). Buildings Material Science, Publication of Indian Academy of Mechanical Properties of Natural Fiber Reinforced Polymer Composites. Sciences 31(5):791–799.

Sudin R, Swamy N (2000). Bamboo and Wood Fiber Cement Composite for Sustainable Infrastructure Regeneration. Forest Research Institute, Malaysia. Wood-Cement Composites in the Asia-Pacific Region: Proceedings of a Workshop held at Rydges Hotel, Canberra, Australia pp. 21–26.

Toledo Filho RD, Scivener K, England GL, Ghavami K (2000). "Durability of alkali sensitive sisal and coconut fibers in cement mortar composites". Cem. Concr. Compos. 22:127-143.

APPENDIX

0% RHA	Years	Composite strength (Mpa)	Flexural strength (Mpa)	Composite strength (KJ/m^2)	Breaking strength (N)	Porosity (%)
0	0.083(28 days)	30.48	6.09	4.9	380	5.13
0	0.5	30.49	6.19	5.32	384	5.28
0	1	30.5	6.2	5.42	390	5.74
0	2	30.4	5.98	5.42	360	6.1
0	4	30.2	5.64	4.6	340	6.1
0	6	30.1	5.52	4	328	6.32
0	8	30	5.44	3.94	315	6.45
10 RHA						
10	0.083(28 days)	28.3	5.9	4.4	360	4.88
10	0.5	28.4	5.9	4.4	362	4.9
10	1	29.3	5.92	4.38	364	4.95
10	2	29.5	5.94	4.32	368	4.98
10	4	29.8	5.98	4.26	370	5
10	6	30	6	4.24	374	5.14
10	8	30.12	6.1	4.2	375	5.22
20 RHA						
20	0.083(28 days)	26.7	5.2	4.2	340	4.8
20	0.5	27.9	5.42	4.28	356	4.74
20	1	29.6	5.87	4.49	378	4.62
20	2	30.42	5.9	4.85	395	4.26
20	4	32.02	6.29	5.05	421	3.94
20	6	33.12	6.43	5.28	442	3.67
20	8	34.6	6.68	5.4	450	3.26
30 RHA						
30	0.083(28 days)	20.2	3.2	2.24	160	4.6
30	0.5	21.4	3.24	2.21	162	4.4
30	1	22	3.43	2.2	164	4.21
30	2	22.69	3.61	2.19	166	4.19
30	4	23.57	3.69	2.14	168	3.52
30	6	23.91	3.72	2.13	170	3.43
30	8	24.2	3.8	2.1	171	3

15

Analysis by homogenisation method of structures in reinforced soil and behaviour interfaces of soil/reinforcement

T. Karech

Université de Batna Dep de Génie-Civil Algérie. E-mail: karech@hotmail.com.

The reinforced soil is a composite material formed by the addition to a non cohesive soil of steel strip reinforcement which is able to withstand important tension forces. Through friction between the soil and the reinforcement, the ground transmits the tension forces, which develop into the mass that cannot be supported, to the steel reinforcement. The stretched reinforcements thus confer to the ground some cohesion along their direction. Therefore providing reinforcements, improves the global mechanical properties of the ground. To oppose lateral expansion of the soil (reinforced ground of artificial filling materials, nailed ground of excavation and slopes) or its movement (blasted columns or micro-piers), the friction "soil-reinforcement" confers to the reinforced ground material an anisotropic cohesion. In a similar manner as with the reinforced and priestesses concrete, the bond between the soil and the reinforcement is an important phenomenon. An interaction analysis involves, separately, the behaviour of two present materials. From such analysis, and contrary to the previous one, appears the real composite behaviour of reinforced soils.

Key words: Homogenisation, interfaces, shearing, finite elements, tension, cohesion, friction.

INTRODUCTION

Many authors were interested in the implementation of a module of homogenisation in a computer code. The application to cases of walls in reinforced ground was undertaken by Cardoso and Carreto (1989), Sawicki (1990). The aforementioned studies made the assumption of perfect adherence between the matrix of the reinforced medium and the elements of reinforcement. Certain authors introduced the possibility of slip between these two materials, this new assumption allowing not over-estimating the resistance of the medium reinforced (De Buhan and Talierco, 1991). Hermann and Al Yassin (1978) on the basis of a computer code based on the finite elements took into account a displacement relating to the interface in the stiffness matrix. They then carried out a comparison with a model or inclusions that are discretized which lead to identical results. The method of homogenisation allows a considerable saving of time in the resolution. Sudret and De Buhan (1999) presented a multiphase model which gives a polar micro description of the reinforced material. Their module not only makes it possible to take into account the relative slip (of elastoplastic type) between the ground and inclusions, but also the effect of the shearing forces and bending moments. Parametric studies were undertaken on networks of intersected piles and inclusions. The principal interest of the implementation of a module of homogenisation lies in the fact that one can take into account, in an axisymmetric configuration, the longitudinal and radial reinforcement (bolting in the tunnels) which makes it possible to avoid carrying a three-dimensional calculation. This makes parametric studies possible considering a short time of resolution of such an approach. In addition to the sophistication of these modules, the study of a displacement relating to the interface soil/reinforcement and even the inflection in inclusions is also possible.

Approach taking into accounts the complete modeling of the ground, inclusion and their interaction

In this technique, the two components (solid mass and reinforcement) are discredited then assembled by means

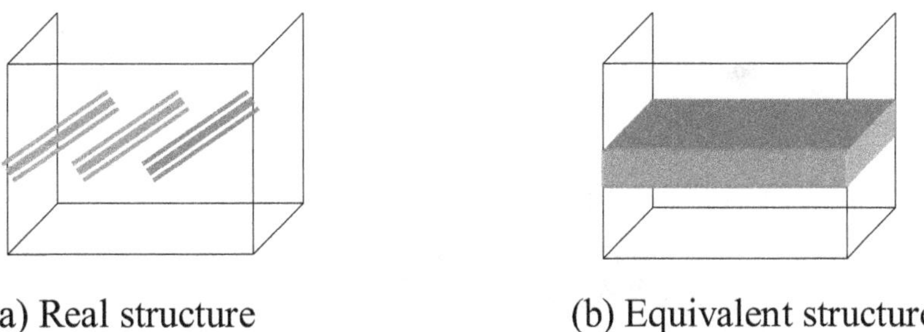

Figure 1. Modeling in 2d with an equivalent plate (Al Hallak, 1999).

of 2d or 3d elements (Chaoui, 1992; Ho and Smith, 1993) or by bar elements. The contributions of these approaches are multiple; they can allow, in particular taking into account the relative displacement of the soil/ reinforcement by the introduction of interface elements and the calculation of the efforts mobilized in the reinforcement. The use of these methods contributes to a better estimate of the contribution of the reinforcement to the limitation of the deformations.

Two-dimensional models

A calculation in plane deformation is a priori acceptable only for the two-dimensional elements of reinforcement (geotextile sheet, mesh wire) which are continuous in their plan on the scale of the work. Two major methods in plane deformations exist for modeling the reinforced solid masses. The first consists of replacing a discontinuous steel sheet by a continuous sheet, whose macroscopic properties are equivalent to those of the real sheet by formulating some assumptions recalled by Chaoui (1992), Unterreiner (1994) for a reinforced solid mass. The composite material "ground + steel" is replaced by a homogeneous plate having properties different from those of the ground and steel (Figure 1).

The second approach consists in studying the section S-S where the ground is not broken when modeling the influence of steels in the section of the ground. Two methods are proposed. The first method "slipping strip analysis" presented by Naylor (1978) is based on the study of a vertical section halfway between two vertical lines of reinforcement. The interaction between the ground and the vertical line of steel is modeled by a vertical zone of interface. With this method the reinforcements are placed out of the section of the studied ground and using a kind of load transfer function to model the interaction between the ground and steels. This approach preserves the vertical continuity of the ground

The second method is proposed by Unterreiner (1994) which considers that it is not necessary to introduce a continuous vertical zone of interface but it is sufficient to model the interaction between the section of ground S-S and each steel by the load transfer function. This one must be calculated in a suitable way or measured starting from wrenching tests on a solid mass.

Method of homogenisation

In the field of the reinforcement of the grounds, the technique of the homogenisation was developed in particular by Buhan and Al (1989). Greuell (1993), Bernaud and Al (1995) and Wong (1997) presented specific approaches for the reinforcement of the grounds. Their models, developed in cases of very simple configurations and boundary conditions, authorize analytical or semi-analytical solutions. From these studies, a model of homogenized behavior of the soil/reinforcement in our computer code is proposed. The possibility of a slip between the reinforcement and the ground is also considered.

Field of validation of the method of homogenisation by the numerical modeling of the reinforced grounds

The homogenisation of a solid mass of reinforced ground consists in replacing the two materials by an equivalent homogeneous material, representing the ground, the reinforcements and their interactions (Figure 2). This approach, however, assumes that various conditions are observed, relating in particular to the periodicity and the density of inclusions.

Representatives of the basic cell

First of all, we will define the concept of the basic cell (Romstad, 1976). This term represents the elementary structure of the composite soil/reinforcement. It is the smallest volume containing the two constitutive materials of the reinforced ground. Figure 2 illustrates in an explicit way a case of reinforcement. The basic cell is composed of two materials

The representatives of this basic cell define the ability of this one to represent the reality the whole reinforced solid mass. While knowing by advance that this condition cannot be strictly met, it is essential nevertheless that inclusion is distributed more or less in a regular way so

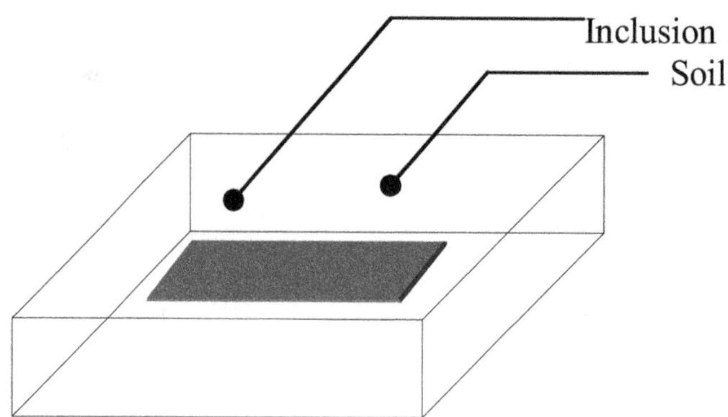

Figure 2. Basic cell representative of the reinforced ground.

that we can model the reinforced ground like a material with periodic structure. It is one of the necessary conditions to the existence of a basic cell representative of the reinforced solid mass.

Nevertheless the use of techniques of discretization in discontinuous elements (finite elements or finite differences) makes it possible to vary, to a certain extent, the density of reinforcement and its orientation in each element, which is impossible in the simplified analytical approaches based on the homogenisation.

Total character of the representation

Contrary to what was presented in numerical calculation taking account of inclusions modeled individually making it possible to locally evaluate the contribution of steels inside an element of ground, the technique of homogenisation allows us to be interested only in the total values inside the cell. In other words, it allows obtaining inside an element of ground only the average force taken by the steel located inside the element since they are also regarded as distributed in the volume of ground. This method thus has an interest only if one is interested in the total sizes (or averages) in the work.

Scale effect

The scale is directly connected to the density of steel D_b in other words the number of inclusions per square meter of wall. This density of reinforcement must be rather high so that the method of homogenisation can be employed (the surface fraction of reinforcement

$$d = \frac{Section\ renforcement}{Section\ cellule}$$ (must be sufficiently weak d<<1)

The observation is based on a comparative study between the experimental results obtained by Siad (1987) on the reinforced earth and theoretical approach by homogenisation carried out by Buhan (1989), which has established a good agreement between their results. Let us specify nevertheless that the scale effect is also related to the size of the studied field, it is thus appropriate not to consider the absolute value of L_b (length of reinforcement) but the relative one compared to the volume of studied ground, i.e. the relative homogeneity of the studied solid mass. Thus, as specified by Jassonnesse (1998), it is appropriate "to consider more objective concepts "than the scale effect as those which we approach under the following conditions.

Mesh smoothness of the numerical model

The finite element or the finite differences method forces to divide the studied continuous medium into a more or less great number of elements representing the mesh. This quantity of elements chosen by the user according to the desired precision defines the smoothness of the mesh. This concept especially dedicated to numerical calculation brings the idea of minimal length on which the digital model provides information. This size must also "be relativized "compared to dimensions of the studied work; while netting very finely. Clearly, it does not seem very useful to go down below the size of the cell but a too loose mesh can lead to a loss of information. Bernaud and Al (1995) propose in the case of a circular tunnel reinforced by radial bolting to keep the same smoothness of mesh as in the case of a non reinforced tunnel.

Period of the basic cell

The last of the conditions to satisfy to homogenize the reinforced solid mass is that the period of reinforcement (the dimension of the basic cell) is small compared to the scale of the work. The dimensions of the basic cell increase with the reduction of the number of the inclusions imbedded in the solid mass until a certain limit that validates the homogenisation method.

If these conditions are satisfied, we can then use the

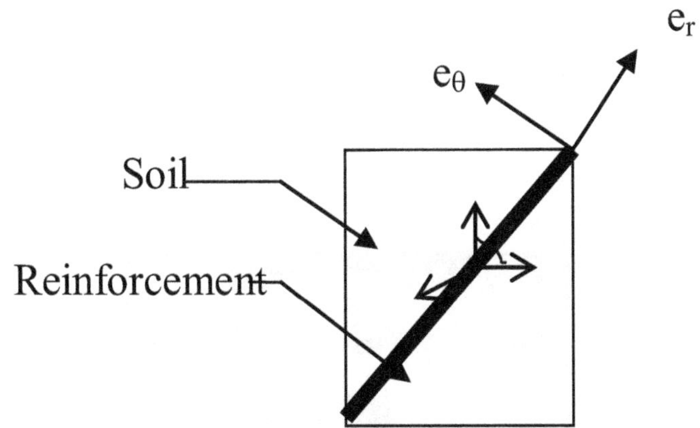

Figure 3. Anisotropic homogenized element.
θ slope of the reinforcement with the horizontal line.

homogenisation of the periodic mediums in order to analyze the behavior of an anisotropy homogeneous composite material having the same geometry, boundary and loading conditions as the reinforced soil.

METHODOLOGY

We now will be interested in the study of the case of a medium to be homogenized in an axisymmetric configuration. Because of the axisymmetric conditions, we are interested, in the same way as Bernaud and Al (1995), only to the radial and axial directions of the reinforcement (Figure 3).

All the configurations of reinforcement are nevertheless possible and, in particular, the association of radial and axial inclusions (case of tunnels) on distinct volumes of ground.

Inclusions are laid out radially around the spherical cavity $0° < \theta < 90°$ of section S_{bs} and density in D_{bs} variable with the length of the reinforcement:

*Radial inclusion: disposed $\alpha\tau\ \theta = 90°$ of section S_{bs} and density in bet D_{bs} variable with the radius:

$$D_{bs}(r) = D_{bs}(R/r)^2 \tag{1}$$

*Axial inclusion: laid out at $\theta = 0°$, of section S_{bs} and of wall constant density D_{bs} for any distance to the wall.

Law of homogenized behavior

We define the behavior of a cell of a homogenized medium from validate relations for each of its basic components, namely the ground and inclusion. We limited our study to the one-way reinforcement by plane inclusions.

Determination of stress fields and deformations in the homogenized medium

In the macroscopic scale of the structure, the reinforced ground can be considered in general as an anisotropic homogenized continuous medium, this in spite the fact that the ground and inclusions are isotropic materials. We can thus substitute the initial heterogeneous by a homogenized material inside which the stress and strain states are defined respectively by the symmetrical tensors $\underline{\Sigma}_{\eta o\mu}$ $\alpha\nu\delta$, $\underline{\varepsilon}_{\eta o\mu}$

The components of these tensors being as follows:

$$[\Sigma_{rr} \quad \Sigma_{\theta\theta} \quad \Sigma_{zz} \quad \sqrt{2}\,\Sigma_{r\theta} \quad \sqrt{2}\,\Sigma_{rz} \quad \sqrt{2}\,\Sigma_{\theta\rho}\ ] \text{ for stress} \tag{2}$$

$$[\ \varepsilon_{\rho\rho} \quad \varepsilon_{\theta\theta} \quad \varepsilon_{\zeta\zeta} \quad \sqrt{2}\varepsilon_{\rho\theta} \quad \sqrt{2}\,\varepsilon_{\rho\zeta} \quad \sqrt{2}\,\varepsilon_{\theta\zeta}\ ] \text{ for strain} \tag{3}$$

In order to be able to simplify the writing of the tensor of the stresses, it is necessary that simultaneously the surface fraction of reinforcement (with S_b section of the reinforcement and, S_{CB} section of base) be very weak d<<1, and that the stiffness of steels is much larger than that of the ground ($E_{steel} >> E_{ground}$). Gruell showed that if these two conditions are joined together, the reinforced material behaves in a macroscopic scale like a transverse isotropic elastic medium around the axis. This demonstration established by using a variationnal approach makes it possible to establish a relation between the tensors $\underline{\Sigma}_{hom}$ and, $\underline{E}_{hom}$. The tensor of the stresses in homogenized material comes from the sum of the contribution of each of two materials:

$$\underline{\Sigma}_{\eta o\mu} = \underline{\Delta}\ \underline{\varepsilon}_{\eta o\mu} + K(\rho)\ \underline{\varepsilon}_{\eta o\mu} \tag{4}$$

$$K(\rho) = \Delta_{\beta\sigma}(\rho)\ \Sigma_\beta\, E_\beta \tag{5}$$

Where: $D_{bs}(R)$ = density of reinforcement (constant when it is axial); S_b = section of the reinforcement and E_b = Young's Modulus of the steel.

Anisotropic field of elasticity

Let us define initially, the behavior of two materials constituting the homogenized material:

1. The isotropic linear elastic ground defined by the Young's modulus E_s and by the Poisson's ratio ν_σ
2. Inclusions: linear elastic bands one-way (direction $\vec{e_r}$) defined by the Young's modulus E_b

Elastoplastic behavior of the homogenized medium

As a criterion of plasticity for the ground, we have adopted that of Mohr-Coulomb. We know that this elastic perfectly plastic criterion is well adapted to the study of the grounds or tender rocks having a coherent / friction behavior.

$$f(\underline{\underline{\sigma}}_s) = (\sigma_1)_s - \frac{1+\sin\varphi}{1-\sin\varphi}(\sigma_3)_s - \frac{2C\cos\varphi}{1-\sin\varphi} \tag{6}$$

Where: $(\sigma_1)_s$ = Major principal stress in the ground and $(\sigma_3)_s$ = Minor principal stress.

The equation $\sigma_b = E_b\varepsilon_{xx} \geq R_b$ defines the acceptable field in the inclusion.

$f(\underline{\underline{\Sigma}}_{hom} - K(x)\varepsilon_{xx}) \leq 0$ defines the elastic range G_S and $f(\underline{\underline{\Sigma}}_{hom} - K(x)\varepsilon_{xx}) = 0$ working limit of stresses in the ground

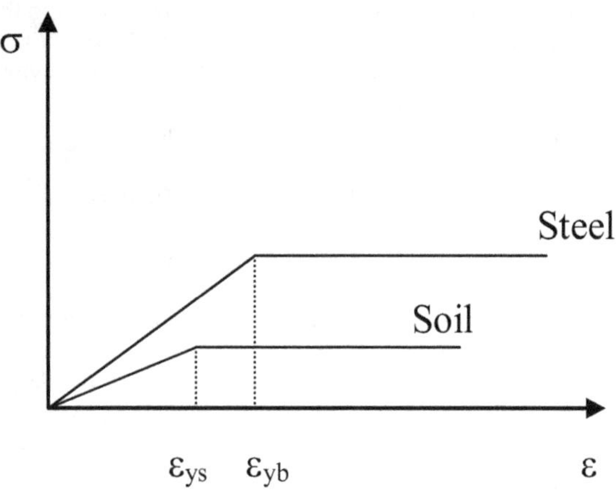

Figure 4. Uni-dimensional behavior of the ground and steels (Wong, 1995).

When a principal direction of the stresses coincides with the direction of reinforcement, the border of the acceptable field for a homogenized material is thus reached when simultaneously the criterion of rupture of the ground is obtained $f(\underline{\underline{\sigma_s}})=0$ and, tensile stress in inclusions reaches the maximum value R_b (Figure 4).

The basis of the limit of load transfer

The concept of limit of transfer of load developed by Jassonnesse (1998) introduces a limitation to σ_β representing in fact a possible slip between inclusion and the ground and thus an imperfect transfer of load from steel to the ground, which limits the resumption of the effort by steels.

The limit of the transfer of load corresponds to the introduction of a rigid-plastic friction/slip law between the ground and the inclusion and, results from the equilibrium of inclusion (Figure 5).

By putting τ friction with the interface soil/inclusion, P_b the perimeter of inclusion, σ_b the stress in the reinforcement and, S_b its section, we obtain the following relations:

$$F(x)=p_b\tau(x)et, T(\xi)=\Sigma_\beta.\ \sigma_\beta(\xi) \quad (7)$$

The inclusion is put into tension by friction τ on the interface inclusion/soil, the equilibrium thus leads to:

$$T(x+dx)-T(x)=F_s(x).dx \Leftrightarrow \frac{d\sigma_b}{dx}=-\frac{p}{S_b}\tau(x) \quad (8)$$

Modeling of the wall in a reinforced soil

Here, we will compare our theoretical results with those obtained by SHAFIEE (1985). These results show the general aspects of the behavior of the wall in a reinforced soil. We treat successively:

1. The evolution of tension T_S in the reinforcements (of the work),
2. The location of the maximum tensions T_{max} in the armed wall,
3. The distribution of displacements of the wall (for various heights of the armed wall).

The work considered is a wall 5 m high, reinforced with 5 beds of reinforcement of 5.1 m in length spaced vertically and laterally (ΔH and e) at 1 m. The facing is in concrete scales of 0.1 m in thickness (Table 1).

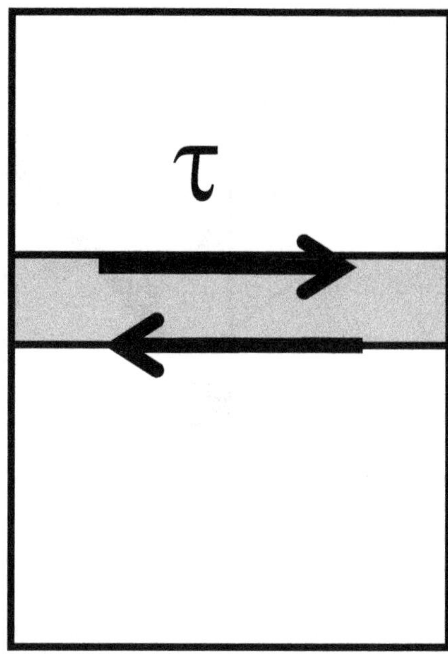

Figure 5. Equilibrium of inclusion.

RESULTS AND DISCUSSION

The evolution of tension in the 5 reinforcements is presented on (Figure 7). This evolution, in conformity with the experimental observations, separates the active zone from the passive one by a line of maximum tension. This vertical line is located at a distance D of the facing. Our results are compared to those of SHAFIEE. Our model slightly underestimates the calculated tensions. This (light) difference is due, in our opinion, to the nature of the calculation (homogenisation, elasto-perfectly plastic with a criterion of Mohr-Coulomb taking into account the effect of the interface). Distribution (non-dimensional) of maximum tension T_{max} * (we standardize all T_{max} by the maximum value of T_{max} calculated) related to the depth Z/H (Figures 6 and 8). The values determined by our program are very close to those obtained by the "CLOUTERRE" program used by SHAFIEE. We, however, note that when we check the usual mode of displacements of a reinforced earth wall (Figure 9), the values determined by our computer code remain slightly lower than the computed values by "CLOUTERRE" (the difference is about 15% in the case of taking into account the effect of the interface and 10% for a perfect adherence).

Conclusions

The behavior of the soils reinforced with linear inclusion

Table 1. Modeling of the wall in a reinforced soil.

Soil (non-cohesive)	Reinforcements in (steel)	Facing
E=10 MPa	E_b=2.10^5 Mpa	E_p=2.10^{4}Mpa
γ= 16 KN/m^3	ν_b=0.25	ν_p=0.25
ν=0.33	(S_b ϕ 50mm)	
ϕ=30°		

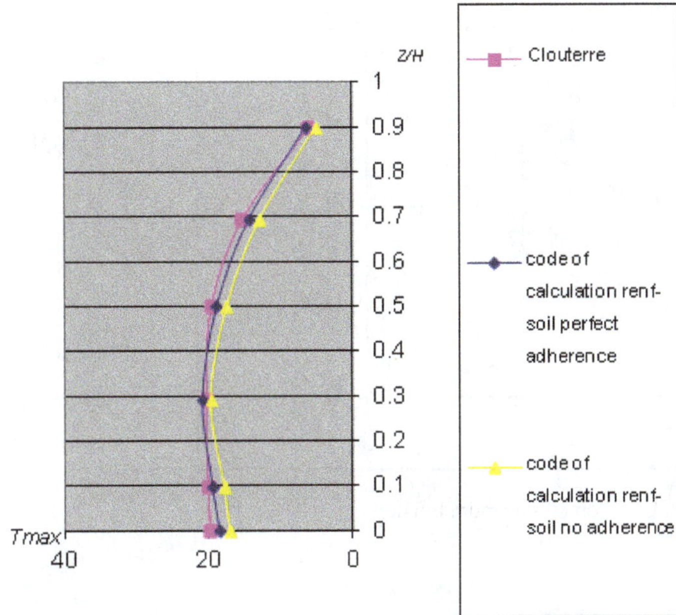

Figure 6. Maximum tension T_{max}.

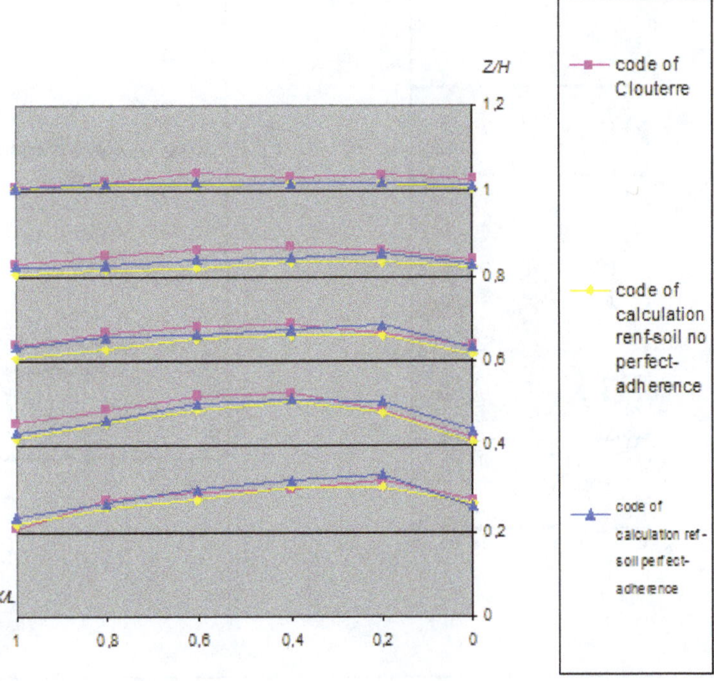

Figure 7. Evolution of tension T_S in the reinforcements.

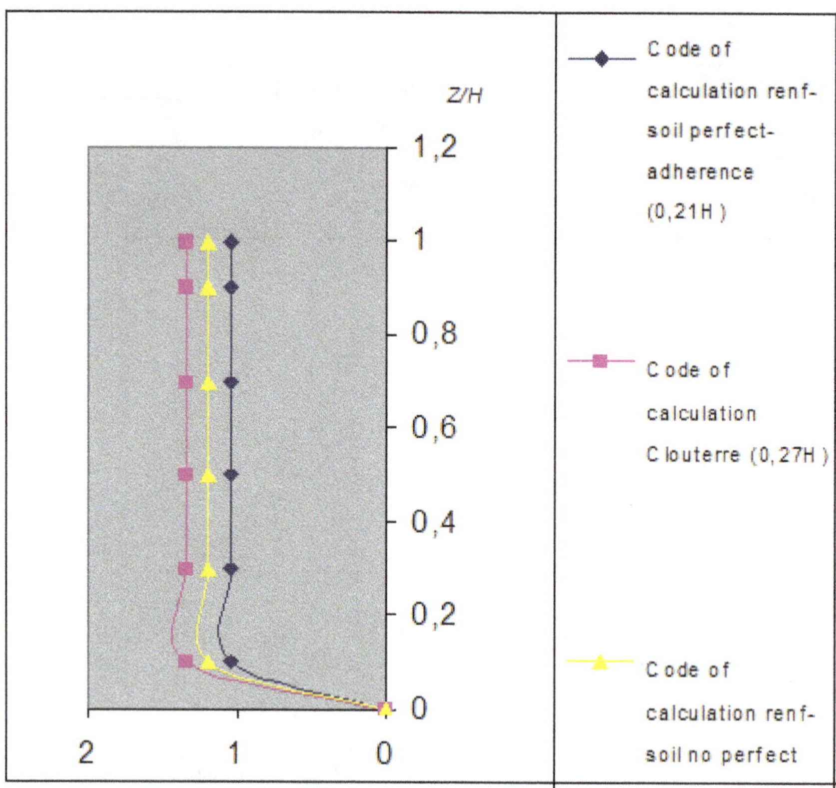

Figure 8. Location of maximum tension.

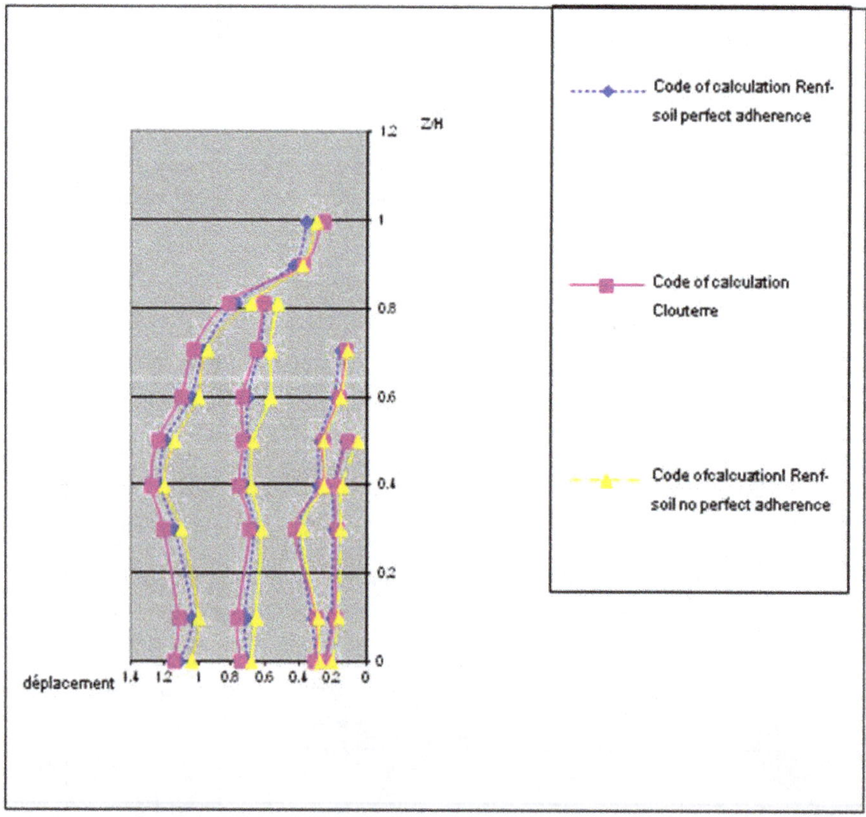

Figure 9. Evolution of displacements of the wall.

is complex and requires taking into account the transfers of the efforts to the interface sol/inclusions. The approaches of the type calculation to rupture aim to determine the equilibrium of the solid mass, but do not allow to evaluate the state of its deformations. The modeling in deformations takes into account the various elements: soil, inclusions and their connection, and leads to two types of approaches; analytical and numerical ones. The homogenisation of the periodic mediums is another approach which makes it possible to consider the composite soil and reinforcement at the macroscopic level as an equivalent material whose global behavior gives an account of that of the soil and the inclusions.

REFERENCES

Bernaud and Al (1995). Numerical simulation of the convergence of a bolt-supported tunnel through a homogenization method, Tnt. J. Num. Anal. Methods Geomech., 19: 267-288.

Chaoui F (1992). Etude tridimensionnelle du comportement des pieux dans les pentes instables, Thèse de doctorat, Ecole Nationale des Ponts et Chaussées, pp. 355.

Greuell NE (1993). Etude du soutènement des tunnels par boulons passifs dans les sols et les roches tendres par une méthode d'homogénéisation, Thèse de doctorat de l'Ecole Polytechnique, Palaiseau. pp. 199

Herman LR, Al Yasin Z (1978). Numerical analyses of reinforced soil systems, Proc. Symp. Earth Reinforced, Pttesburgh, pp. 428-570

Jassonnesse C (1998). Controle de la déformation du massif renforcé par boulonnage au front de taile d'un tunnel, Thèse de doctorat, INSA de Lyon, pp. 234

Shafiee S (1986). Simulation numérique du comportement des sols cloués interaction sol-renforcement et comportement de l'ouvrage , Thèse DDI, ENPC. pp. 278

Sawicki A (1990). Development of failure in reinforced soil structures, Performance of reinforced soil structures, Glasgow, pp. 31-40.

Sudret B, de Buhan P (1999). Modélisation multiphasique de matériaux renforcés par inclusions linéaires, C. R. Acad. Sci., , Paris, t. 327, Série Hb, pp. 7-12.

Unterreiner Ph (1994). Contribution à l'étude et à la modélisation numérique des sols clones : application au calcul en déformation des ouvrages de soutènement, thèse de doctorat, ENPC., 2 : 499.

Effect of fines mineralogy on the oedometric compressional behavior of sandy soils

K. Lupogo

Department of Geology, College of Natural and Applied Sciences, University of Dar es Salaam, P. O. Box 35052, Dar es Salaam Tanzania.

Oedometer tests have been performed on reconstituted samples prepared to study the consolidation behavior of sandy soils. The influence of fines content and type can be determined with oedometer test by utilizing intergranular void ratio. Result shows that, up to transition fines content, compression behavior of the mixtures is mainly controlled by the sand grains. When concentration of fines exceeds transition fines content, fines controls the compression. The transition fines content varies between 15 and 35% regardless of fines mineralogy. This range of fines content is also consistent with various values reported in literature regarding the strength alteration. It can be conclude that the presence of fines on sand does not increase the compression of the sand but it is improving it. Therefore for mixtures with fines content above 10% between 15 to 35% are useful for reclamation activities because of their low oedometer compression.

Key words: Silty sand, fines, fines content, oedometer, compression.

INTRODUCTION

Fines are often encountered in dredging reclamation works. The presence of fines in sandy soil is generally recognized as a problem in geotechnical engineering as it is believed in dredging practice and tendering that soil containing fines is more compressible and increases tendency to creep.

The occurrence of fines in sandy soil may be due to degradation of material during dredging processes and due to an increase shortage of clean sandy soils near the project locations the use of mixed sand, which includes silty sand, clayey sand, sandy clay and sandy silt in reclamations works will be a viable option.

It has been pointed out by Mitchell (1993) that the fines content in sandy soils have a major influence on engineering properties such as strength and stiffness, resistance to liquefaction, volume change behavior and hydraulic conductivity.

In dredging reclamation works, sandy soils with different fines contents are often encountered. The amount of fines in the fill material depend on dredged material, applied loading method, transport method and placement method. Therefore, the fines contents of the fill material can vary widely from a project to project and even within a single reclamation work. The influence of the variation of amount fines content on the engineering properties of sandy soil is not well understood. As practice in dredging and reclamation works, mostly the fills are constructed with material, which differ fundamentally from clean sand. The presence of fines in sandy soil is assumed to increase the compressibility of fill material.

It has been noticed that volume change behavior of these types of materials, especially compressibility problems have not been studied systematically in the laboratory level (Monkul et al., 2007). This study will focus more on compressibility of sandy soil as influenced by type of fines and variation of fines contents.

The parameters influencing the mechanical behaviour of sandy soils under stress conditions have been extensively studied by Amini and Qi (2000), Naeini and Baziar (2004), Della et al. (2009), Sharafi and Baziar (2010) and Belkhatir et al. (2012). The influence of laboratory initial conditions such as the relative density, the degree of saturation, the sample preparation method, the overconsolidation ratio and the stress ratio are well understood. However, the influence of other parameters such as the minarealogy of fines, the structure, size and shape of the grains are incomplete and requires further investigation.

Martins et al. (2001) noted that the presence of fines avoids obtaining a unique compression line for a particular coarse grained soil. Moreover, they stated that a new framework is needed for sandy soil which do not behave in accordance with the general compression behavior describe for other soils in literature. Fukue et al. (1986) studied the consolidation behaviour of mixture of sand and clay. They described that the compressibility decreases strongly if the void ratio of sand reaches a given value, defined as the threshold sand void ratio. During consolidation, the clayey part dominates the consolidation properties of the mixture until the sand content reaches the point where sand grains come into contact with each other. If the mixtures are compacted beyond the threshold void ratio, the soil has very low compressibility. The void ratio of the sand skeleton in this state was mentioned to vary between 1.25 and 1.4. The mixtures will have low compressibility and high strength because the friction resistance and cohesion are expected to increase during consolidation or shearing in this range of void ratios. Generally, mixtures or soils behave like sand with properties of fines in this range of void ratio.

Fukue et al. (1986) mentioned the advantages of using sand–clay mixtures for reclamation of embankments if the proper mixture can be found. These advantages are: 1) a cohesion can be expected and may reduce the liquefaction potential, 2) in terms of settlements, compressibility may be as low as that of sand or gravel, 3) most of the displaced soft sediments can be used in mixtures, 4) soil containing certain amounts of clay can also be used for reclamation if they satisfy the new standard for materials.

Monkul et al. (2006) studied the compression behavior of clayey sand and transitional fines contents. He concluded that the compressional behavior of clayey sand can be determined by oedometer test and by utilizing the intergranular void ratio rather than the global void ratio. For a given initial condition and predetermined stress, one dimensional compression of clayey sand is controlled by the coarse grain matrix up to the transitional fines content. For soils containing fines greater than transitional fines content, the compression is controlled by fine grain matrix. He described that the transition fines content is independent of the minimum void ratio.

Transition fines content can be defined to be between 19 to 34% .These studies can be considered as pioneering studies for the consolidation behaviour of sand–clay mixtures and for classification of such materials.

The objective of the present study is to systematically investigate the compressional behavior of sandy soil under the influence of different clay types from the perspective of Equivalent grain void ratio concept by defining the transition fines content and granular compressional index parameters. Influence of fine content and stress condition has been studied by means of oedometer test on reconstituted fines-sand mixtures.

Void ratio and sandy soils texture

The evaluation of the initial soil state is very important for studying the mechanical behavior of granular soils. There are several possibilities to define the initial state of granular soils, the most common parameters are the intergranular void ratio (), the interfine void ratio (), the global void ratio () and the relative density ().Void ratio and relative density are common but others are specific for sand with fines.

Mitchell (1993) introduced the use of the skeleton void ratio or granular void ratio, whereas Thevanayagam et al. (2000) introduced the use of the intergranular void ratio to study the behavior of soil that contains coarse and fine particles. The parameters intergranular void ratio, granular void ratio, or skeleton void ratios are based on the same concepts (Yang, 2004; Monkul et al. 2007; Rahman et al., 2008). These parameters are calculated by considering the fines as voids.

The intergranular void ratio is a more representative packing index to correlate sand-like properties. A formulation of intergranular void ratio by Thevanayagam (2000) can be represented as:

$$e_s = \frac{e + fc}{1 - fc} \tag{1}$$

Where e is global void ratio, - fine content. The use of the intergranular void ratio is limited to soils with fines content below the threshold fines content (TFC) which is between 20 to 30% as pointed out by different authors (Thevanayagam, 1998; Ni et al., 2004; Yang, 2004; Rahman et al., 2011). At higher fines content (relative to the void space), the fines begin to participate in the force structure. Therefore, Thevanayagam et al. (2002) proposed the use of equivalent granular state parameter, e_s defined by Equation (2) below as a better alternative state to *e*:

$$D_r = \frac{e_{\max} - e}{e_{\max} - e_{\min}} \tag{2}$$

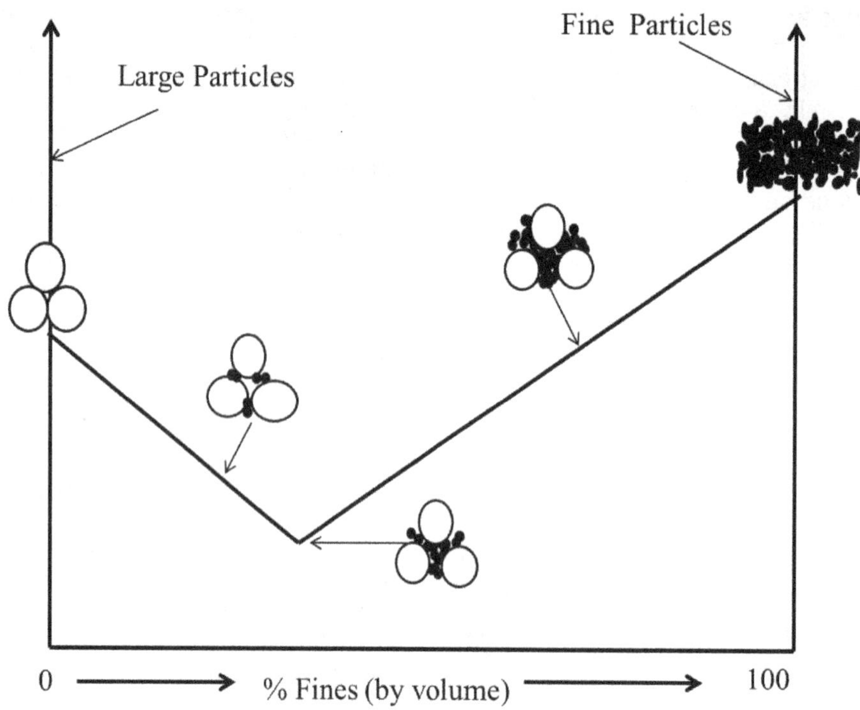

Figure 1. Schematic explanation of silt sand mixture (Lade and Yamamuro, 1998).

where, b represents the fraction of fines that are active in force structure. The basic assumption of behind Equation (2) requires. The e_s was also being referred to as, equivalent inter-granular contact void ratio (Thevanayagam et al., 2002), corrected intergranular void ratio (Yang et al., 2006b), equivalent inter-granular contact index void ratio (Thevanayagam, 2007), equivalent granular void ratio (Rahman and Lo, 2008b; Rahman et al., 2008, 2011).

In Figure 1, the evolution of the void ratio is shown for different fines contents. There are three zones between the two extremes for sand silt mixtures. When the silt content is about 10 to 20% the sand behavior is dominant. The second zone applies for approximately 25 to 45% of silt for which the silt skeleton is replacing the sand skeleton. In this case the silt particles fill the sand voids, which leads to a considerable decrease of the void ratio and neither the sand nor the silt can play its own role as they do alone. The third zone applies for a higher percentage of silt content for which the silt behavior dominates (Bahadori, 2008). Thevanayagam and Mohan (2000) worked on quartz sand with different contents of plastic fines, and they concluded that a fines content between 20 and 30% in a coarse grain matrix defines the transition between the two types of behavior. For a fines content of 10% or lower, the coarse–grained matrix dominates the compressibility and for fines contents of 40% or higher the fines dominate the behavior. It can be concluded that, the transition behavior of the mixed sand-fines mixture is somewhere between 25 and 45%. It is expected that at this range of fines contents the mixture will have a low void ratio and a low compressibility.

MATERIALS AND METHODS

To get undisturbed mixed samples from field for laboratory testing is difficult. An alternative approach is to test reconstituted samples obtained by mixing sand with various quantities of fines. Four types of materials were used. The host sand consists of a blend of two types of sand. Coarse quartzic sand from the river Maas was mixed with artificially (crushed) quartz sand produced by Sibelco LM25 in a proportion of 2 to 1. In the first series of tests (series A), the filler material was a quartz flour also produced by Sibelco by iron-free grinding of selected quartz sand with a high silica-content in ball- or vibration mills and control of particle size. The flour is commercialized under the name of M10. In the second series of tests, fines were sampled above a sand quarry in Bierbeek nearby Leuven, Belgium. The Belgium silt belongs to an Eocene loess deposit that lies above the sand of Brussels formation.

The index parameters of sand and fines mixtures shown in Table 1 were determined by means of sieve, hydrometer, specific gravity and consistency limit tests. These tests were performed in accordance with British standards (BS-1377, 1990). Maximum void ratio (e max) of sand was obtained by pluviation of sand grains into a mold of known volume filled with water.

Sand – fines mixtures were prepared on dry weight basis using oven dried sand and fines which were thoroughly mixed. Therefore, the fines content (FC) refers to the percentage of fine grains in total weight of solids. The sand – fines mixtures were mixed manually in dry state. The mixing process continued for 10 to 15 min, until the mixtures are observed to be visually homogeneous. Then the mixtures were inundated to desired water content soaking period of 24 h. After soaking period, wet mixing was performed for 15 min in such a manner that sand grains were dispersed in mixture

Table 1. Characteristics of the mixtures of artificial silt and Belgium silt.

Material fines content (%)	Specific gravity (g/cm^3)	Coefficient of uniformity	D10 (mm)	D30 (mm)	D50 (mm)
		Artificial silt			
Host sand	2.65	1.21	0.75	0.85	0.9
7	2.67	100.00	0.018	0.7	1.23
10	2.67	175.00	0.01	0.5	1.22
15	2.67	269.23	0.0065	0.25	1.21
20	2.67	309.09	0.0055	0.08	1.2
25	2.67	340.00	0.005	0.025	1.2
30	2.67	326.53	0.0049	0.01	0.9
50	2.67	111.11	0.0045	0.005	0.01
Host silt	2.67	10.00	0.002	0.006	0.0105
		Belgium silt			
Host sand	2.65	1.21	0.75	0.85	0.9
7	2.67	16.67	0.108	1	1.23
10	2.67	17.50	0.1	0.8	1.22
15	2.67	116.67	0.015	0.6	1.21
20	2.67	154.55	0.011	0.35	1.2
25	2.67	157.41	0.0108	0.2	1.2
30	2.67	158.42	0.0101	0.1	0.9
50	2.67	71.43	0.007	0.025	0.1
Host silt	2.67	6.67	0.006	0.02	0.032

uniformly and segregation was prevented.

Oedometer samples were tested in 7.5 cm diameter rings. Care was taken in order to avoid air entrapment during placing sample into rings. For all samples the degree of saturation is 0.95. Loadings were initiated from 3.5 KPa and terminated at 1108.3 kPa. It was found in preliminary tests that primary consolidation had been already completed in 2 h. Therefore, loading duration was set at 2 h in accordance with ASTM standard D 2435-96 method.

The initial water content, initial global void ratio and initial intergranular void ratio of the samples for both sets are given in Table 2.

RESULTS

Relationship between effective stress and vertical strains

The relationship between effective stress and vertical strains for the different materials tested at different relative densities and fines contents is shown in Figure 2. The artificial silt mixtures were tested at an assumed relative density of 50% the slope of their virgin compression curves do not show any trend as function of fines content. The Belgium silt mixtures were tested at an assumed relative density of 50%; the compression line steepness increases with an increase in fines content. Figure 3 shows the relationship between void ratio and vertical effective stress. The initial void ratios are as expected, considering the targeted void ratios.

DISCUSSION

The main objective of this study was to understand the effect of fines mineralogy and fines content on the compressibility of sandy soil and including the change on the void ratio and compressive index.

Equivalent grain void ratio concepts

Intergranular void ratio concept employ the concept of using the void ratio created by the granular material and considers the fines as voids. Based on this concept the influence of fines on the compressibility of sand silt mixture can be depicted.

Figure 4 shows variation of Equivalent grain void ratio with vertical effective stress. These curve have similar features as those plotted for vertical effective stress as function of void ratio. However, using intergranular void ratio the curves are shifted in position but the nature of deformation of the sample remain the same.

Monkul and Ozden (2005) introduced the concept of transition fines content as fines content at which contact between grains occurs. In this concept, it is assumed that direct grains contacts of coarse grains can be initiated when intergranular void ratio of the mixture become equal to the maximum void ratio of the host granular material. Transition fines content is determined by intersection of

Table 2. The initial water content, initial global void ratio and initial intergranular void ratio of artificial silt and Belgium silt.

Sample	Silt (%)	Minimum void ratio	Maximum void ratio	water content	Initial density (kg/m^3)	Dry density (kg/m^3)	Initial void ratio	Relative density (%)
				Artificial silt				
A10	10	0.318	1.046	0.20	1932.9	1610.8	0.651	54
A11	10	0.318	1.046	0.20	1752.2	1642.7	0.619	59
A12	15	0.276	0.867	0.21	2065.6	1700.4	0.564	51
A13	20	0.201	0.734	0.18	2175.0	1843.2	0.443	55
A31	20	0.201	0.734	0.17	2120.7	1812.6	0.468	50
A14	25	0.171	0.781	0.20	2147.8	1789.8	0.486	48
A32	25	0.171	0.781	0.18	2126.4	1802.1	0.476	50
A15	30	0.16	0.783	0.17	2143.5	1786.2	0.489	47
A16	30	0.16	0.783	0.17	2143.5	1786.2	0.489	47
A33	30	0.16	0.783	0.18	2115.2	1792.6	0.484	50
A17	50	0.233	1.323	0.20	1901.5	1584.6	0.679	59
A18	50	0.233	1.323	0.20	1901.5	1584.6	0.679	59
A34	50	0.233	1.323	0.29	1929.6	1495.8	0.778	50
				Belgium silt				
B1	7	0.318	1.046	0.20	1932.9	1610.8	0.651	54
B2	10	0.318	1.046	0.20	1752.2	1642.7	0.619	59
B3	15	0.276	0.867	0.21	2065.6	1700.4	0.564	51
B4	20	0.201	0.734	0.17	2120.7	1812.6	0.468	50
B5	25	0.171	0.781	0.18	2126.4	1802.1	0.476	50
B6	30	0.16	0.783	0.18	2115.2	1792.6	0.484	50
B7	50	0.233	1.323	0.29	1929.6	1495.8	0.778	50

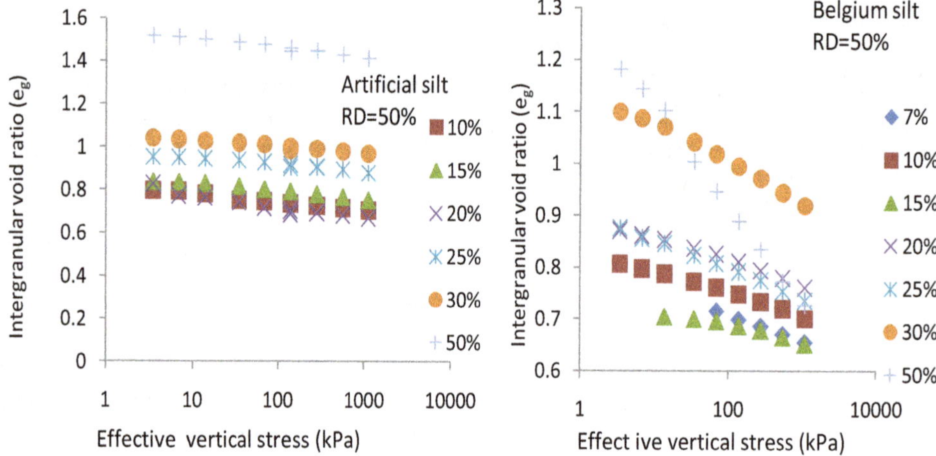

Figure 2. Relationship between vertical strain and effective vertical stress.

maximum void ratio line with intergranular void ratio against effective vertical stress curves.

Granular compression index (C_{c-s})

The granular compression index (C_{c-s}) was introduced by Monkul and Odzen (2005). The importance of this parameter is on depicting the compressional characteristics of the silt sand mixture. The definition of C_{c-s} is similar to the definition of compression index Cc and is expressed based on the decrease of Equivalent grain void ratio with effective stress increment as in Equation 3.

$$C_{c-s} = \frac{\Delta e}{\Delta \log \sigma'} \quad (3)$$

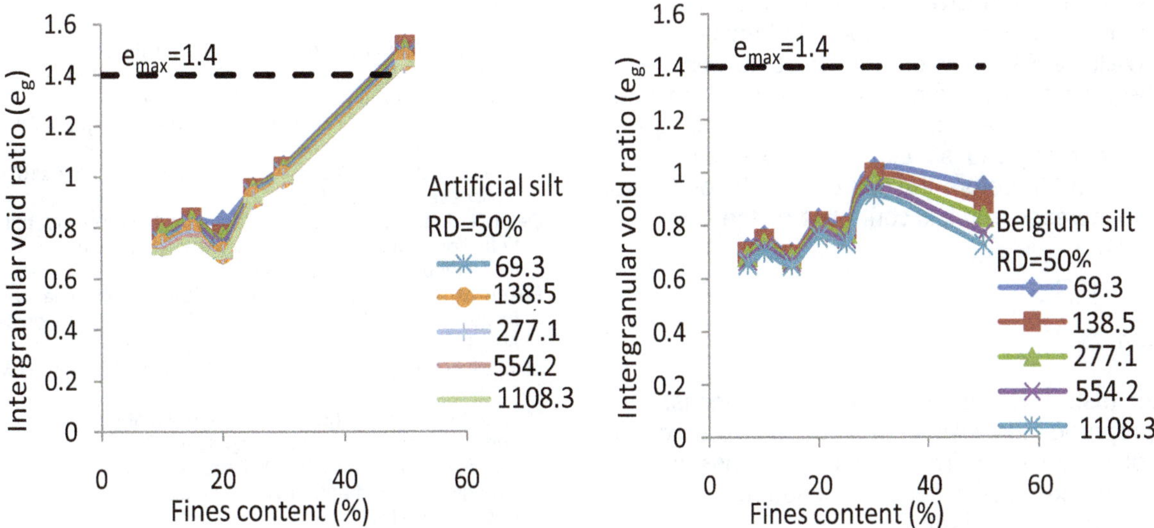

Figure 3. Relationship between void ratio and effective vertical stress.

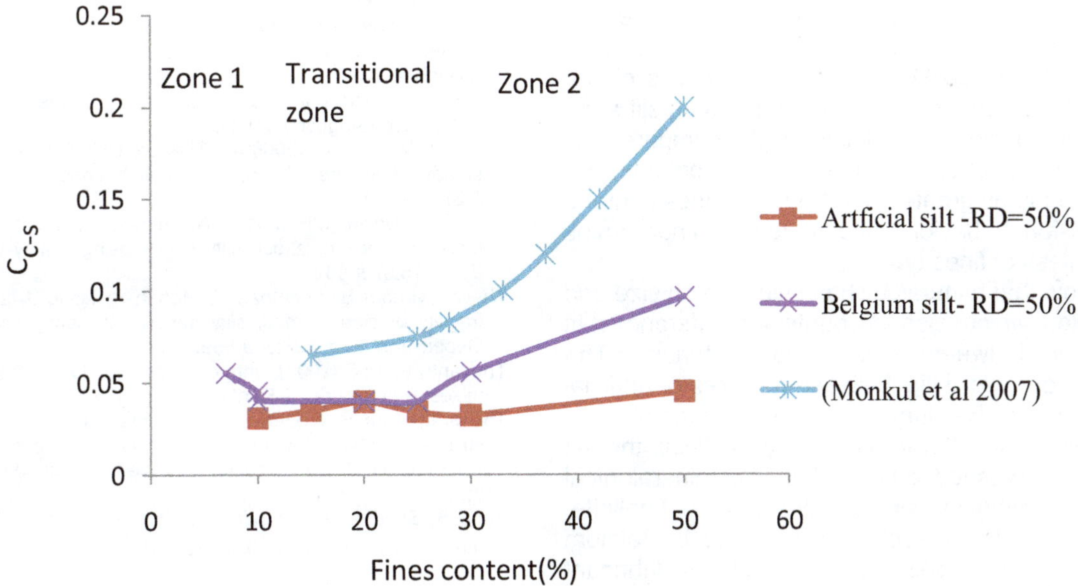

Figure 4. Relationship between equivalent grain void ratio and effective vertical stress.

Figure 4, shows relationship between $C_{c\text{-}s}$ and fines content.The trend observed in these curve is the same as those shown in the figure showing relationship between Cc and fines content. Relationship between $C_{c\text{-}s}$ and fines contents it shows clear the influnce of fines n compressibility as one can see from Figure 4. Mixtures with fines content below 10% Cc-s decrease with increase fines content while for mixture between 10 and 25% have low $C_{c\text{-}s}$ values while for mixture above 25% $C_{c\text{-}s}$ values increases with and increase in fines content. Tested material show to have low Cc-s compared to material tested by Monkul et al. (2007) as it can be observed in the Figure 4. This can be due to the difference in mineralogy and grain sizes in the mixtures.Relationship between Cc-s and fines contents does shows clear differences at higher fines content for Artifical silt and belgium silt.

There are three zones can be observed in Figure 4 (that is, zone 1, transitional zone and zone 2). In zone 1, the coarse grain matrix can be assumed to have almost a continuous framework with grain to grain contacts and fines are mostly located in the intergranular voids and in the contacts between the grains, as load applied the fines at the contacts tends to move and fill the intergranular voids, hence compressive behavior decrease with increase in fines contents in this zone. In transition zone, all fines available fill the intergranular voids and hence forming the continuous framework with grains to grain contact. In zone 2, as fines increases, sand grains become more dispersed so that there exist almost no grain contact and therefore the compressibility of the soil continues to increase and it is controlled by the finer grain matrix.

Conclusion

Oedometer tests have been performed on reconstituted samples prepared in laboratory to study the consolidation behavior of sandy soils. The influence of fines content and type can be determined with oedometer test by utilizing intergranular void ratio.

For a given initial condition and predetermined stress, one dimensional compression behavior of sandy soils is mainly controlled by coarser matrix up to transition fine content. This behavior does not depend on the type of fines and it is almost the same for both plastic fine and non-plastic fines. For both type of fines , at transitional fines, content (15 to 30%) the packing density is higher resulting to lower compression and for Belgium silt sand mixtures compression is relative high compared to artificial silt sand because its mineralogical contents. For soils containing fines greater than transition fines content, the compression behavior is controlled by finer grains matrix regardless of fines types.

Results show that mineralogy has influence on size and shape of fines which also accounts for differences in compressibility between these two materials. The difference in compressibility between mixtures of artificial silt and mixtures of Belgium is that due to mineralogical compositional of the silts. In mixtures of artificial, the silt used was pure crushed quartz while the mineralogical composition of Belgium silt contains quartz, kaolinite, mica and calcite. The presence of the mica in Belgium mixtures for higher fines contents act as lubricant because of its flexible sheet structure which deform easily under loading. As a result, Belgium silt is more compressible as fines content increases above transitional fines content.

ACKNOWLEDGEMENTS

The authors greatly appreciate the support of TUDelft and Boskalis Dredging Company for partially funding this study. Dr. Dominic Ngan–Tillard and Co. are acknowledged for their valuable suggestions, continuous concern and steady assistance. The author also acknowledges Mr. Wim Vaal for his assistance in the experimental program.

REFERENCES

Bahadori H, Abbas G, Towhata I (2008). Effect of non-plastic silt on the anisotropic behavior of sand. Soils Found. 48:531-545.

Belkhatir M, Arab A, Della N, Missoum H, Schanz T (2012). Experimental Study of Undrained Shear Strength of Silty Sand: Effect of Fines and Gradation. Geotechnical and Geological Engineering. Available at: http://rd.springer.com/article/10.1007/s10706-012-9526-1 [Accessed August 12, 2012]

Della N, Arab A, Belkhatir M, Missoum H (2009). Identification of the behavior of the Chlef sand to static liquefaction. J. Comp. Rendus Me´canique (CRAS) 337:282–290.

Fukue M, Okusa S, Nakamura T (1986). Consolidation of sand-clay mixtures. Consolidation of soils, ASTM Special technical Publication 892, R.N Yong and F.C Townsend, pp. 627-641.

Mitchell J (1993). Fundamentals of Soil Behavior. New York: John Wiley & Sons, Inc.

Monkul M, Ozden G (2007). Compressional behaviour of clayey sand and transition fines content. Eng. Geol. 89:195-205.

Naeini SA, Baziar MH (2004). Effect of fines content on steady state strength of mixed and layered samples of a sand. Soil Dyn. Earthq. Eng. 24:181-187.

Ni Q, Tan T, Dasari G, Hight D (2004). Contribution of fines to the compressive strength of mixed soils. Geotechnique 54:561-569.

Rahman M, Lo S, Gnanendran C (2008). On equivalent granular void ratio and steady state behaviour of loose sand with fines. Can. Geotechn. J. 45:1439-1456.

Rahman M, Robert LO (2011). Effects of fines and fines type on undrained behaviour of sandy soils under critical state soil mechanics framework. In World Scientific Publishing Co. Pte. Ltd., pp. 403-408. Available at: http://www.worldscientific.com/doi/abs/10.1142/9789814365161_0050 [Accessed August 5, 2012].

Rahman M, Lo SR (2008b). "The prediction of equivalent granular steady state line of loose sand with fines." Geomech. Geoeng. 3(3):179-190.

Sharafi H, Baziar MH (2010). A laboratory study on the liquefaction resistance of Firouzkooh silty sands using hollow torsional system. EJGE 15:973–982.

Thevanayagam S, Shenthan T, Mohan S, Liang J (2002). Undrained fragility of clean sands, silty sands and sandy silts. J. Geotechn. Geoenviron. Eng. 128:849-859.

Thevanayagam S (2007). "Intergrain contact density indices for granular mixes: Framework." J. Earthq. Eng. Eng. Vib. June.

Thevanayagam S, Mohan S (2000). Intergranular state variables and stress-strain behaviour of silty sands. Geotechnique 50(1):1–23.

Yang S (2004). Characterization of properties of sand-silty mixture. PhD thesis.

Yang S, Sandven R, Grande L (2006). Steady-state lines of sand-silt mixture. Can. Geotech. J. 43:1213-1219.

Effects of curing conditions on high strength concrete

Abalaka A. E.[1]* and Okoli O. G.[2]

[1]Department of Building, Federal University of Technology, Minna, Nigeria.
[2]Department of Building, Ahmadu Bello University, Zaria, Nigeria.

Compressive strength of high strength concrete (HSC) cubes cured in water and air (uncured) were investigated at different ages and free water/cement (w/c) ratio. Tensile strength of the HSC cylinders was also determined at 90 days for uncured and water cured specimens at different w/c ratios. Compressive strength decreases were recorded for all the ages tested for uncured specimens compared to water cured specimens: the maximum compressive strength decrease of 35.99% was recorded for uncured cubes at 90 days at w/c ratio of 0.35. Tensile strength of the cylinders at 90 days increased with water curing at all the w/c ratios. Water cured specimens had improved durability properties at 90 days compared to uncured specimens of the same age.

Key words: High strength concrete, strength, curing.

INTRODUCTION

High strength concrete (HSC) has a more uniform and compact internal structure compared to the more porous internal structure usually associated with normal strength concrete (NSC) Giaccio et al. (2007). HSC mixes have cement content higher than 400 kg/m^3, with compressive strength greater than 50 Mpa at 28 days (Long, 2008). High compressive strength usually associated with HSC enables the construction of slim concrete sections that would otherwise be impossible with NSC. The internal structures and properties of HSC are so different from normal concrete that they are given separate classification by ACI Committee 363 (1998). The improvements in properties of HSC facilitate the construction of structures with long service lifespan and low life-cycle maintenance costs; it is used mainly in the construction of bridges, high rise buildings, pre-stressed concrete and high performance structures.

Curing of concrete is the process of maintaining satisfactory moisture content and temperature in the concrete for a definite period of time (Mamlouk and Zaniewski, 2011; Neville, 1981). If curing stops for some time and then resumes again, then strength gain will also stop and reactivate (Bushlaibi and Alshamsi, 2002); though the detrimental effect of early improper curing are irreversible (ACI, 1991). Concrete curing starts soon after concrete hardens and the aim is to control temperature and moisture movement from and into the concrete; it essentially promotes cement hydration (Neville and Brooks, 2008). Since cement hydration takes place only in water filled capillaries, it is important to keep concrete saturated or as nearly saturated as possible until the water filled pores in the fresh cement paste is filled to the desired extent by hydrating calcium silicate hydrate gels and other solid hydration products (Neville, 1981; Taylor, 2000). Hydration of cement can be defined as the combination of all chemical and physical processes that take place after contact of the anhydrous solid with water (Stark, 2011).

Earlier work by Powers (1949) shows that increased hydration leads to increase in gel/space ratio for w/c ratio in the range of 0.10 to 0.95; as hydration products form, capillary water volume reduces. The degree of hydration of cement had been shown to be dependent on the vapour pressure in concrete and maximum hydration rate occurs under saturation conditions (Neville, 1981). The work of Powers (1947) also shows that the degree of hydration is negligible at a vapour pressure below 0.3 of the saturation pressure. Water is known to improve hydration of cement (Zhutovsky and Kovler, 2012).

Earlier researches by Gonnerman and Shuman (1928)

*Corresponding author. E-mail: aabalaka@gmail.com.

Table 1. Composition of OPC by mass.

Composition (%)													
SiO_2	Al_2O_3	Fe_2O_3	CaO	MgO	SO_3	K_2O	Na_2O	Mn_2O_3	P_2O_5	TiO_2	Cl-	SR	AR
24.79	6.35	0.92	58.50	2.87	4.91	0.80	0.65	0.0	0.15	0.06	0	3.41	6.88

SR: silica ratio=SiO_2/ (Al_2O_3+Fe_2O_3), AR=alumina ratio= Al_2O_3/Fe_2O_3.

Table 2. Particle size distribution of aggregates as percentage by weight passing sieve sizes.

Aggregate	Sieve size (mm)							
	20	10	5	2.36	1.18	0.60	0.30	0.15
Fine aggregates	-	-	92.4	81.6	61	38.3	14.5	5.3
Coarse aggregates	95.00	40.62	0.80	-	-	-	-	-

Table 3. Concrete mix proportions.

Cement content	Sand	Coarse aggregates	Free w/c ratio
530 kg/m^3	458 kg/m^3	1,302 kg/m^3	0.30 - 0.55

and Price (1951), show that concrete cured in air had lower compressive strength compared to water cured concrete at all the ages tested. The work of Soroka and Baum (1994) showed that at 28 days, compressive strength of concrete cube specimens continuously wet cured was 40% higher than those uncured and at 90 days specimens continuously moisture cured had compressive strength 20% higher than those of uncured cubes.

Guneyisi et al. (2005) reported compressive strength loss of 10 to 20% of concrete cubes that were ambient air cured compared to concrete cubes that were wet cured; compressive strength losses were recorded at curing ages of 28, 90 and 180 days for cube specimens that were air cured compared to wet cured cubes. Wet curing was reported by Guneyisi et al. (2005) to be more effective in improving compressive strength at later ages for higher w/c ratio specimen than lower w/c ratio specimen. Alizadeh et al. (2008) reported compressive strength increases of concrete cubes cured in water compared to air cured cubes at 7 and 28 days using Portland cement at cement content of 400 kg/m^3.

The characteristic strength of concrete at 28 days is generally accepted as standard for design and construction and concrete specifications are usually based on 28 days strength of concrete cubes continuously cured in water. However, in practice site concrete rarely have the benefit of the continuous wet curing method used in the laboratory due to the technical difficulties in providing continuous wet curing for concrete elements.

The aim of this study was to determine the compressive strength development of HSC specimens continuously cured in water and specimens stored in the open on laboratory floor after de-molding in a tropical environment. In technical terms, the cubes stored in the open air are actually uncured specimens. The ambient air storage of the cube specimens after de-molding simulates practical concrete production since site concrete elements are not subjected to continuous wet curing. This is particularly the case for vertical structural concrete elements; in site concrete production active curing nearly always ceases long before the maximum possible hydration has taken place (Neville and Brooks, 2008).

MATERIALS AND METHODS

A commercial brand of ordinary Portland cement (OPC) available in Nigeria was used for this study. The compositions of the OPC used, determined by X-ray florescence (XRF) are given in Table 1.

Crushed granite of 20 mm maximum size with specific gravity of 2.63 was used as coarse aggregates; natural river bed quartzite sand with specific gravity of 2.73 was used as fine aggregates. The results of the sieve analysis of the aggregates are given in Table 2. The particle size distribution of the fine aggregates correspond to zone 2 sand by the British Standards Institution, BS 882 (1983) classification. The concrete mix proportions used in this study are given in Table 3.

The concrete was mixed in a tilting drum mixer for 2 min, and manually compacted in two layers in 100 mm steel moulds. Forty cubes and three cylinders (150 mm × 300 m) were cast for each concrete mix. A chloride free lignosulphonate based plasticizer (Fosroc's *Conplast P505*) complying with British Standards Institution, BS EN 934 (2001) was used to increase the slump of lower w/c ratio mixes. After 24 h in the moulds, the cubes to be continuously cured in water were de-molded and cured in water in compliance to British Standards Institution, BS 1881 (1997)

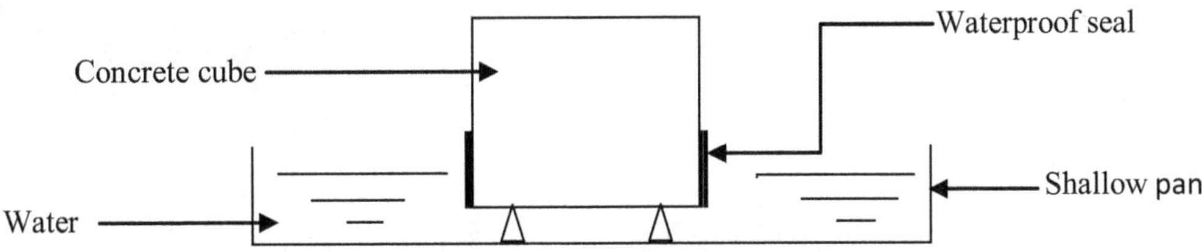

Figure 1. Coefficient of water absorption and sorptivity test.

standard. The cubes that were designated uncured were removed from the moulds after 24 h and stored in the open air on the laboratory floor. The compressive strength of the cubes were determined in compliance to British Standards Institution, BS 1881 (1970) using *ELE ADR 3000* digital compression machine at a loading rate of 3.00 kN/s; split tensile strength of concrete cylinders were determined in compliance to British Standards Institution, British Standards Institution, BS 1881 (1983) using the same machine at a loading rate of 2.10 kN/s.

Concrete cubes cured in water were used as control. Three samples were tested for each parameter investigated and the results represented are an average of three test results. Coefficient of water absorption and sorptivity of the concrete cubes were measured at 90 days using the experimental setup as shown in Figure 1.

Coefficient of water absorption

Coefficient of water absorption is a measure of permeability of concrete (Ganesan, 2008; Giannotti da Silva, 2008). This is determined by measuring water uptake in dry concrete in a time of 1 h. The concrete specimens were heated in an oven at 98°C until a constant weight was attained at ten days and the cubes were allowed to cool gradually to room temperature for 24 h. Four sides of 100 mm cube samples were sealed with 1 mm thick silicone sealant to a height of 30 mm to allow water absorption on only one surface of the cube. The samples were immersed to a depth of 10 mm in water as shown in Figure 1. After immersion in water for one hour, the cubes were taken out and the wet surface was wiped of excess water and weighed. The coefficient of water absorption of the specimens at 90 days was calculated from the formula,

$$K_a = \left[\frac{Q}{A}\right]^2 \times \frac{1}{t}$$

where K_a is the coefficient of water absorption (m^2/s), Q is the quantity of water absorbed (m^3) by the oven dry specimen in the time (t), t=3600 s and A is the surface area (m^2) through which water was absorbed (Ganesan et al., 2008).

Sorptivity

Sorptivity is a measure of the capillary forces exerted by the pore structure causing fluids to be drawn into the body of the material (Ganesan et al., 2008; Hall, 1989). The concrete specimens were heated in an oven at 98°C until a constant weight was attained at ten days and then allowed to cool to room temperature for 24 h. The sides of the cubes were coated with silicone sealant to allow the flow of water on only one surface of the cube specimen. The cube specimens were immersed to a depth of 10 mm in water on only one surface. The initial mass of the cube was taken at time 0 and at time intervals of 1, 2, 4, 8, 10, 20, 30, 60 and 90 min, the samples were removed from water and excess water blotted off and the sample weighed. It was then placed back in water and the process repeated at the same selected time intervals. The sorptivity value of the specimens at 90 days were calculated using the formula

$$i = S/\sqrt{t}$$

where i is the cumulative water absorption per unit area of the surface (m^3/m^2); S is the sorptivity ($m/\sqrt{t}$) and t is the elapsed time (s) (Stanish et al., 1997).

RESULTS

The effects of curing conditions on compressive strength of concrete specimens at different ages are given in Table 4. Table 5 shows the effect of curing on tensile strength of concrete cylinders, sorptivity and coefficient of water absorption of concrete cubes at 90 days at different w/c ratio. Percentage compressive and tensile strength losses of uncured specimens compared to control are shown in Table 6. Table 7 shows the percentage increase in durability properties of coefficient of water absorption (***Ka***) and sorptivity (s) of uncured specimens compared to control at the age of 90 days.

Compressive strength

From the results in Table 4, at a w/c ratio of 0.30, cube specimens uncured and those continuously cured in water show progressive compressive strength increase with age. The results in Table 4 show appreciable strength reductions with increase in w/c ratio; in HSC mixes it is important to have high cement content and low w/c ratio in order to develop high compressive strength at 28 days. The results of compressive strength tests at w/c ratio of 0.30 at 28 days and 90 days were less than the strength recorded at w/c ratio of 0.35 because of the high dose of plasticizer used for the mix.

Compressive strength losses were recorded for all the uncured specimens compared to water cured specimens at all the ages tested. Percentage compressive strength

Table 4. Effects of curing conditions on compressive strength of HSC.

Free w/c	Plasticizer (l/m^3)	Slump (mm)	Uncured compressive strength (N/mm^2)					Wet cured compressive strength (N/mm^2)				
			3 days	7 days	14 days	28 days	90days	3 days	7 days	14 days	28 days	90 days
0.30	6.8	30	29.43	30.85	31.91	32.84	32.95	36.04	40.02	47.38	49.18	49.75
0.35	5.7	90	29.03	34.24	36.64	37.63	39.50	35.42	41.68	45.73	52.71	61.71
0.40	2.4	120	28.79	30.36	33.60	34.78	35.01	32.49	34.95	40.42	44.53	49.96
0.45	1.1	200	22.13	29.96	31.00	33.40	33.34	23.82	27.68	33.69	33.94	40.11
0.50	0	200	19.19	20.45	25.54	25.79	26.75	20.89	24.18	28.13	30.39	34.85
0.55	0	200	13.29	16.25	18.56	20.39	21.21	15.20	16.99	21.01	21.34	26.37

Table 5. Effect of curing conditions on tensile strength and durability properties.

Free w/c	Age (days)	Uncured						Water cured					
		0.30	**0.35**	**0.40**	**0.45**	**0.50**	**0.55**	**0.30**	**0.35**	**0.40**	**0.45**	**0.50**	**0.55**
Tensile strength (N/mm^2)	90	3.074	3.153	2.911	2.994	2.238	1.845	4.276	3.978	3.557	3.760	3.516	2.953
Coefficient of water absorption Ka (m^2/s) $\times 10^{-20}$	90	8.837	7.140	20.332	21.501	46.883	78.407	4.016	3.134	4.786	9.036	11.950	13.499
Sorptivity $S\ (m \times \sqrt{t})$ $\times 10^{-6}$	90	1.555	1.338	2.209	2.319	3.019	5.515	0.924	0.767	1.274	1.657	1.951	2.393

losses are shown in Table 6 for uncured specimens compared to the control. The maximum value of 34.03% compressive strength loss was recorded for uncured cubes recorded at w/c ratio of 0.30 at 90 days. The results in Table 6 further show percentage compressive strength losses increased with age. Similar pattern of progressive compressive strength losses with age were recorded for uncured cubes at the other w/c ratios.

Tensile strength

The results of tensile strength tests of cylinder specimens at 90 days for uncured and water cured specimens at 90 days at different w/c ratios are shown in Table 5. The results show that for both uncured and water cured specimens, tensile strength reductions were recorded as w/c ratio increased. Percentage tensile strength losses of uncured cylinders compared to the control at different w/c ratios are given in Table 6, with a maximum loss of 37.52% at w/c ratio of 0.55.

Coefficient of water absorption and sorptivity

The results of tests on cube specimens for durability properties of coefficient of water absorption and sorptivity at 90 days for both uncured and water cured cubes are given in Table 5. For uncured specimens, the coefficient of water absorption increased as the w/c ratio increased; similarly for water cured specimens the coefficient of water absorption increased as the water/cement ratio increased. However, the values coefficients of water absorption of water cured specimens are less than that of uncured specimens at any w/c ratio. The percentage increases of coefficient of water absorption of uncured specimens over control are given in Table 7.

The results of sorptivity tests on cube specimens at 90 days are given in Table 5 for

Table 6. Percentage strength losses of uncured specimens.

Free w/c	Compressive strength losses of uncured cubes relative to control (%)					Tensile strength losses of uncured cylinders (%)
	3 days	7 days	14 days	28 days	90 days	90 days
0.30	-18.34	-22.91	-32.65	-33.22	-34.03	-28.11
0.35	-18.04	-17.85	-19.88	-26.28	-35.99	-20.74
0.40	-11.39	-13.13	-16.87	-19.07	-29.92	-18.16
0.45	-7.09	+8.24	-7.98	-1.59	-16.88	-20.37
0.50	-8.14	-15.43	-9.21	-15.14	-23.24	-36.35
0.55	-12.57	-4.36	-11.66	-4.45	-19.57	-37.52

Table 7. Percentage increase in durability properties of uncured specimens compared to control.

Free w/c ratio	0.30	0.35	0.40	0.45	0.50	0.55
Coefficient of water absorption (Ka)	120	128	325	138	292	481
Sorptivity (s)	68	74	73	40	55	130

uncured and control specimens. As expected the sorptivity of values increased as the w/c ratio increased for both uncured and water cured specimens. The percentage sorptivity increases of uncured specimens above control are given in Table 7.

DISCUSSION

Compressive strength

Low w/c ratio concretes are known to be more susceptible to moisture loss (Bentz et al., 2012). Water plays an important role in the hydration of cement and at low w/c ratio, the effect of loss of moisture from the concrete means that less water is available for hydration. Self desiccations would also result in lower pore vapour pressure; and since the degree of hydration at vapour pressure below 0.8 is low, strength reduction would result. This appears to account for the continuous increase in compressive strength losses with age in the low w/c ratio mixes.

The w/c ratio of 0.45 does appear to have the least effect on compressive strength loss for uncured cube specimens. At a w/c ratio of 0.45, the maximum compressive strength loss of uncured specimens recorded at 90 days was 16.88%. An understanding of this effect is related to hydration dynamics in concrete. Hydration of cement takes place in concrete only in water filled capillaries and the maximum hydration proceeds at saturation vapour pressure. In addition, proper curing of concrete in theory requires that water filled pores in the fresh cement paste be occupied by hydration products. Though self-desiccation was evident in the results of compressive strength at 90 days for uncured specimens, the concrete at this w/c ratio appears to have the lowest compressive strength loss at 90 days and the other ages. It does appear that this w/c ratio represents a balance between optimum hydration resulting in strength gain and self-desiccations that results in strength reduction.

Uncured cube specimens at w/c of 0.50 had a progressive increase in compressive strength losses with age, reaching a maximum value of 23.24% at 90 days. These losses could be attributed to increased porosity of the concrete as the w/c ratio increased.

At w/c of 0.55, compressive strength increases with age were recorded for both uncured and water cured cubes at all ages. An increase of w/c ratio from 0.35 to 0.55 resulted in compressive strength reduction of 57.27% at 90 days for water cured cubes and strength reduction of 46.30% for uncured cubes at 90 days. As opined by Stark (2011), much of the mixing water in concrete is used in chemical reactions with cement and the remainder still being in the liquid state is enclosed in the microstructure of the concrete. This remaining water is directly responsible for porosity of the hardened cement paste that affects mechanical and transport properties of concrete. It is known that increase in w/c ratio leads to increases in porosity of the transition zone in concrete that result in strength reduction (Elrahman et al., 2011; Prokopski and Langier, 2000). Higher w/c ratio has been associated with larger pore size in cement hydration (Friedemann et al., 2006). The maximum compressive strength loss recorded at 90 days for uncured cubes at this w/c ratio was 19.57%. The lower reduction of compressive strength loss with age of uncured specimens at higher w/c ratio mixes compared to the high losses recorded at lower w/c ratio mixes may be attributed to additional water available for hydration at a higher w/c ratio mix to compensate for moisture loss in the concrete. However the use of higher w/c in concrete leads to lower compressive strength and reduction in durability properties.

The pattern of results in Tables 4 and 5 shows that the low w/c ratio mixes were more susceptible to moisture losses that could adversely affect compressive strength at 90 days. Increase in compressive strength of water cured cubes relative to uncured cubes is due to continuous hydration resulting from the saturation of concrete pores with water. By saturating the concrete pores with water by continuous water curing, the growth of calcium-silicate-hydrate (CSH) gels that are responsible for strength gain of concrete would be promoted (Goñi et al., 2012); this effectively improves the gel/space ratio.

Tensile strength

The growth of more CSH gels in water cured specimens resulted in improved tensile strength of cylinders cured in water at 90 days as shown in Table 6. The results further show reductions in tensile strength as w/c ratio increased, attributable to increased porosity resulting from higher water content. Though CSH gel is mainly responsible for compressive strength of concrete, it is also known to be weak in tensile strength; hence the tensile strength pattern shown in these results. Though the reasons for this is not certain, Murray et al. (2010) suggest that bond break in silicate chains at the atomic scale is responsible for the low tensile strength in concrete.

Coefficient of water absorption and sorptivity

As more CSH gels grow as a result of improved hydration in water cured cubes, the pore structure of the concrete improved. This resulted in increased coefficient of water absorption recorded for uncured cubes compared to water cured cubes, and reductions in coefficient of water absorption as the w/c reduced for both water cured and uncured cubes as shown in Table 5. Increase in extra water not used in hydration has been associated with development of pores in concrete, and as w/c increases this extra water increased thus leading to increased sorptivity and coefficient of water absorption for both uncured and water cured specimens. The results also show increase in sorptivity for uncured cubes compared to control, and sorptivity increase with increase in w/c. For uncured specimens, the developments of solid products of hydration are less compared to water cured specimens; thus higher sorptivity and coefficient of water absorption for uncured specimens. The exceptions were at w/c of 0.35, where the values of sorptivity and coefficient of water absorption reduced compared to the values at w/c of 0.30. This was due to the high dose of the plasticizer at the low w/c ratio of 0.30 that equally resulted in compressive strength reductions compared to the strengths at w/c ratio of 0.35. Proper curing reduces the rate of moisture loss and provides a continuous source of moisture required for the hydration that reduces the porosity and provides a fine pore size distribution in concrete (Alamri, 1988). Since curing promotes cement hydration, it leads to the development of hydration products that ultimately contributes to the promotion of a more compact microstructure of the concrete pores.

Conclusions

The results of this study show that low w/c is very important in achieving high compressive strength in concrete mixes. It also shows that uncontrolled moisture loss would result in strength losses. The results show that continuous water curing promotes the hydration of cement and the development of better pore structures in concrete that improves durability properties. Low w/c mixes have been shown to be particularly vulnerable to moisture losses; these mixes have the highest compressive strength losses at all the ages investigated. The results show the need to provide adequate curing, particularly in low w/c concrete. Without adequate curing, concrete losses strength but the extent to which losses can occur has been established.

REFERENCES

ACI Committee 305 (1991). Hot Weather Concreting, American Concrete Institute, Detroit, MI.

ACI Committee 363 (1998). State-of-art report on high strength concrete. (ACI 363 R-92/97). In: Man. Con. Prac., ACI p. 55.

Alamri AM (1988). Influence of curing on the properties of concrete and mortars in hot climates, PhD thesis, Leeds University, U.K.

Alizadeh R, Ghods P, Chini M, Hoseini M, Ghalibafian M, Shekarchi M (2008). Effect of curing conditions on the service life design of RC structures in the Persian Gulf region. J. Mater. Civ. Eng. 1(2):0899-1561.

Bentz DP, Snyder KA, Stutzman PE (2012). Hydration of Portland cement: The effects of curing conditions. Available at http://fire.nist.gov/bfrlpubs/build97/PDF/b97002.pdf [Accessed 16 Feb. 2012]

British Standards Institution, BS 1881 (1970). Methods of testing concrete for strength. London. Part. 4.

British Standards Institution, BS 1881 (1983). Splitting tensile (indirect) strength of cylindrical concrete specimens. London. Part 117.

British Standards Institution, BS 822 (1983). Specification for aggregates from natural sources for concrete. London.

British Standards Institution, BS 1881 (1997). Methods of normal curing of test specimens. (20°C method). London, Part.111.

British Standards Institution, BS EN 934 (2001). Admixtures for concrete, mortar and grout. Concrete admixtures, Definitions, requirements, conformity, marking and labeling. London, Part 2.

Bushlaibi AH, Alshamsi AM (2002). Efficiency of curing on partially exposed high-strength concrete in hot climate. Cem. Concr. Res. 32(6):949-953.

Elrahman MAA, Imam MA, Reheem AHA, Tahwia AM (2011). Production and properties of superplasticized concrete. J. Civ. Eng. Arch. 5(4):341-352.

Friedemann K, Stallmach F, Kärger J (2006). NMR diffusion and relaxation studies during cement hydration - A non-destructive approach for clarification of the mechanism of internal post curing of cementitious materials. Cem. Concr. Res. 36(5):817-826.

Ganesan K, Rajagopal K, Thangavel K (2008). Rice husk ash blended cement: Assessment of optimal level of replacement for strength and permeability properties of concrete. Constr. Build. Mater. 22(8):1675-1683.

Giaccio G, Rodrı´guez de Sensale G, Zerbino R (2007). Failure mechanism of normal and high-strength concrete with rice-husk ash. Cem. Concr. Compos. 29(7):566-574.

Giannotti da Silva F, Liborio JBL, Helene P (2008). Improvement of physical and chemical properties of concrete with Brazilian silica rice husk (SRH). Revista Ingeniería de Construcción, Available at www.ing.puc.cl/ric 2008 [Accessed 2011 February 12]. 23(1):18-25.

Goñi S, Frias M, Vegas I, García R, Vigil de la Villa R (2012). Effect of ternary cements containing thermally activated paper sludge and fly ash on the texture of C–S–H gel. Concr. Build. Mater. 30:381–388.

Gonnerman HF, Shuman EC (1928). Flexure and tension tests of plain concrete. Major series 171, 209 and 210. Report of the Director of Research. Port. Cem. Assoc. pp. 149-163.

Guneyisi E, Ozturan T, Gesoglu M (2005). A study on reinforcement corrosion and related properties of plain and blended cement concretes under different curing conditions. Cem. Concr. Compos. 27(4):449-461.

Hall C (1989). Water sorptivity of mortars and concretes: A review. Mag. Concr. Res. 41(14):51-61.

Long TP (2008). High strength concrete at high temperature- an overview. Available from http://fire.nist.gov/bfrlpubs/build02/pdf/b02171.pdf [Accessed 2012 January 20].

Mamlouk MS, Zaniewski JP (2011). Materials for Civil and Construction Engineers. Third Edition. Pearson Education, Inc New Jersey. pp. 292-297.

Murray JS, Subramani VJ, Selvam RP, Hall KD (2010). Molecular dynamics to understand the mechanical behavior of cement paste. Tran. Res. Rec.: J. Trans. Res. Bd. 21(42):75-82.

Neville AM (1981). Properties of concrete. Third edition. Longman Scientific & Technical, England.

Neville AM, Brooks JJ (2008). Concrete technology. Pearson educational Ltd. England.

Powers TC (1947). A discussion of cement hydration in relation to the curing of concrete. Proc. Highw. Res. Board 27:177-188.

Powers TC (1949). The non-evaporable water content of hardened Portland cement paste: Its significance for concrete research and its method of determination. ASTM Bull. 158:68-76.

Price HW (1951). Factors influencing concrete strength. J. Am. Concr. Inst. 47:417-432.

Prokopski G, Langier B (2000). Effect of water/cement ratio and silica fume addition on the fracture toughness and morphology of fractured surfaces of gravel concretes. Cem. Concr. Res. 30(9):1427-1433.

Soroka I, Baum H (1994). Influence of specimens size on effect of curing regime on concrete compressive strength. J. Mater. Civ. Eng. ASCE 6(1):15-22.

Stanish KD, Hooton RD, Thomas MDA (1997). Testing the chloride penetration resistance of concrete: A literature review. FHWA contract DTFH61 1997, Department of Civil Engineering, University of Toronto, Canada pp. 19-22.

Stark J (2011). Recent advances in the field of cement hydration and microstructure analysis. Cem. Concr. Res. 41(7):666-678.

Taylor GD (2000). Materials in construction, an introduction. Third edition. Pearson Education limiTed. England. U.K. pp. 50-51.

Zhutovsky S, Kovler K (2012). Effect of internal curing on durability-related properties of high performance concrete. Cem. Concr. Res. 42(1):20-28.

18

Sorption characteristics of cement composite reinforced with some locally available lignocellulosic materials in Nigeria

Omoniyi, T. E , Olorunnisola A.O. and Akinyemi B.A

Department of Agricultural and Environmental Engineering, University of Ibadan, Nigeria.

The aim of this study was to investigate the sorption property of wood cement composite produced from bagasse (*Saccharum officinarum*), bamboo (*Bambusa vulgaris*) and coir (*Cocos nucifera* L). The mass of the fibre varies from 1 to 6% of the mass of cement. The result indicated that the mass fraction has significant effect on the sorption properties of the composites. Water absorption (W.A.) rate increases with increase in the fibre content of the composites. Thickness swelling (T.S.) in all the composites was less than 1.7% at 24 h water immersion at room temperature. There was linear correlation between mass fraction, water absorption and thickness swelling of the composites. The relatively low W.A. capacity and T.S. at content less than 3% of mass of cement suggests that they can be employed in outdoor situations and at this level they are dimensionally stable but beyond this level it is not advisable.

Key words: Water absorption, thickness swelling, bamboo, bagasse and coir fibres.

INTRODUCTION

The use of technologically by-product agricultural wastes in various segments of the construction and building industry is increasing continuously. Among the different types of fibres used in cement-based composites, non woody materials offer distinct advantages such as availability, renewability, low cost, and current manufacturing technologies. Fibre-cement composites exhibit improved toughness, ductility, flexural capacity, and crack resistance as compared to non- reinforced cement-based materials. Due to their hygroscopic nature, fibre-cement composites are sensitive to moisture changes in the material itself and in the ambient environment. Generally, flexural strength and stiffness tend to decrease as the moisture content increases. It has been reported (Mai et al., 1983; Coutts and Kightly, 1984; Coutts, 1987) that the decrease in stiffness when wet and the resulting ductility gained changes both the behavior of the fibres as well as the interfacial characteristics between the cement matrix and the fibres. These changes in the properties and in the cement matrix interface leads to changes in the mode of failure of the fibres. In the wet state, it is believed that the bond between the cement matrix is weakened. On the other hand, in the dry state, the bond strength is increased (Mohr et al., 2003).

About 75 genera and 1250 species of bamboo are found in different countries of the world. *Bambusa vulgaris* is the best known and most widely used species in Asia. For building, *Guadua angustitifolia Kunth* is also used and is a common plant in Latin America, especially in Colombia, Peru and Ecuador (Agopyan, 1988; Hidalgo-Lopez, 2003). One of the main shortcomings of bamboo is water absorption when it is used as a reinforcement and/or permanent shutter form with concrete. The dimensional variation of untreated bamboo due to water absorption can cause micro or even macro cracks in cured concrete (Ghavani, 2005). Coconut is a tall cylindrical-stalked palm tree, reaching 30 m in height and 60 to 70 cm in diameter. It is a tropical plant for low altitudes. Coconut fibres (coir) can be extracted from either immature or mature fruits. They are lignocelluloses fibres obtained from the mesocarp of the coconut fruit, which constitutes about 25% of the nuts. They are one of the least expensive of the various natural fibres available in the world. They are not brittle like glass fibres; they are responsive to chemical modification and are non-toxic. However, the waste from their disposal causes environmental problems. Tomczak (2007) and Joana et al. (2011) studied coir/cement composites and stated that the composites produced with NaOH presented low dimensional stability because of high values of thickness swelling and water absorption. This is as a result of the alkaline treatment which modified the surface and increased the roughness and the new additional surface area may be used as a new pathway for water absorption. Bagasse is a ligno-cellulosic material left after the removal of sugar and moisture from sugarcane. With an average yield of 80 to 100 tons cane per hectare, the annual yield of sugarcane in Nigeria is two to three million tons 45% of which ends up as bagasse (Wada et al., 2004).

Water absorption is one of the most important characteristics of natural reinforced concrete when exposed to environmental conditions. The availability of moisture is a necessity for decay to occur in a material. A problem associated with natural fibres in composites is their high moisture absorption and dimensional instability (swelling) (Sayyed et al., 2011). Swelling of fibres leads to micro cracking of the composites and eventual degradation of its mechanical properties. Matoke et al. (2012) worked on bamboo fibres and stated that water absorption is a disadvantage in composites. Water absorption in composites influences dimensional stability. The higher the fibre content, the higher the water uptake and vice versa. They also reported that water uptake increases with increase of the filler content. Since lignocellulose fibres are hydrophilic in nature, the increased amount of fibres used as filler in the composite showed significant effect on the water absorption. The percentage water absorption of the composites is expected to achieve equilibrium. As the filler loading increases, the formation of agglomerations increases hence it is difficult to achieve homogeneous dispersion of a filler of high filler loading. This agglomeration of the filler in composite increases the water absorption of the composites.

Dimensional stability of composite is important since construction materials should have the ability to withstand the stresses of shrinkage or swelling due to the changes of temperature and moisture. The initial drying state of the fibres also affects its dimensional stability during subsequent wetting and drying. Fibres which have been dried once prior to introduction to a matrix material are expected to swell less upon rewetting, as compared to fibres which have not been previously dried (Mohr et al., 2003). Results by Mohr et al. (2003) showed that composites produced with fibres which have been dried exhibited superior dimensional stability compared to composites produced with fibres which had never been dried.

Thickness swelling (T.S) is also another important factor that affects dimensional stability and it's highly correlated with the cement- ratio. In general, the higher cement content of a composite, the lower the T.S. In order to minimize T.S., with the negative side effect of decreased MOR, reducing cement-wood ratio is necessary so that water absorption can decrease. (Meneeis et al., 2007). Also, pre-treatment of the fibres has an effect on T.S. Addition of chemicals can decrease T.S. by using $CaCl_2$ as an accelerator. This improvement is caused by better contact as a result of improved bonding ability with cement. Besides $CaCl_2$, hot water soaking of these fibres can effectively reduce T.S. (Semple and Evans, 2004). T.S. is highly dependent on particle geometry. There is an increase in T.S. with increasing particle thickness and decreasing particle length. In summary, lowest T.S. as exhibited by some studied natural fibres-reinforced concrete arises because of sufficient encapsulation of the particles at high cement-ratios and the minimal swelling of the small particles (Frybort et al., 2008). This work was carried out to ascertain the sorption characteristics of three major non-woody products in Nigeria to determine the order of suitability for use in terms of dimensional stability in cement bonded composite boards. The work is limited to evaluation of sorption properties at different fiber contents while density, mixing ratio, particle sizes, ratio of water to cement, type of cement, accelerator, curing time and production methods were constant. Strength characteristics were not considered in this research.

MATERIALS AND METHODS

Material preparation

The following three materials were used for this research and their preparations for composite boards are highlighted below:

1. Bagasse was obtained from Bodija in Ibadan, Oyo State. The raw bagasse was received at about 30% moisture content. It was sun-dried for two weeks, manually depithed and further sun-dried for two weeks to a moisture content range of 7 to 10%. Bagasse was hammer-milled to produce bagasse particles. The hammer-milled particles were passed through sieves of sizes 2.4 mm and 850 μm. Particles that passed through 2.4 mm but were retained on the 850 μm sieve were categorized as coarse particles were used.

2. Coconut husk was procured from Badagry, Lagos State, reduced by grinding in the hammer mill to produce particle fibres of different sizes. The particle fibres were then sieved with sizes listed for bagasse fibres for uniformity purpose.
3. Bamboo specie used is from the family of *B. Vulgaris.* Field trips were made to four locations within the University of Ibadan, Nigeria to obtain bamboo culm samples. Identification was carried out in the herbarium of the Department of Botany, University of Ibadan. The wet bamboo was allowed to dry for four weeks and then sliced to about 60 cm length by 3 cm width with the slicing machine and further reduced to 15 cm length by 1cm width with the knife before it was crushed by an hammer mill. The different fibres were sieved with the same procedure stated above.

The binder used for the study is the Portland cement. It was purchased locally from dealers in standard bags of 50 kg weight. The type of cement available in the country complies with the British Standard BS 12:1958. Additives are used as chemical pre-treatments to improve the compatibility of the non woody fibrous materials with cement. Calcium chloride was used as an additive at a concentration of 3% of the cement weight in the boards. Water source was from the University of Ibadan supply.

Method for the design of the composite board

The only variables used for the composite board production were the mixing ratio and the five non-woody fibrous materials. The fibre content varies from 1 to 6% of the mass of cement. The mixing ratio gives an indication of the proportion of cement to non-wood per oven dry weight of the board. To make the board, the quantity of non-wood required was measured and put in a plastic bowl. To it was added the required amount of water containing the needed amount of additive. The chemical solution was then mixed with the non-wood aggregates. Quantity of cement needed was added to the wet non wood s and mixed thoroughly until a homogenous non-wood -cement mix was formed as reported by Badejo (1988). The composites were produced by vibration. The cement and particles were mixed in dry form until a high level of uniformity was achieved. Water was added to attain a level of plasticity to permit the shaping of the mixture. The slurry was then placed on interphase sheets spread on the surface of the vibrating table and vibrated for about 40 s. The mixture was later transferred to a plastic mould of 75 × 50 × 10 mm. After this, it was cured for 28 days. Five replications of each fibre percentage were made. Thirty boards were manufactured altogether. All the experimental panels were made with the following standard specifications:

Board Type: Homogenous and 1 layered
Board Dimension: 75 mm × 50 mm × 10 mm
Board Density: 1,200 kg/m^3 based on oven dry weight and volume of board.
Binder: Portland cement purchased in standard bag of 50 kg.
Additive: Powdered $CaCl_2$ applied at concentration of 3% of the cement weight in the board.

Water absorption

Water absorption is used to determine the amount of water absorbed by a composite. The water absorption test followed ASTM standard test method D570. The test specimen was in the form of a bar 75 mm long, 50 mm wide and 10mm thick. Before the measurement, the sample was dried in an air oven at 50°C for 24 h, cooled in a desiccator and immediately weighed to the nearest 0.001 g which is then taken as the dry initial weight of the sample. Then the specimen was immersed in distilled water maintained at a temperature of 23 ± 1°C for 24 h. After 24 h, the specimen was removed from water and placed on blotting paper to remove excess water before weighing to the nearest 0.001 g. For each composite, five sub samples were measured. The water absorption of the sample was calculated as percent weight change (w %) as follows:

$$WA = \frac{M2-M1}{M1} \times 100\%$$

Where M_1 = weight of dry piece (g)
M_2 = weight of wet piece (g)
WA = water absorption (%)

Thickness swelling

This test, like water absorption was important in ascertaining dimensional changes. The thickness swelling samples were 75 mm × 50 mm × 10 mm. Five specimens for each material and ratio were tested. The samples were soaked in distilled water for 24 h. The immersed samples were taken out and wiped by dry cloth to remove water from the surface. The thickness was measured using a vernier caliper to the nearest 0.01 along the length at room temperature and average results recorded. The thickness swellings of the samples were calculated according to ASTM standards D1037-03.

RESULTS AND DISCUSSION

Thickness swelling

Table 1 and Figure 1 summarize the data obtained. The range of thickness swelling observed for all the boards from bagasse, bamboo and coir fibres ranged between 0.3 to 1.6% after 24 h of submersion in water. The result further showed that thickness swelling of the boards after immersion in water and chemical treatment had least value for bamboo at 1% fibre content while the largest swelling value was obtained from coir at 6% fibre content. The less the fibre content, the lower the thickness swelling. Therefore, thickness swelling increases with increase in fibre content of the composite. Also, the greater the cement content of the composite, the lower the thickness swelling which means more cement coating on the fibres may have restrained the boards from swelling. It can be observed that bamboo cement composites boards are relatively more dimensionally stable than bagasse and coir cement composites when exposed to water at room temperature. Regression equations were also developed to relate thickness swelling for the three fibres used as shown below:

$$T.S_{\text{bamboo 24 h}} = 0.22 + 0.108_{\text{b.f.c.}} \quad R^2 = 0.937 \tag{1}$$

$$T.S._{\text{bagasse 24 h}} = 0.233 + 0.2_{\text{ba.f.c.}} \quad R^2 = 0.929 \tag{2}$$

$$T.S._{\text{coir 24 h}} = 0.193 + 0.225_{\text{c.f.c}} \quad R^2 = 0.940 \tag{3}$$

Where $T.S_{\text{bamboo 24 h}}$ represents percentage thickness swelling for bamboo after 24 h, $T.S._{\text{bagasse 24 h}}$ represents

Table 1. Mean thickness swelling.

Content (%)	Mean thickness swelling (%)		
	Bagasse 24 h	Bamboo 24 h	Coir 24 h
1	0.4	0.3	0.5
2	0.6	0.5	0.5
3	0.8	0.5	1.0
4	1.2	0.7	1.2
5	1.3	0.7	1.2
6	1.3	0.9	1.6

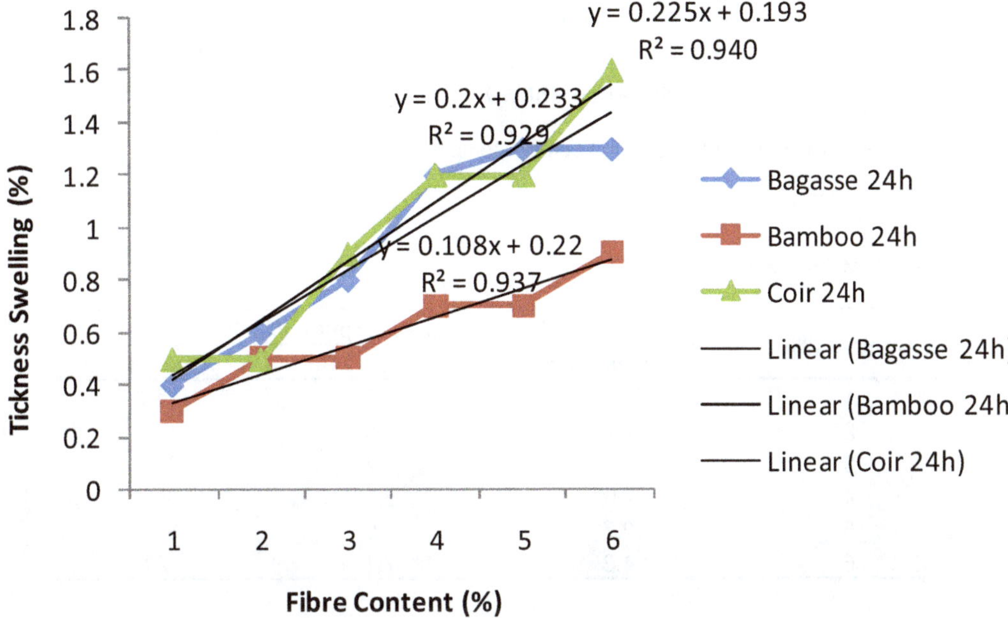

Figure 1. Thickness swelling of some -cement composites.

percentage thickness swelling for bagasse, $T.S._{coir\ 24\ h}$ represents percentage thickness swelling for coir, b.f.c represents bamboo fibre content, ba.f.c. represents bagasse fibre content and c.f.c represents coir fibre content respectively. As shown in the equations above, there is a strong linear relationship between thickness swelling and fibre content after 24 h of immersion for the three different fibres used. The thickness swellings values obtained compare favourably with the range of values from 0.2 to 0.7% as reported by Olorunnisola (2004).

The mixing ratio of cement: sand: water is 1:3:0.5 is constant for all the homogenous composites produced.

Water absorption

The mixing ratio of cement: sand: water is 1:3:0.5 is constant for all the homogenous composites produced. Water Absorption varied from 3.5 to 7.5% for bamboo, 6.7 to 14.8% for bagasse and 11.2 to 70% for coirs. After 24 h of immersion in water at room temperature the composition at 5% content or below has acceptable water absorption for materials suitable for interior use. Composites manufactured from bagasse and bamboo generally absorb less water than that of coir as shown in Figure 2 and Table 2. Water Absorption showed a positive correlation (R^2 =0.8 for bamboo cement, R^2 =0.7 for bagasse and R^2 =0.9 at 24 h of immersion respectively). This result suggests that coir has more affinity for water and also is a hygroscopic material. As the coconut coir content increases ability to absorb water also increases. Water absorption rate is a primary indicator of the durability of cement component. According to Olorunnisola (2004), the presence of water can cause cracking (associated with swelling and shrinkable phenomena), biodegradation of wood aggregates and the dissolution of the composites. The

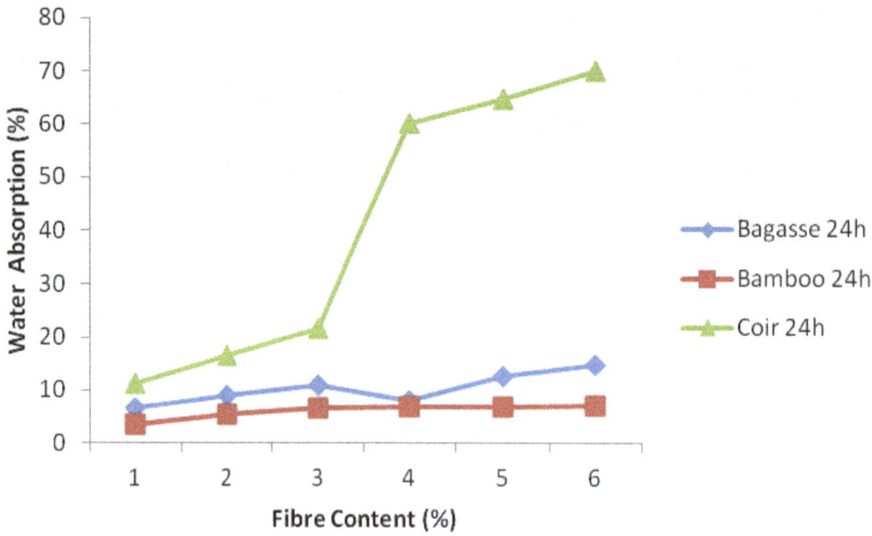

Figure 2. Water absorption of wood-cement composites.

Table 2. Mean water absorption.

Content (%)	Mean water absorption (%)		
	Bagasse 24 h	Bamboo 24 h	Coir 24 h
1	6.7	3.5	11.2
2	9.0	5.5	16.7
3	11.0	6.7	21.7
4	8.1	6.9	60.1
5	12.7	7.0	64.9
6	14.8	7.1	70

relatively high water absorption rate of the coir composites corroborates the findings of Olorunnisola (2004) who noted that the water absorption of vegetables was usually high sometimes reaching over 100% in only 1hr of immersion in water. One way to address this weakness is through the water repellent treatment.

Regression equations were also developed to relate water absorption for the three fibres used as shown below:

$$W.A._{bamboo\ 24\ h} = 3.613 + 0.748_{b.f.c.} \quad R^2 = 0.837 \tag{4}$$

$$W.A._{bagasse\ 24\ h} = 5.513 + 1.391_{ba.f.c.} \quad R^2 = 0.737 \tag{5}$$

$$W.A._{coir\ 24\ h} = -6.933 + 13.62_{c.f.c} \quad R^2 = 0.896 \tag{6}$$

Where $W.A._{bamboo\ 24\ h}$ represents percentage water absorption for bamboo after 24 h, $W.A._{bagasse\ 24\ h}$ represents percentage water absorption for bagasse, $W.A._{coir\ 24\ h}$ represents percentage water absorption for coir, b.f.c represents bamboo fibre content, ba.f.c. represents bagasse fibre content and c.f.c represents coir fibre content respectively. As shown in the equations above, there is a strong linear relationship between water absorption and fibre content after 24 h of immersion for the three different fibres used.

Conclusion

Composite boards were produced using 3 lignocellulose materials at different ratios in cement composite. From the result, the following conclusions were drawn;

1) The W.A. capacity of boards produced by vibrating table was relatively low but the best performance was observed in bamboo composite which meant that they could be employed in outdoor situation provided that the mass fraction is not more than 3% provided other properties are taken into consideration.
2) The sorption properties were relatively low when compared with some other lignocellulose materials reported in the literature. The reason for this is that most of the reported results were produced with the application

of pressure hence relatively high strain recovery due to external pressure during manufacture whereas the composite boards used in this research were produced by vibrating table with no external pressure.

(3) There is linear relationship and positive correlation between thickness swelling, water absorption and percentage of fibre used after 24 h of immersion.

REFERENCES

Agopyan V (1988). 'Vegetable reinforced building materials – Developments in Brazil and other Latin American countries.' in Swamy R N (Ed.), Natural Reinforced Cement and Concrete (Concrete Technology and Design. Glasgow, Blackie, pp. 208–242.

American Society for Testing and Materials (ASTM) (1991). Standard Methods of Evaluating the Properties of Wood-Based Fibre and Panel Materials. ASTM D 1037-1091. Annual book of ASTM Standards, 04.09 Wood, Philadelphia, PA. pp. 169–191.

Badeji OO (1988). Dimensional Stability of Cement Bonded Particle board from Eight Tropical Hardwoods Grown in Nigeria." Nig. J. For. 16:1, 2

Coutts RSP (1987). "Matrix interface in air-cured wood-pulp -cement composites." J. Mater. Sci. Lett. 6:140-142.

Coutts RSP, Kightly P (1984). "Bonding in wood -cement composites." J. Mater. Sci. 19:3355-3359

D1037 - *12* Standard Test Methods for Evaluating Properties of Wood-Base.

D570 - 98(2010). Standard Test Method *for* Water Absorption of Plastics, absorption, immersion, plastics, water, Immersion--plastics, Water analysis—plastics.

Frybort S, Mauritz R, Alfred T, Ulrich M (2008). "Cement Bonded Composites – A Mechanical Review." BioResources 3(2):602 -626.

Hidalgo-Lopez O (2003). Bamboo: The Gift of the Gods, Bogota, Oscar Hidalgo-Lopez.

Joana MF, Claudio HS, Del M, Divino ET, Sabrina AM (2011). Effects of Treatment of Coir and Cement/ ratio on properties of cement bonded composites. BioResources 6(3):3481–3492.

Khosrow G (2005). Bamboo as reinforcement in structural concrete elements. Cem. Concrete Compos. 27(2005):637–649.

Mai YW, Hakeem MI, Cotterell B (1983). "Effects of water and bleaching on the mechanical properties *of* cellulose cements." J. Mater. Sci. 18:2156-2162.

Matoke GM, Owido SF, Nyaanga DM (2012). Effect of Production Methods and Material Ratios on Physical Properties of the Composites. Am. Int. J. Contemp. Res. 2(2), February.

Meneeis CHS, Castro VG, Souza MR (2007). "Production and Properties of a medium density Wood-Cement Boards produced with oriented strands and silica *fumes."* Maderas: Ciencia y tecnologia 9(2):105–116.

Mohr BJ, Nanko H, Kurtis KE (2003). "Durability of pulp -cement composites to wet/dry cycling." Submitted to Cement and Concrete Composites, June 2003.

Olorunisola AO (2004). Compressive strength and water resistance behaviour of cement composites from Rattan Cane and Coconut Husk. J. Trop. For. Res. 20(2):1-13.

Sayyed KH, Mahdi M, Vahidreza S, Behzad K (2011).Decay resistance, hardness, water absorption and tthickness swelling of a bagasse /plastics composites. Bioresources 6(3):3289–3299.

Semple KE, Evans PD (2004).'Wood-Cement Composites – Suitability of Western Australian Mallee Eucalypt, Blue gum and Melaleucas." RIRDC/Land and Water Australia.

Tomczak F, Sydenstricker THD, Satyanarayana KG (2007). 'Studies on lignocellulosics of Brazil. Part II: Morphology and properties of Brazilian coconut *s'.* Compos. Part A: Appl. Sci. Manuf. 38(7):1710–1721.

Wada AC, Ishaq MN, Agboire S (2004). Evaluation and Characterization of Sugarcane Germplasm Accessions for their Breeding Values in Nigeria. Retrieved 26, November, 2004 http://www.ipgri.cgiar.org/pgrnewsletter/article.asp., pp. 1-3.

19

Mechanical and hardening properties of accelerator on the cement concrete composites containing metallic reinforcements

A. Sivakumar* and V. M. Sounthararajan

Structural Engineering Division, School of Mechanical and Building Sciences, VIT University, India.

Research efforts in the past have been very successful in obtaining high strength concrete using various supplementary cementitious materials in concrete; however, the restriction on adding these mineral admixtures is primarily due to delayed reactivity with the hydration product of cement and the decelerated strength gain compared to plain cement concrete. Research efforts in the past have focused on the various aspects of improving the matrix properties and make it suitable for improving the mechanical and durability properties of concrete. Matrix strengthening envisages the improved mechanical properties and accelerators improve the accelerated rate of hardening when mixed with cement. The present study is aimed at to study the influence of incorporation of accelerators and metallic reinforcements in concrete. The effect of accelerators on the compressive strength gain properties of different concrete mixes were studied for different w/c ratio, fine to aggregate ratio, cement to total ratio and metallic fibre contents. The results showed that for lower water cement ratio (0.3) and for higher f/c ratio (0.8) the strength was higher than all other concrete mixes. Similar increase in strength was noted for higher water cement ratio (0.4) at lower finer to coarse aggregate ratio. The rate of increase in strength gain was higher for lower water cement ratio at all fine to coarse aggregate ratio.

Key words: Metallic reinforcements, accelerators, compressive strength, ultrasonic pulse.

INTRODUCTION

The improved concrete properties have become an increasing concern in recent years due to increasing interest towards sustainable development. A good amount of work has been carried out in recent years for the large scale utilization of fibre reinforcements in concrete. The beneficial properties of matrix strengthening can be realized in terms of the improved mechanical properties in concrete after longer curing period (Romualdi and Batson, 1963). However, the early age setting properties of cement concrete is greatly affected when replaced with mineral admixtures. This leads to negative effects on the use of concrete for fast track concreting such as concrete pavement applications, (ACI-544, 1993; Balaguru and Shah, 1992).There is an increasing demand to study the engineering properties of metallic matrix reinforced concrete with accelerators and produce a high toughened concrete (Bentur and Mindess, 1990). The rapid development in the production of high quality concretes was made possible with the advent of innovative techniques that can alter the mechanical properties of concrete. The production of high strength concrete incorporating accelerators and metallic reinforcements had not been widely studied (Hanant, 1978). The high volume metallic reinforcement addition in concrete has not gained good attention. However, the negative effects on the setting properties and deceleration in the rate of strength gain restricts the maximum replacement levels of metallic reinforcements in cement concrete. Processing of concrete is one of such promising technique that can completely offset the negative aspects of metallic reinforcements on the rate of

*Corresponding author. E-mail: sivakumara@vit.ac.in.

Table 1. Mixture proportions of various concrete mixtures.

Mix Id	Cement (Kg)	Fine aggregate (Kg)	Coarse aggregate (Kg)	F/C ratio	W/C ratio	Cement to total aggregate ratio	MMP (% by weight of cement plus fine aggregate)
M1A19	5.32	7.562	12.51	0.8	0.3	0.27	1
M1A20	5.32	7.562	12.51	0.8	0.3	0.27	2
M1A21	5.32	7.562	12.51	0.8	0.3	0.27	3
M2B1	5.32	8.7	12.75	0.6	0.3	0.25	1
M2B2	5.32	8.7	12.75	0.6	0.3	0.25	2
M2B3	5.32	8.7	12.75	0.6	0.3	0.25	3
M2B25	5.32	8.7	12.75	0.8	0.4	0.25	1
M2B26	5.32	8.7	12.75	0.8	0.4	0.25	2
M2B27	5.32	8.7	12.75	0.8	0.4	0.25	3
M3C1	5.32	10	12.5	0.8	0.4	0.23	1
M3C2	5.32	10	12.5	0.8	0.4	0.23	2
M3C3	5.32	10	12.5	0.8	0.4	0.23	3
M3C25	5.32	10	12.5	0.8	0.4	0.24	1
M3C26	5.32	10	12.5	0.8	0.4	0.24	2
M3C27	5.32	10	12.5	0.8	0.4	0.24	3

strength gain in concrete. Processing envisages the early setting properties and improved micro structural properties. This can guarantee the accelerated strength gain and long term durability properties of metallic reinforced concrete. There were extensive studies carried out which reflects the toughness enhancements of concrete using metallic reinforcements (Zollo, 1975). Cement replacement with finer metallic reinforcements showed the strength of concrete being slightly lower than that of controlled concrete in the initial ages, but the development of strength was superior after 28 days. The replacements of cement by fine metallic reinforcements on equal mass basis were 1, 2 and 3% and the results confirmed that with high fineness of metallic reinforcements at higher replacements levels the compressive strength of concrete was increased by more than 20%. Physical properties like shape, length and distribution of reinforcement affects the matrix properties. Several investigators reported the effect of metallic reinforcements in concrete through a comparison of the compressive strength of metallic reinforcements with the normal concretes (Naaman, 1992; Parimi and Rao, 1971). At higher replacements of metallic reinforcements, a highly durable concrete was obtained. The test results also confirmed that the high volume metallic reinforcements in concrete showed high toughness, low permeability as well as good resistance to freezing and thawing. Concrete made with metallic reinforcements exhibited high compressive toughness compared to that of ordinary Portland cement concrete. It is also realized that the strength of processed metallic reinforcement in concrete showed higher compressive strength of 20% more than the strength of the controlled concrete (Kar and Pal, 1972). Also, it was noted that the porosity and permeability of the processed metallic reinforced concrete mixtures was lower than that of the controlled mix.

Metallic reinforcements have little bonding efficiency at the early ages and rather act as filler materials at later ages (Soroushian and Lee, 1990; ACI-544, 1993). Metallic reinforcements do not have the same properties for different fractions and the effect of metallic reinforcements on mortar strength is a combined effect of its size fractions (Shah and Rangan, 1971). A good amount of research studies has been carried out for the metallic reinforcement utilization in concrete. The present study is aimed at to accelerate the hardening and compressive properties of concrete. The present study can provide a major thrust towards the large scale utilization of metallic matrix composites in construction industry since the replacement levels of metallic reinforcements in cement still remains low due to poor workability and poor early age hardened concrete properties

MATERIALS AND METHODS

Ordinary Portland cement conforming to IS 12269 of 53 grade was used for producing concrete. The specific gravity of cement was found to be 3.37. River sand with a specific gravity of 2.69 and fineness modulus of 2.55 was used as fine aggregate. Coarse aggregates of size 12 mm and down and of specific gravity 2.75 were used. An accelerator was used to obtain a high early strength concrete and metallic reinforcements were added at different dosages of 1, 2 and 3% by weight of concrete. The concrete mixture proportions used in the study are shown in Table 1. A total of 18 different concrete mixtures were proportioned based on the cement to total aggregate ratio (C/TA) (0.25, 0.27), water to cement ratio (W/C) (0.3, 0.4), fine to coarse aggregate ratio (F/C) (0.6, 0.8) and metallic reinforcements (MMP) (1, 2 and 3%). The concrete

Table 2. Compressive strengths of various concrete mixtures.

Mix Id	W/C	F/C	C/TA	% of metallic reinforcements	Accelerator dosage (%)	Average compressive strength (MPa)		Average split tensile strength
						7 days	28 days	(28 days) (MPa)
M1A19	0.3	0.8	0.27	1	1	40.5	46.1	3.31
M1A20	0.3	0.8	0.27	2	1	38.9	50.8	3.28
M1A21	0.3	0.8	0.27	3	1	40.9	46.3	3.56
M2B1	0.3	0.6	0.25	1	1	40.1	45.9	3.76
M2B2	0.3	0.6	0.25	2	1	35.3	44.3	3.57
M2B3	0.3	0.6	0.25	3	1	38.1	44.9	3.82
M2B25	0.4	0.8	0.25	1	1	27.5	33	2.48
M2B26	0.4	0.8	0.25	2	1	29.1	39.1	2.78
M2B27	0.4	0.8	0.25	3	1	34.2	42	3.09
M3C1	0.4	0.8	0.23	1	1	37.8	47.8	3.27
M3C2	0.4	0.8	0.23	2	1	43.3	47.4	3.41
M3C3	0.4	0.8	0.23	3	1	43	47.9	3.37
M3C25	0.4	0.8	0.24	1	1	30.1	35.6	2.58
M3C26	0.4	0.8	0.24	2	1	37.4	39.6	3.28
M3C27	0.4	0.8	0.24	3	1	36.8	38.9	3.61

Table 3. Compressive strengths of control concrete mixtures for 7 & 28 days.

Mix Id	W/C	F/C	C/TA	Compressive strength for 7 days (MPa)	Compressive strength for 28 days (MPa)
R1	0.3	0.8	0.27	47.4	53.7
R2	0.3	0.6	0.25	37.7	45.4
R3	0.4	0.6	0.27	46.9	55.1

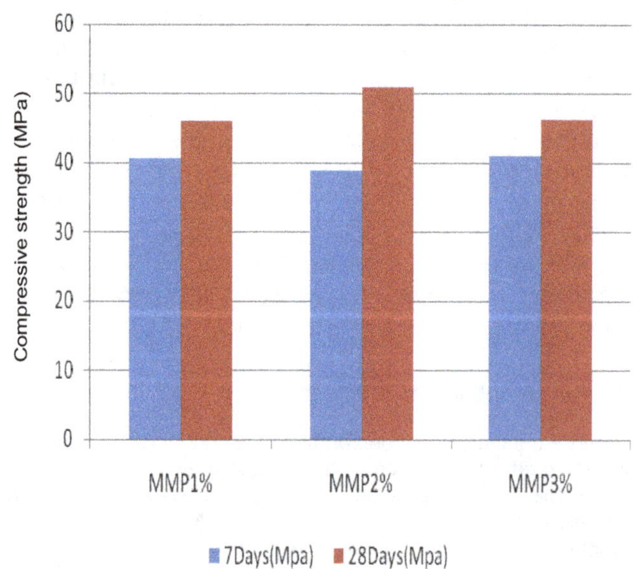

Figure 1. Compressive strength of concrete mixes for different W/C ratio 0.3, F/C ratio 0.8 and C/TA ratio 0.27.

mixer and casted in standard cube (100 ×100 ×100 mm) moulds and cylinders (100 mm diameter × 200 mm height). The specimens mixtures were mixed using a 40 liters capacity vertical mixing drum were tested after sufficient curing in water ponding and tested in universal testing machine for determining compressive and split tensile properties. Ultrasonic pulse velocity measurements were recorded for all the concrete specimens immediately after demoulding for the first few hours and continued after 3, 7 and 28 days.

RESULTS AND DISCUSSION

The compressive test results of different concrete mixtures are shown in Tables 1, 2 and 3, and Figures 1 to 6. It can be observed that compared to control concrete all the metallic matrix concrete composites showed higher strength. It can also be noted that the variables such as cement to aggregate ratio and fine to coarse aggregate ratio affected the compressive properties greatly when the W/C ratio is 0.3 as shown in Figure 1. Similarly, it is shown in Figures 2 and 4 that compared to 0.3 W/C ratios, 0.4 W/C showed higher strength due to higher F/C ratio, whereas the C/TA was lesser than the lower W/C ratio. Also, the same trend(as observed in Figure 3 and 5) was observed for the controlled concrete mix which resulted in higher compressive strength for higher W/C and C/TA ratio. Also, it is well noted that at higher F/C ratio used in concrete results in higher

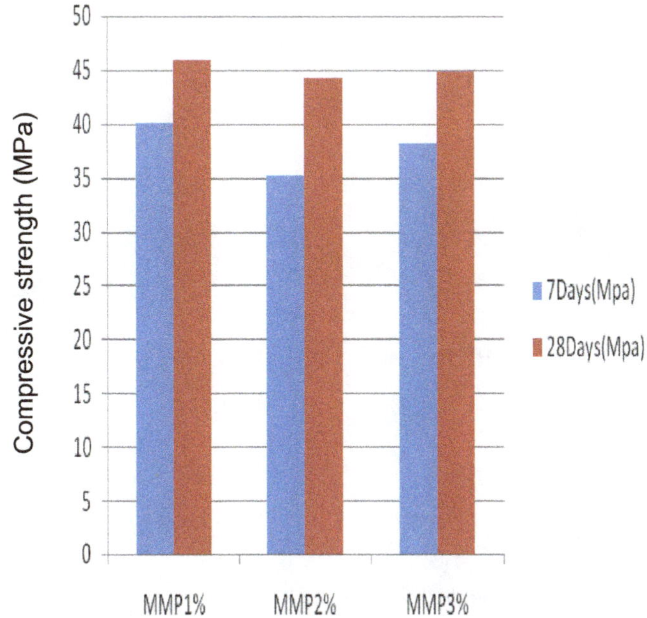

Figure 2. Compressive strength of concrete mixes for different W/C ratio 0.3, F/C ratio 0.6 and C/TA ratio 0.25.

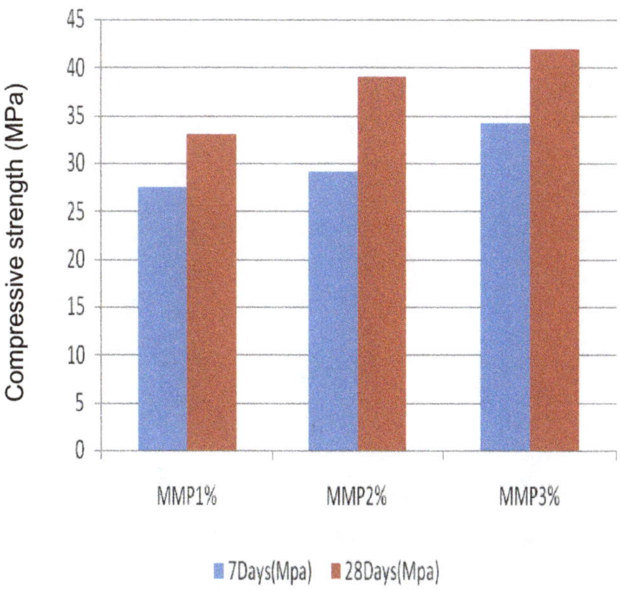

Figure 3. Compressive strength of concrete mixes for different W/C ratio 0.4, F/C ratio 0.8 and C/TA ratio 0.25.

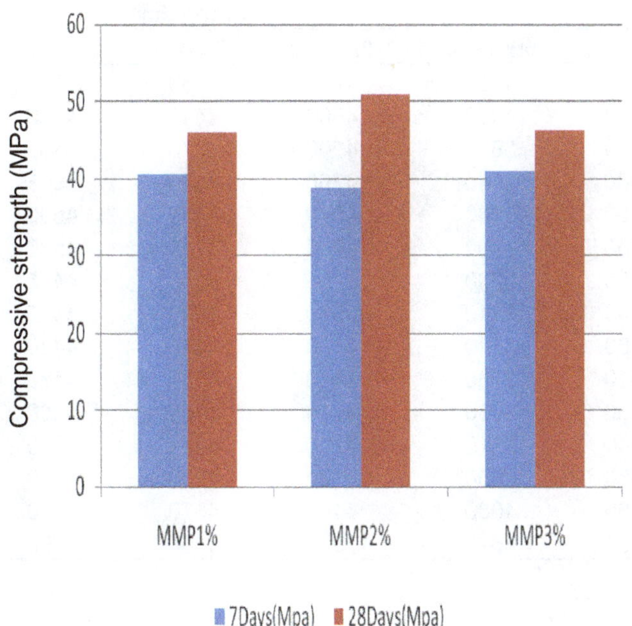

Figure 4. Compressive strength of concrete mixes for different W/C ratio 0.4, F/C ratio 0.8 and C/TA ratio 0.23.

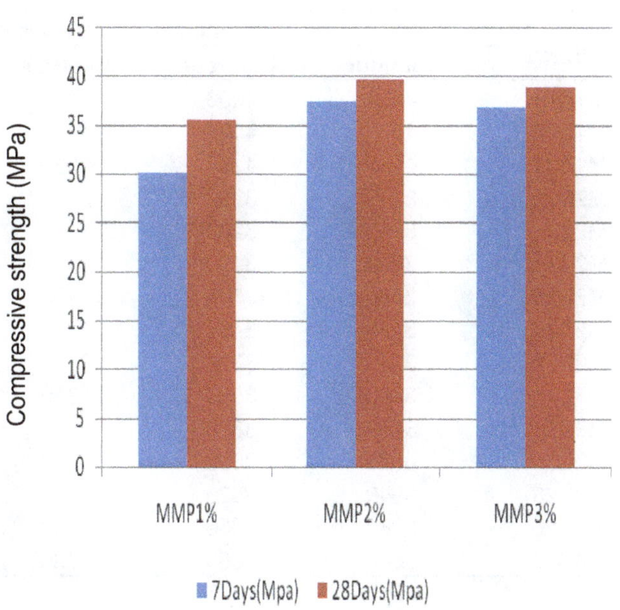

Figure 5. Compressive strength of concrete mixes for different W/C ratio 0.4, F/C ratio 0.8 and C/TA ratio 0.24.

strength (as shown in Figure 6). This can be justified based on the fact that both the parameters such as C/TA and F/C ratio have significant effect on the improvement in strength due to effective binder content for the volume of aggregate and higher fractions of fine particles. This can lead to delayed cracking in concrete upon loading.

The split tensile properties were further showing similar contrasting properties as that of compressive strength. The test results on the early age hardening in few hours after demoulding the concrete specimens were recorded and shown in Table 4. All the concrete specimens showed an increased strength immediately after demoulding. However, soon after few hours the ultrasonic value was found to decrease and after a day the ultrasonic pulse velocity was higher as shown in

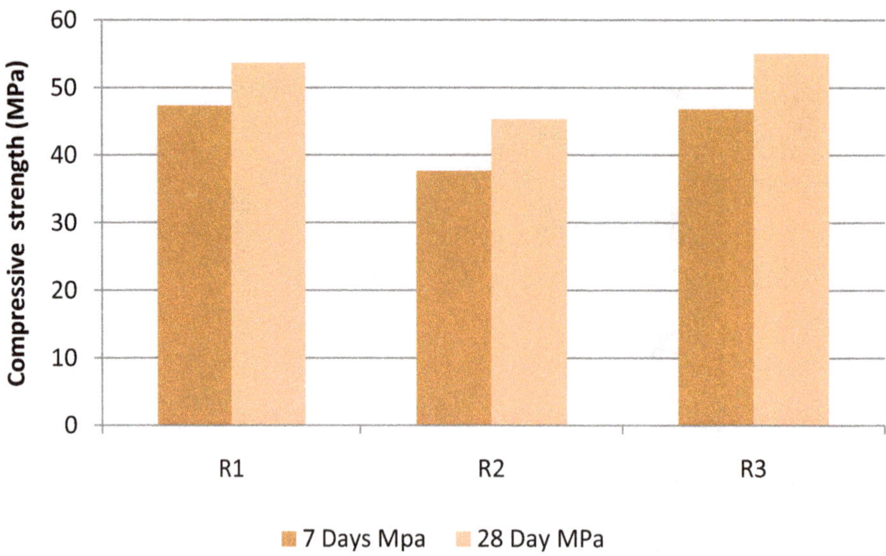

Figure 6. Compressive strength for different mix proportion of controlled concretes.

Table 4. Ultrasonic pulse velocity (m/s) measurements for various concrete mixtures.

Mix Id	Immediately after demoulding					Curing days		
	1st hour	2nd hour	3nd hour	4th hour	5th hour	3rd day	7th day	28th day
M1A19	4390	4200	4140	4050	4050	4400	4530	4550
M1A20	4200	4120	4100	4120	4080	4600	4440	4580
M1A21	4150	4150	3920	3870	3820	4200	4070	4460
M2B1	3830	3250	3090	3060	3040	4350	4510	4605
M2B2	3980	3200	3240	3200	3100	4040	4430	4590
M2B3	3940	3190	3340	3300	3300	4000	4410	4520
M2B25	3390	3300	3150	3120	3200	3400	4130	4455
M2B26	3420	3400	3250	3300	3140	3530	4220	4415
M2B27	3340	3325	3140	3130	3340	4180	4375	4440
M3C1	3695	3550	3750	3410	3730	4440	4560	4570
M3C2	3750	3800	3830	3730	3730	4370	4450	4500
M3C3	3675	3800	3720	3640	3580	4100	4470	4510
M3C25	3940	3920	3880	3880	3890	4100	4200	4450
M3C26	3860	4030	4050	4080	4080	4100	4370	4510
M3C27	3680	3650	3730	3620	3640	3920	4000	4250

Figures 7 to 12. This can be substantiated and the accelerator has phenomenal influence in the first few hours after mixing water and saturates thereafter. While the early setting property is visibly seen in first few hours for all the concrete mixtures; however, a deceleration in the strength gain was observed for the concrete specimens during first few hours after demoulding. This can be evident that the strength gain for concrete is reduced due to volumetric shrinkage occurring immediately after the surface is exposed to air drying. As shown in Figures 7 to 12 that in strength gain after 3 days, there was a good increase in pulse velocity and satisfies the Indian standard requirements (IS 13311, 1992).

Conclusion

The comprehensive analysis on the various experimental test results showed that accelerated setting properties of concrete can be more useful in high early fast track concreting in pavement applications. Also, the addition of metallic reinforcements has significant effect on the compressive properties of concrete as well as split tensile

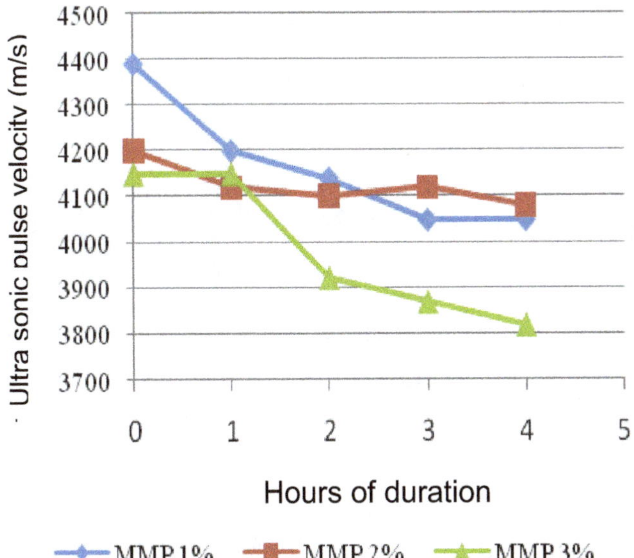

Figure 7. Ultrasonic pulse velocity for different W/C ratio 0.3, F/C ratio 0.8 and C/TA ratio 0.27 – day one

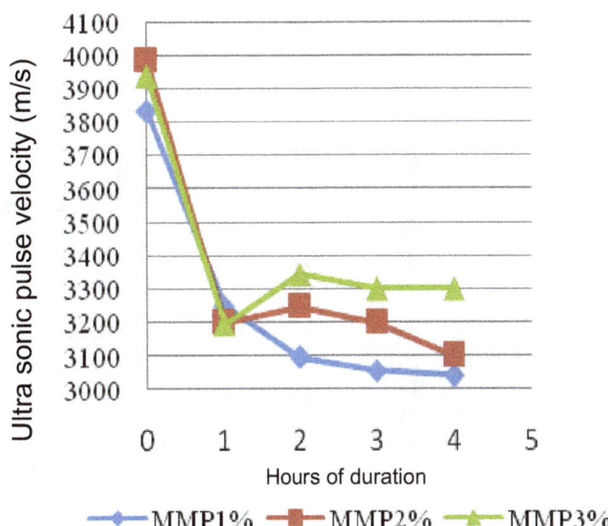

Figure 8. Ultrasonic pulse velocity for different W/C ratio 0.3, F/C ratio 0.6 and C/TA ratio 0.25 – day one

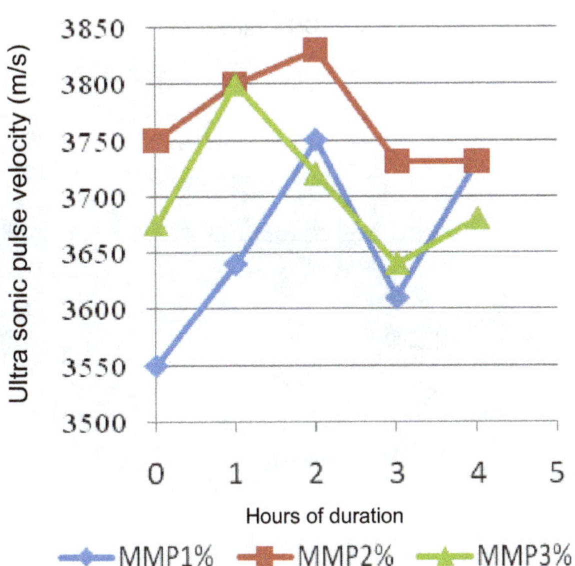

Figure 9. Ultrasonic pulse velocity for different W/C ratio 0.4, F/C ratio 0.8 and C/TA ratio 0.25 – day one.

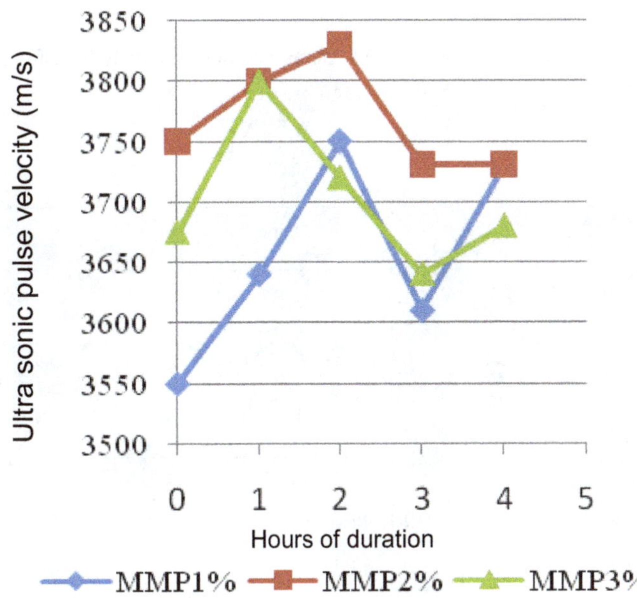

Figure 10. Ultrasonic pulse velocity for different W/C ratio 0.4, F/C ratio 0.8 and C/TA ratio 0.23 – day one.

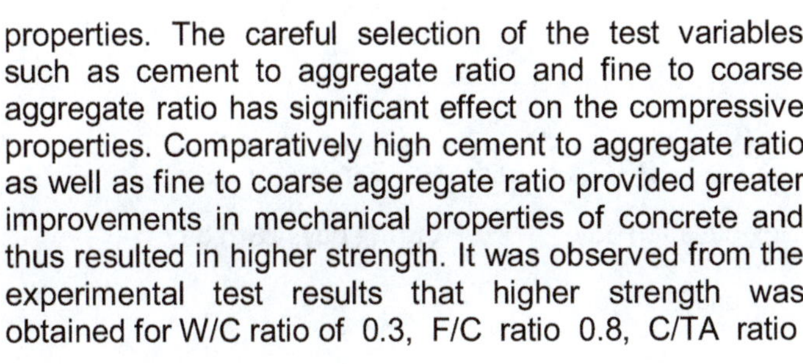
properties. The careful selection of the test variables such as cement to aggregate ratio and fine to coarse aggregate ratio has significant effect on the compressive properties. Comparatively high cement to aggregate ratio as well as fine to coarse aggregate ratio provided greater improvements in mechanical properties of concrete and thus resulted in higher strength. It was observed from the experimental test results that higher strength was obtained for W/C ratio of 0.3, F/C ratio 0.8, C/TA ratio 0.27 with metallic reinforcements 2% and the compressive strength at 7 and 28 days were 38.9 and 50.8MPa, respectively and a similar trend was noted for higher W/C ratio of 0.4., F/C ratio 0.8, C/TA ratio 0.23 with metallic reinforcements 3% the compressive strength at 7 and 28 days 43 and 47.9MPa, respectively. This is evident that higher fraction of aggregates and higher cement content envisage higher strength. The addition of accelerator has direct effect on the early strength gain

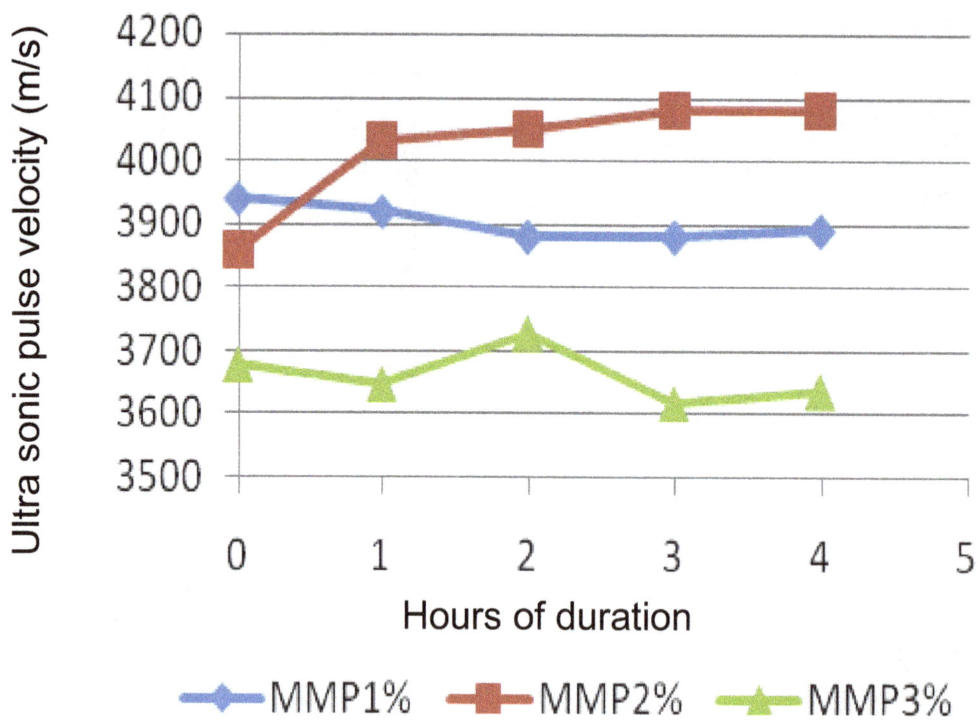

Figure 11. Ultrasonic pulse velocity for different W/C ratio 0.4, F/C ratio 0.8 and C/TA ratio 0.24 – day one.

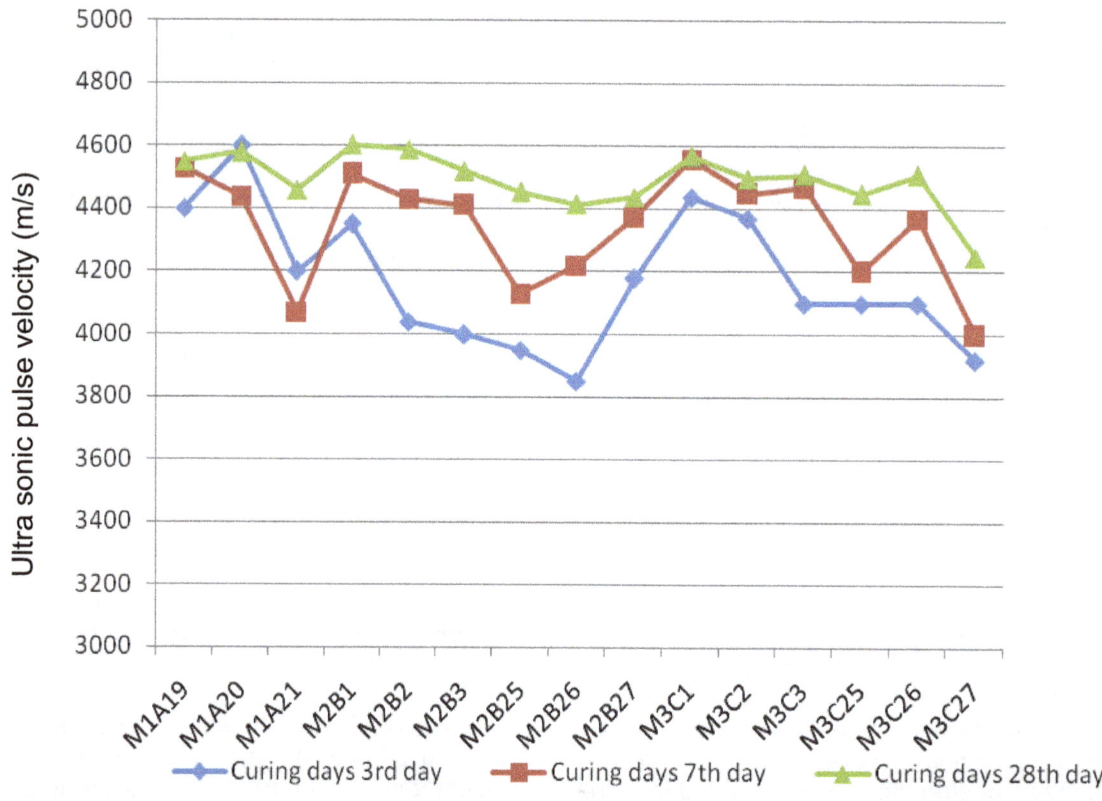

Figure 12. Ultra sonic pulse velocity for differenet mixes at 3, 7 and 28 days.

and resulted in attaining the 28 days in a short duration of 14 days. The potential use of accelerators can be realized in high strength concrete incorporating mineral admixtures of which the delayed strength gain can be offset and early strength gain can be achieved.

REFERENCES

ACI-544 (1993). Guide for Specifying, Proportioning, Mixing, Placing and Finishing Steel Fiber Reinforced Concrete, 544.3R-93., American Concrete Institute, Detroit, MI, pp. 241-253.

Balaguru PN, Shah SP(1992). Fiber Reinforced Cement Composites. McGraw Hill, Inc., New York. pp. 156-167.

Bentur A, Mindess S (1990). Fiber Reinforced Cementi-'eous Composites. Elsevier Applied Science, New York. pp. 29-38

Hanant DJ (1978). Fibre Cements and Fibre Concretes. John Wiley and Sons, Chichester, pp. 219-225.

IS 13311 (1992). Non-destructive testing of concrete: Part 1 Ultrasonic pulse velocity.

Kar JN, Pal AK (1972). Strength of fibre-reinforced concrete. X Struct. Div., Proc. ASCE, ST5, pp. 11-20.

Naaman AE(992). Sifcon: tailored properties for structural performance. In High Performance Fiber Reinforced Cement Composites, Proc. Int. RILEM AC1 Workshop. E and FN Span/Chapman and Hall, London, 1.

Parimi SR, Rao JKS (1971). Effectiveness of random fibres in fibre-reinforced concrete. In Mechanical Behaviour of Materials, Proc. Int. Conf., pp. 15-26.

Romualdi JP, Batson GB(1963). Mechanics of crack arrest in concrete. Proc. ASCE, 89 EM3, pp. 147-168.

Shah SP, Rangan BV (1971). Fiber reinforced concrete properties. ACI Mater: J., 68(2): 126-135.

Soroushian P, Lee Cha-Don (1990). Distribution and orientation of fibers in steel fiber reinforced concrete. ACI Mater: J., 87(5):433-439.

Zollo RF (1975). Fiber reinforced concrete extrusions. Proc. ASCE, IOI(ST12),pp. 2573-2583.

20

Utilization of natural and industrial mineral admixtures as cement substitutes for concrete production in Jordan

Omer Nawaf Maaitah[1], **Nafeth A. Abdel Hadi**[2] **and Monther Abdelhadi**[3]

[1]Faculty of Engineering, Mutah University, Karak, Jordan.
[2]Balq'a Applied University, Jordan.
[3]Department of Civil Engineering, Al-Ahliyya Amman University, Jordan.

Several materials such as tuff and Tripoli which is naturally occurring and industrial by-products wastes as high calcium ash and slag were investigated as cement substitutes in this paper. Compressive strength of various standard mortar samples have been tested at 7 and 28 days. The obtained results show that these materials have improved the properties Ordinary Portland Cement (OPC) concrete. It was found that an increased compressive strength of 22% were attained at an optimum of 10% addition of ash. Meanwhile, the compressive strength increases by 30% when the content of Tripoli is of 10%. The slag and Tuff have adverse effect on the strength. The reason may be attributed to the mode and original of crystals. The finding of this research work also show that using these materials will reduce the use of OPC. This reduction will reduce some of the adverse environmental effect due to OPC production and consequently, reduces the consumption of energy in Jordan.

Key words: Ordinary Portland Cement (OPC), cement substitutes, tuff, Tripoli, ash, environmental.

INTRODUCTION

Mineral admixtures are materials used as an ingredient of concrete or mortar and added to the batch immediately before or during mixing mainly to improve or modify several properties of Portland cement concrete PCC. Mineral admixtures are used in conjunction with Portland or blended cements as a supplementary cementing material (SCM) through hydraulic or pozzolanic activity or both. PCC mixtures are multiphased, particle-reinforced composites that consist of irregularly shaped and randomly oriented aggregate particles embedded in an inelastic matrix. PCC mixtures generally exhibit complicated mechanical behavior and multiple modes of damage. Although precise identification and prediction of the inelastic damage modes of the PCC is extremely difficult, it is important to seek out simpler approaches of predicting mechanical behavior including damage characteristics of the mixture in place of expensive and time-consuming laboratory experiments where possible.

Table 1. Classification of class C and F compared with Karak ash (Analysis was done using XRF technique).

Oxide (%)	Ferguson et al. (1999)		Ordinary Portland Cement*	Karak ash
	Class C ash	Class F ash		
SiO_2	54.9	39.9	19.94	23.82
Al_2O_3	25.8	16.7	5.37	5.34
Fe_2O_3	6.9	5.8	3.18	1.94
CaO	8.7	24.3	63.65	52.84
MgO	1.8	4.6	2.59	0.9
SO_3	0.6	3.3	2.88	8.71

*Neville (1995).

Recently, the use of reinforced concrete in multi store structure and industrial plants in the vicinity of the capital Amman and surrounding towns is continuously increasing in the last decade. There is shortage in production and huge demand on OPC which the local cement factories cannot satisfy the local market needs. This has led to an increase in the demand energy. On the other hand, there is a negative environmental impact of cement production. For instance the production of one ton of OPC releases one ton of CO_2 and these issues are considered the most important challenges facing construction sector.

Some of the local cement factories in Jordan have investigated the use of bituminous limestone ash as a source of raw material after direct combustion and reducing its energy in the kiln system. This may be considered as a good approach for energy saving, but it will lead to complicated environmental problems due to the evolved SO_x and CO_x gases.

To achieve sustainability of the OPC, changes in working practices are required. This means that low energy, low carbon, and low waste techniques have to be developed to replace more intensive techniques. For example, *in situ* remediation of contaminated land may become a more attractive than a ‘dig-and-dump’ approach, where the material is removed and disposed of to landfill and then replaced with new material. These techniques are not only high in energy consumption and produce large amounts of contaminated waste, but also highly expensive. Alternative methods however, have been recently developed to reuse several types of industrial waste materials in civil engineering construction by many researchers such as Tay (1987), Churchill (1994), Perez et al. (1996), Stefanov (1986), Pereira et al. (2000), Pavlova (1996) and Maaitah (2012).

In this paper, some materials such as Tripoli, ground tuff, high calcium ash and air cooled slag will be added to the mortar. These materials have been investigated as cement substitutes. The aim of this work is to investigate the enhancement of mechanical properties of the PCC by the addition of mineral admixtures to the concrete mixtures as cement substitutes. The optimum percent or fixation value for each additive will be investigated. The weight of cement content decreases but the strength is increased that is required to produce a certain class of reinforced concrete with improved quality and durability through adding self cementitious materials as mineral admixtures. This may reduce the consumption of OPC which in turn may help reduce all the adverse effect to the cement production.

EXPERIMENTAL PROCEDURE

Materials

The naturally occurring material such as ash and Tripoli are abundant in Jordan. Bituminous limestone as a source of ash and Tripoli are available as millions of tones in the vicinity of Al-Karak city about 120 km to the south of the capital Amman. Slag is available at many steel factories located at and around Amman. The material that produces waste as high calcium ash and slag were used as cement substitutes.

In the present work the Karak ash (from various outcrops from El-Lajjun) was obtained by direct combustion of Karak bituminous limestone at a temperature of (900 to 1000)°C. The sample was allowed to cool down to the ambient temperature which can be considered as fast cooling. This means that the crystal is micro crystalline. Then, the sample was ground under dry conditions to obtain the possible minimum grain size. Small ball mills and Los Angles machine were used. The sample was crushed using a jaw crusher to obtain bituminous limestone aggregates of 9 mm nominal size particles.

Ash could be the solid waste product of possible utilization of the Karak bituminous limestone. The ash used in this study is a high calcium ash that has been produced by direct combustion of the bituminous limestone at 950°C (Maaitah, 2012). The chemical properties of Karak ash are summarized in Table 1. The combusted bituminous rocks in Karak/Jordan have indicated the presence of two groups of minerals; high temperature which is equivalent to clinker cement (Khoury, 1993; Al-Hamaiedh, 2010) and low temperature which is similar to the hydrated cement products (Khoury and Nasser, 1982). The low temperature mineral group has a similar composition to the hydrated cement products and has been precipitated from high alkaline circulating water (pH > 12.5). This naturally occurring alkaline water is analogous to the cement percolating water (Khoury, 1993).

Tuff can be obtained from the northern province of Jordan. Tuff is a natural volcanic material that is characterized by very high porosity, low density, rich in SiO_2, and has a considerable content of Al_2O_3 in addition to Fe_2O_3. Huge quantities of this material are available in the eastern and north eastern provinces of the Kingdom; these natural recourses are utilized in various

Table 2. Chemical composition of slag, tuff, and Tripoli by using XRF technique. (Tests were carried out at the Natural resources authority (NRA) in Amman).

Oxide Wt. %	Tuff	Ash	Slag
SiO_2	50.6	25.30	16.55
Al_2O_3	15.2	2.35	7.8
Fe_2O_3	11.2	1.37	18,33
CaO	9.0	45.21	7.12
MgO	5.8	1.63	7.83
P_2O_5	4.6	5.47	------
Na_2O	----	0.85	2.83
TiO_2	---	0.14	------
MnO	2.5	0.02	4.52

Table 3. Chemical composition and physical properties of Tripoli.

Area in Karak	SiO_2	CaO	Al_2O_3	Na_2O	Fe_2O_3	MgO	L.O.I	Specific gravity
Shahabiyeh*	94.50	0.44	0.33	0.45	0.16	0.23	3.10	2.43
Adnanieh*	90.64	2.16	0.11	0.94	0.13	0.37	4.77	2.6

* both village in west of Karak.

engineering and agricultural aspects.

About 20 kg of reddish crushed sand size was sampled from Amani quarry at Tall Hassan in the vicinity of Al-Azraq area. The sample was ground to get the maximum possible passing #200 sieve fraction, the Loss Angles abrasion machine was used for this purpose to produce a bulk fine tuff sample that will be used later to investigate the cement-tuff mortars. The chemical composition of each material was determined using the XRF technique as shown in Table 2.

Slag can be obtained from the United Iron and Steel MFG. Co., 30 km to the south of Amman. Huge quantities of blackish, tough aggregation stockpiles of iron slag wastes are available as by-product of iron production at different steel factories in the vicinity of Amman and surrounding areas. The sample was obtained from Al Manaseer Steel Factory about 30 km south of Amman. The sample was ground to fine powder and passing sieve No. 200 was used.

The chemical composition of each material was determined using the XRF technique as shown in Table 2. Conplast C_3O as a super plasticizer was used in constant dosage in all trials and with Specific Gravity of 3.3. The unit weight is 1750 kg/m^3 and the Absorption is up to 2.32. The slag crystal is not compatible, some crystal is glassy because of the cooling is very fast in the surface and some fine. The fine crystal is constituted because it is cooled slowly somehow at the down of the stack. Therefore, the production is out of control and the composition varies from time to time. It is difficult to find similar sample among the slag stack after manufacturing.

Tripoli is microcrystalline silica. It is a form of silica, earthy, light colored, light weight, friable, very fine grained chalcedonic and opaline silica of cryptocrystalline, and it is commercially classified as "Soft Silica" or "Amorphous Silica", having a 90 to 94% Silica content with high purity and high melting point. The Tripoli can be used as filler material in paints and rubber, plastic industries, mild abrasive, insecticides. The occurrences are found in Karak District, 168 km Southwest of Amman, in the following areas:

i) Around thirteen million metric ton in El-Adnanieh (8 km South of Karak);
ii) Around seven million metric ton in El-Shahabiyeh, 4 km Southwest of Karak;
iii) Undetermined huge millions metic tonnage in Wadi Rakin and Wadi Ben Hamad 4.5 km Northwest of Karak;
iv) Undetermined millions of metic tonnage in Wadi Falqa, 5.5 km Southwest of Karak;
v) Undetermined huge millions metic tonnage in Ainun, 5.5 km South-Southwest of Karak along the west bank of Wadi Daba.

The chemical composition of Tripoli was determined using the XRF technique as shown in Table 3. The Adnanieh Tripoli were used in this work.

Approach

The combusted limestone ash, slag, tuff and Tripoli were ground using Loss Angles machine for one and half an hour each, followed by sieving the ground material on No. 200 sieve. The fraction passing #200 sieve was collected in tight plastic bags. The silica sand was sieved to prepare standard sand for mortar preparation. A reference mortar mix composed of 1500 gr of the standard silica sand was mixed with 500 gr of OPC at a w/c ratio of 0.485. The mortar was mixed well and cast into 5x5x5 cm cubes, two layers into each cube and 10 blows on each layer using a standard rod with 2x5 cm cross sectional area. Six cubes were prepared and de-molded after 24 h. The samples were cured in water until testing at 7 and 28 days. Compressive strengths presented are the averages for 3 cubes at each of the (7 and 28 days). The same procedures were repeated for the other mortars- admixtures with substituting cement by various percents of that admixture. ASTM C109 was followed strictly for the whole procedure. The compressive strength of various standard mortar samples have been tested at 7 and 28 days.

Mortar composition

The proportions of materials for the standard mortar shall be one part of cement to 2.75 parts of graded standard sand by weight.

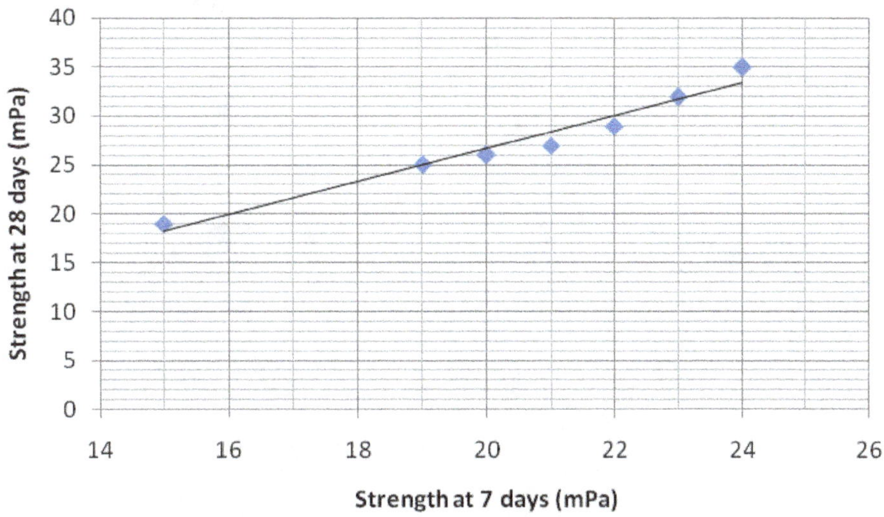

Figure 1. Comparison between strength at 7 and 28 days for Tripoli.

The water/cement ratio of 0.485 was used for all Portland cements. The water/cement ratio for other than Portland and air entraining Portland cements shall be such as to produce a flow of 110 +/- 5.

The specimen mold preparation was conducted using Mortar Mixing Procedure in accordance with (ASTM C305). This was performed by applying a thin coating of mold release to the interior surfaces of the molds and base plates. Surfaces were wiped with a cloth to remove any excess, with dry paddle and bowl placed in the mixing position of the mixer. The strength for native sample (without any additives) after 7 days is 21 kg/cm^2 and after 28 days is 27 kg/cm^2. An increase in strength of 28.5%, possibly due to curing time.

RESULTS AND DISCUSSION

The bond strength between cement substitutes and concrete materials is related to the interface properties, interface fracture mechanics and most likely to the crystal. The mode of crystal (that is, phanertic, aphanertic and glassy) play a significant role in determining the success of the substitutes. The loadings will impart both tensile and compressive normal and shear stresses at the interface, and thus failure will be under multi-component stresses.

The development of the self cementaceous properties and strength are controlled by the additive content and the curing period. The additive is similar to a great extent to the Portland cement. The results are obtained through the standard mixtures of each sample (Table 4). The mixtures are prepared and cured under the same standard procedures and conditions. Figure 1 show the effect of curing time on Tripoli which improve that the Tripoli reacts with OPC. This is because it has micro crystalline original and a high content of SiO_2 and CaO reacts with cement (Table 5).

The strength of mortar increases as the Tripoli content increases up to specific value (that is, 20%) and then it has no effect on strength up to 40% content. The content of Tripoli more than 40% has adverse effect because the strength decreases as shown in Figure 2. At fixation point (10% of Tripoli content) the strength improved 30% from native sample after 28 days. On the other hand, the strength increases by 14% after 7 days for the same Tripoli content. This is an improvement on the fact that the Tripoli reacts with time. The well known reaction between the elite (C_3S) and water is

$$3CaO.SiO_2 + H_2O \rightarrow 2CaO.H2O + Ca(OH)_2 + 120\ cal/gr$$

The extra Portlandite $Ca(OH)_2$ will react with SiO_2 from Tripoli to produce Belit which will cause an increase in the strength by 30% as shown in Figure 2.

$$Ca(OH)_2 + ((SiO_2 + CaO)\ (from\ Tripoli)) \rightarrow 2CaOSiO_2\ (Belit)$$

The extra Belit ($2CaOSiO_2$) which is due to the addtion of Tripoli is appearantly resposible for the increase in compressive strength.

From the obtained results in Figure 3, it is clear that the mortar strength has increased by 22% of ash fixation point after 28 days, whilst, at 7 days an increase of 14% were observed. The fixation point is at 10% ash content.

The ash and Portland cement are essentially composed of lime (CaO). Silica (SiO_2) and alumina (Al_2O_3) are present at higher concentrations in the Portland cement and react with CaO at about 1425°C to form alite. Heat treatment of the bituminous rocks and Portland cement raw material involves dehydration, thermal decomposition of clay minerals (300 to 650°C), decomposition of calcite (greater than 800°C), the formation of belite (C_2S), tricalcium aluminate (C_3A), and tetracalcium alumina ferrite (C_4AF). The liquid phase and

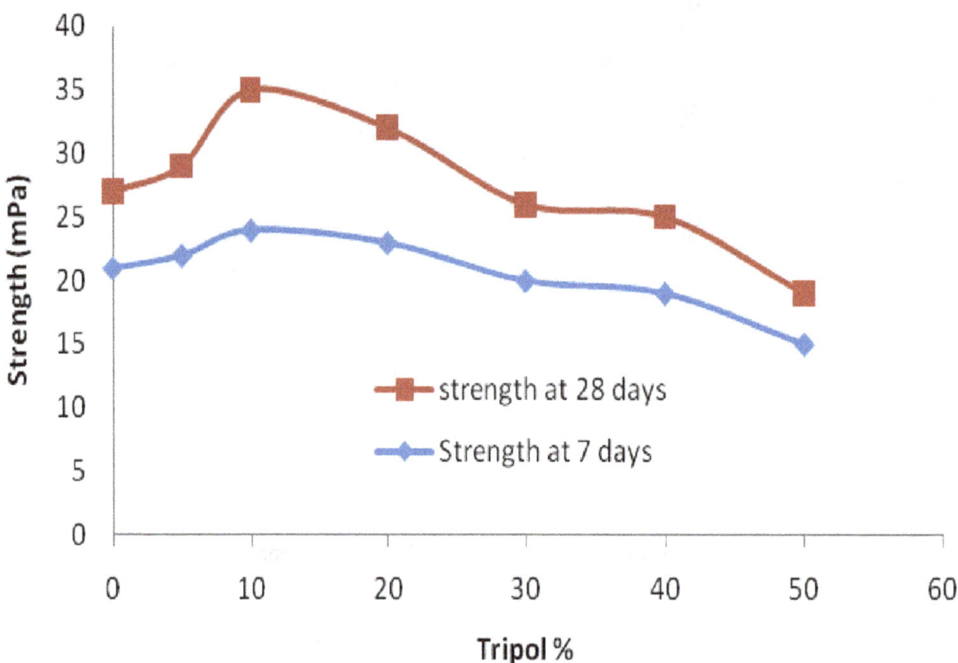

Figure 2. Mortar strength versus Tripoli content.

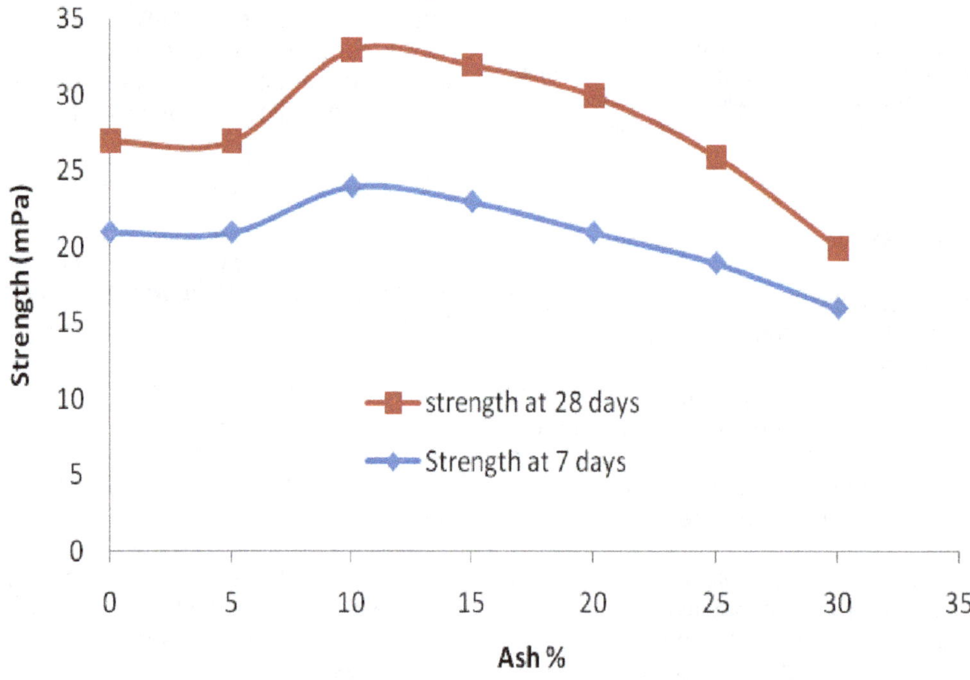

Figure 3. Mortar strength versus Ash content.

sintering at about 1425°C form alite (C_3S) which is responsible for the strength of concrete (Khoury, 1993).

The variable content of SO_3, the alkali CaO, and pozzolanic content (SiO_2 +Al_2O_3+ Fe_2O_3), are found in both the ash samples and OPC raw material. The strength build up in all the samples is related to the setting reactions of lime (CaO) with the pozzolanic constituents to produce calcium silicate hydrate (CSH) and calcium aluminate hydrate (CAH). High pH solution is highly reactive with amorphous Al-Si rich phases at

Table 4. The following batch component, which is sufficient for 6 samples.

Material	Amount (g)	w/c
Portland Cement	500	
Silica Sand	1375	
Distilled Water	242	Portland (w/c=0.485)

Table 5. Summary of the results.

Additives	% of additives	%Strength at 7 days/strength for native sample	%Strength at 28 days/strength for native sample
Tripoli	20	14	30
	50	-28.5	-30
Ash	10	14	22
	20	1	11
Slag	5	-10	-4
	50	-76	-70
Tuff	10	-5	-8
	30	-23	-24

normal room temperature. Sulfate minerals as ettringite are expected to form because of the availability content of SO_3. The CaO content plays an important role in the alkali–pozzolanic reaction.

The reaction products leave the kiln as a clinker. The clinker leaves the kiln here to be cooled, mixed with gypsum, and then ground into a fine powder (cement). The setting of cement involves a number of stages at different rates. It is generally known that a complex series of reactions do take place as the cement reacts with water. Setting of C_2S involves slow hydration reactions and the formation of Portlandite Ca (OH_2) and calcium silicate hydrate.

Portlandite Ca $(OH)_2$ plays an important role in the setting reaction. Portlandite reacts with silicates and aluminum rich phases to form insoluble compounds which contribute to the strength formation (pozzolanic reactions). Excess portlandite reacts with atmospheric CO_2 to precipitate calcium carbonate that helps in strengthening the product after aging.

The combusted bituminous rocks in central Jordan have indicated the presence of two groups of minerals; high temperature which is equivalent to clinker cement (Khoury, 1993) and low temperature that is similar to the hydrated cement products (Khoury and Nasser, 1982). The low temperature mineral group has a similar composition to the hydrated cement products and has been precipitated from high alkaline circulating water (pH > 12.5). This naturally occurring alkaline water is analogous to the cement percolating water (Khoury, 1993).

Figure 4 and Table 5 show that the tuff has an adverse effect on strength. This is possibly, because the Tuff is poured rock and the powder has high surface area. The Tuff powder absorbs extra water, in turn affecting the water/cement ratio for reaction. This could be related to the kind of crystal of tuff that shows no reaction with OPC. Figure 4 and Table 5, also, show that the slag has an adverse effect on strength. The slag production is out of control. The slag contains strange materials and impurities. The properties of slag vary from sample to sample. The slag durability may be low due to the rust.

The effect of curing time can be seen Figure 5 and Table 5. Curing period of 28 days and more has influenced the compressive strength results. High compressive strength values obtained for intact samples indicate no disintegration features under fully saturated conditions. All hydrated samples have shown a similar behavior to the hydrated OPC products but with lower compressive strength.

Conclusion

Table 5 summaries the results of this research and illustrates that an addition of 20 and 10% Tripoli and ash, respectively, lead to an improvement of the strength with respect to native sample by 19 and 22%. It is also, apparent that any increases in the additive content more than the fixation point will cause a reduction in the compressive strength. The obtained results suggest that the presence of Tuff and slag additives have determined

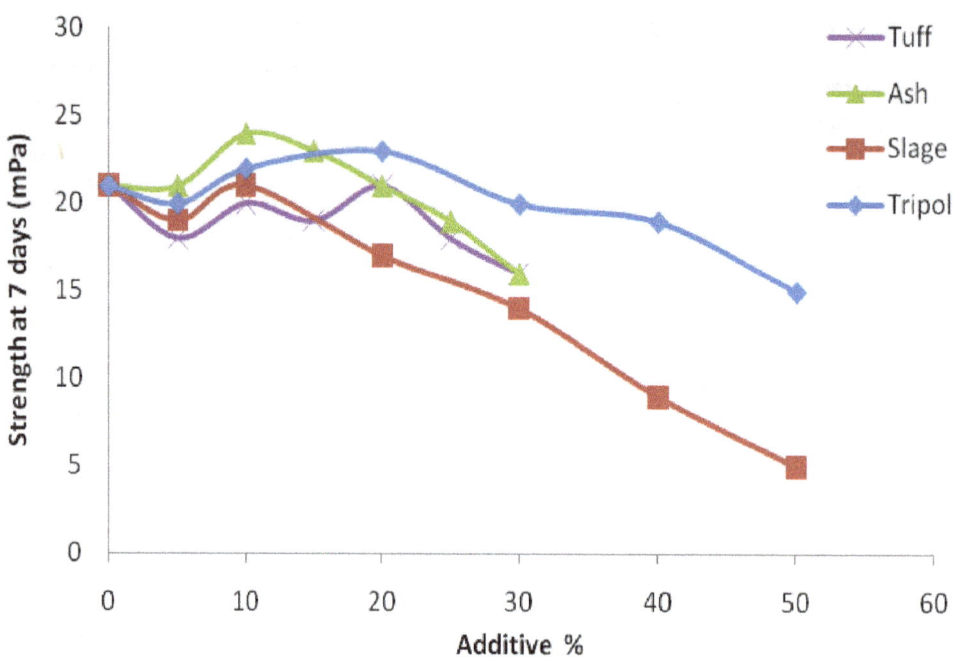

Figure 4. Strength versus percent of additives after 7 days.

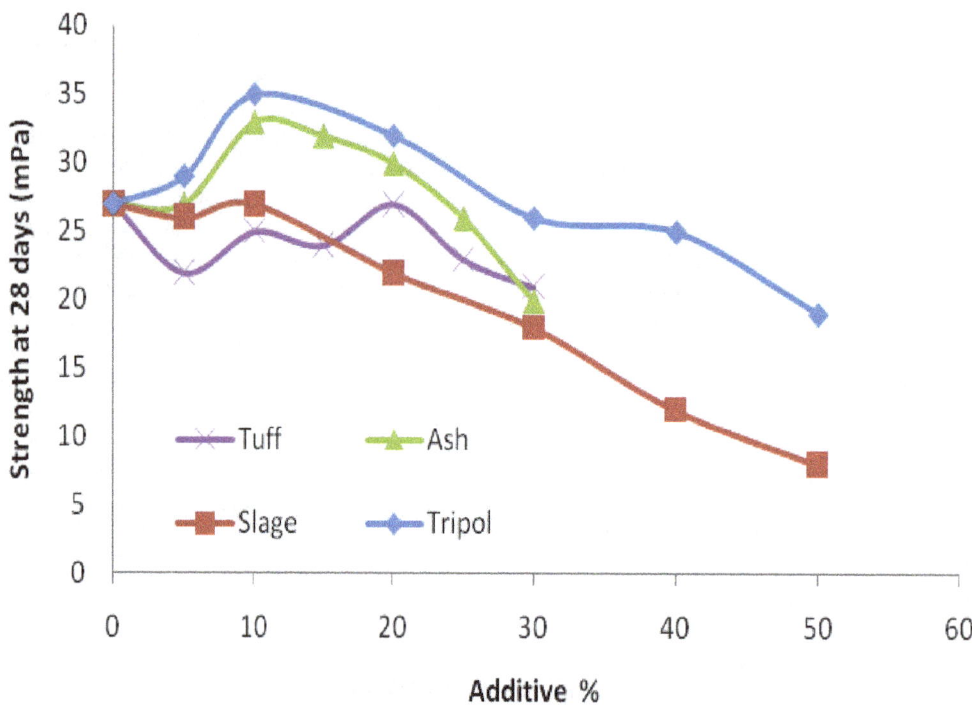

Figure 5. Strength versus percent of additives after 28 days.

effect on the compressive strength, which can be explained as a result of the mode and original of crystal (that is, phanertic, aphanertic and glassy). The crystalline plays a significant role in determining the success of the substitutes. Curing period of 28 days and more has a good influence on the compressive strength results.

Conflict of Interest

The authors have not declared any conflict of interests.

REFERENCES

Churchill M (1994). Aspects of sewage sludge slime utilization and its impact on brickmaking, Global Ceram. Rev. 1:18.

Khoury H (1993). Mineralogy and Isotopic Composition of the Metamorphic Rocks in the Bituminous Limestone of the Maqarin Area, Jordan, Dirasat 20B(2).

Khoury H, Nasser S (1982). A discussion of on the origin of Daba-Siwaqa marble, Dirasat. 9:55-66.

Maaitah ON (2012). Evaluation of Al-Karak Ash for Stabilization of Marl Clayey Soil, 17[2012], Bund. G, EJGE.

Pavlova L (1996). Use of industrial waste in brick manufacture, Tile and Brick Int. 12:224.

Pereira DA, Couto DM, Labrincha JA (2000). Incorporation of aluminum-rich residues in refractory bricks, CFI – Ceramic Forum International, 77:21.

Perez JA, Terradas R, Manent MR, Seijas M, Martinez S (1996). Inertization of industrial wastes in ceramic materials. Ind. Ceram. 16(7):571-584.

Stefanov S (1986). Use of industrial wastes in the brick and tile industry, Ziegelindustrie Int. 3:137.

Tay JH (1987). Bricks manufacture from sludge slime. J. Environ. Eng. 113:278. http://dx.doi.org/10.1061/(ASCE)0733-9372(1987)113:2(278)

21

Strength properties of groundnut shell ash (GSA) blended concrete

Raheem, S. B.[1], Oladiran, G. F.[1], Olutoge, F. A.[2] and Odewumi, T. O.[1]

[1]Civil Engineering Department, the Polytechnic, Ibadan, Oyo State, Nigeria.
[2]Civil Engineering Department, University of Ibadan, Oyo State, Nigeria.

This research work detailed the report of an experimental study into the strength of modified concrete produced from mixes containing partial replacements of Ordinary Portland Cement (OPC) with groundnut shell ash (GSA). The experiments were designed to include two main mixes (with variations in the water/cement ratios) with different percentages by weight of OPC to GSA in the order of 100:0, 95:5, 90:10, 85:15 and 100:0, 90:10, 80:20 for mixes 1:2:4 and 1:2.3:2.6 respectively. For the ratio 1:2:4 mix, a total of 32 concrete cubes of sizes 150 × 150 × 150 mm and 32 cylindrical concrete specimens (100 mm diameter and 200 mm long) were cast and tested. Also, for the 1:2.3:2.6 mix, 24 concrete cubes and 24 cylindrical concrete specimens, with the same sizes as above, were cast and tested at 7, 14, 21, 28 days of curing. Compressive and splitting tensile tests were conducted to assess the strength of concrete. Generally, strengths of modified concrete increased with curing period but decreased with increased GSA percentage. For mix ratio 1:2:4, the highest compressive and tensile strengths were 24.06 (2.67) and 21.34 (2.11 N/mm^2) at 28 days for 0 and 10% GSA respectively. While mix ratio of 1:2.3:2.6 gave the highest compressive and tensile strengths of 35.11 (4.21) and 27.33 (4.01 N/mm^2) at 28 days for 0 and 10% GSA respectively. It was observed that 10% GSA replacement was appropriate for both mixes. GSA therefore seems to be a promising and local partial replacement material for cement in concrete making.

Key words: Concrete, pozzolana, partial replacement, groundnut shell ash (GSA), concrete strength.

INTRODUCTION

Due to increasing industrial and agricultural activities, tones of waste materials are deposited in the environment with little effective method of waste managing/recycling. Some of these deposits are not easily decomposed and the accumulation is a threat to the environment and people at large. Some of these waste materials are rice husks, maize combs, snail shells, palm-kennel shell, coconut shell, saw dust, groundnut shell etc. Global pollution coupled with resource depletion has challenged many researchers and engineers to seek locally available materials with a view to investigating their usefulness wholly as a construction material or partly as a substitute for conventional ones in concrete making. In search for new materials which address the issues aforementioned and which are cost effective and more efficient, pozzolans attract much interest. Malhotra and Mehta (1996) define "pozzolan" as "a siliceous or siliceous and aluminous material, which in itself possesses little or no cementing property, but will in a finely divided form – an in the presence of moisture -

chemically react with calcium hydroxide at ordinary temperatures to form compounds possessing cementitious properties." Numerous achievements have been made in these regards and the subject is attracting attention due to its functional benefit of waste reusability and sustainable development, reduction in construction costs and its indigenous technology and equipment requirements are added advantages. Among others, Alabadan et al. (2005) investigated the potentials of Groundnut Shell Ash (GSA) as a partial replacement for ordinary portland cement in concrete. In the study, it was generally reported that the strength of the control was higher and concluded that the replacement of cement with ash up to 30% gave promising results over others. This research intends to investigate the pozolanic activity and usefulness of one of these agricultural waste materials (groundnut shell ash) as a partial substitute for cement in concrete making. If found useful, it will promote waste management/recycling at little cost, reduce pollution by the waste and increase the economic base of the farmer.

Admixture is defined as a material, other than cement, water and aggregates that is used as an ingredient of concrete and is added to the batch immediately before or during mixing (Shetty, 2005)

Blended cement is obtained by adding mineral admixtures like fly- ash, slag and silica fumes to OPC. There are a number of systems that are used to make blended cements. Some systems are capable of "on-demand" blending, while others may blend the materials in a fixed percentage into a storage silo. All of the systems meter the constituent products in the desired proportions, and then blend them to a uniform mixture. In most cases, proportions can be adjusted to produce blends that optimize the desired properties in concrete.

MATERIALS AND METHODS

The research was carried out in stages; the first stage involved the sourcing for and preparation of material (groundnut shell), at the second stage, preliminary tests on the groundnut shell ash (calcined at 600°C) was conducted at the Department of Agronomy, University of Ibadan, Ibadan, Nigeria. At the next stage, concrete cubes (150 × 150 × 150 mm) and cylindrical concrete specimens (100 mm diameter, 200 mm length) were cast using groundnut shell ash (GSA) as partial replacement for cement and subsequent tests were carried out on them. The class of specimens is seen in Tables 1 and 2. The concrete comprised of ordinary portland cement (OPC), fine aggregate and coarse aggregate, water and groundnut shell ash (GSA). The ordinary portland cement packaged by Dangote Group was used. It was stored under dry condition and free from lump. The coarse aggregate used was granite stone. It was of high quality and free of deleterious organic matter and only the ones retained on sieve 3.75 mm were used. Also, the fine aggregate used was white sand obtained from river. Groundnut shell was obtained from marketers at Bodija, Ibadan, Oyo- State, Nigeria. About 8 kg of the shells was obtained and burnt to ash completely at temperature 600°C in a furnace at Fine Art Department, The Polytechnic, Ibadan. The ash was then sieved through British Standard sieve of 312 μ after grounding. The portion passing the sieve was reported to the required degree of fineness that is 312 μ and below while the ash retained on the sieve was reground and sieved again.

In this research, a mix ratio of 1:2:4 (cement:fine aggregate:coarse aggregate) by mass was adopted and OPC/GSA ratio of 100:0, 95:5, 90:10, 85:15 percentages by mass were used. Also, high strength concrete (40 Mpa) was designed and OPC/GSA ratio of 100:0, 100:10, 100:20 percentages by mass were used to investigate the effect of GSA replacement on HSC. 32 concrete cubes (150 × 150 × 150 mm) and 56 cylindrical specimens (100 mm diameter, 200 mm length) were cast and cured in the curing tank containing clean water. The compressive and splitting tensile strengths of the cubes and cylindrical specimens respectively were obtained from the crushing and splitting tensile tests at ages 7, 14, 21, 28 days of curing.

Class of specimens

Tables 1 and 2 provide class of specimens.

Preliminary test

This includes tests conducted on the constituent materials used in the production of the specimens.

Chemical analysis of GSA

Chemical analysis of GSA was carried out at Laboratory of the Department of Agronomy, University of Ibadan, Ibadan, Nigeria and the result is presented in Table 3.

Sieve analysis of GSA, fine aggregate and coarse aggregate

Sieve analysis was conducted on the GSA, fine and coarse aggregates used at the Material Laboratory of the Oyo State Secretariat, Ibadan. The sieve were mounted into a frame and shaken in a mechanical sieve shaker for 10 min. The apparatus and materials used were set of sieve (7, 14, 25, 36, 72, 200, 200 for GSA and fine aggregate and ¾", ½", 3/8" for granite), balance sensitive to 0.1 g, brush (for cleaning sieves), mechanical shaker, sample of soil, large pan. The results of the sieve analysis of the GSA, fine aggregate and coarse aggregate are given in Figure 1.

Aggregate impact value (AIV)

According to BS 812: Part 112 (1990), two procedures are available for the determination of AIV, one in which the aggregate is tested in a dry condition, and the other in a soaked condition. The former was adopted in this research. Aggregates passing a 14.0 mm test sieve and retained on a 10.0 mm test sieve were used for the test as stated in BS 812:Part 112 (1990). The test specimen was compacted, in a standardized manner, into an open steel cup. The specimen was tamped and then subjected to 25 numbers of standard impacts from a dropping weight in 3 layers. This action broke the aggregate to a degree which is dependent on the impact resistance of the material. This degree was assessed by a sieving test on the impacted specimen with the use of 2.36 mm sieve. Weights of retain and passing were measured. The result of the test is given in Table 4. AIV was calculated from the equation as follows:

$$AIV = M1/M2 \times 100$$

Where M_1 = the mass of the test specimen (in gram); M_2 = the mass

Table 1. Normal concrete (1:2:4 mix), for compressive test.

Sample	GSA%	7 days	14 days	21 days	28 days
CN_0	0	2	2	2	2
CN_5	5	2	2	2	2
CN_{10}	10	2	2	2	2
CN_{15}	15	2	2	2	2

CN_0 = Compressive test for normal concrete at 0%GSA.

Table 2. Normal concrete (1:2:4 mix), for splitting tensile test.

Sample	GSA%	7 days	14 days	21 days	28 days
TN_0	0	2	2	2	2
TN_5	5	2	2	2	2
TN_{10}	10	2	2	2	2
TN_{15}	15	2	2	2	2

TN_0 = Tensile test for normal concrete at 0%GSA.

Table 3. Chemical properties of GSA.

Constituent	% By weight (g) test 1	% by weight (g) test 2	% by weight (g) average
ZnO	2.56	2.61	2.59
C_uO	1.86	1.89	1.88
Fe_2O_3	2.24	2.27	2.26
MnO_2	3.48	3.53	3.51
MgO	4.86	4.86	4.86
SiO_2	32.96	33.05	33.01
Al_2O_3	7.06	7.06	7.06
K_2O	9.75	9.78	9.77
CaO	11.23	11.18	11.21
Na_2O	6.53	6.55	6.54
Loss on ignition	8.76	8.84	8.80
Others	8.71	8.38	8.55
Total	100	100	100

of the material passing the 2.36 mm test sieve (in gram).

Aggregate crushing value (ACV)

The test specimen was compacted in a standardized manner into a steel cylinder fitted with a freely moving plunger in 3 layers. The specimen was then subjected to a standard load of 400 kN applied through the plunger for 10 min. This action crushed the aggregate to a degree which is dependent on the crushing resistance of the material. The degree was then assessed by a sieving test on the crushed specimen with the use of 2.36 mm sieve. Weights of retain and passing were measured. The result of the test is given in Table 4. ACV was calculated from the equation as follows:

AIV = M1/M2 × 100

Where M_1 = the mass of the test specimen (in gram); M_2 = the mass of the material passing the 2.36 mm test sieve (in gram).

Specific gravity determination

Specific gravity bottles, weighing balance, distilled water and a drying cloth were used in the determination of specific gravity of GSA. Empty, clean and dry specific gravity bottle with its stopper was weighed (W1). The bottle was filled up to one-third full with the GSA sample and reweighed (W2). A small amount of distilled water was then added and the bottle contents shaken to remove entrapped air. Shaken continued and more water added until the bottle was full. The stopper was inserted and excess water cleaned on bottle and weighed (W3). The bottle thereafter was emptied, thoroughly washed and wiped dry and then filled with distilled water and the stopper inserted and excess water cleaned and weighed (W4). These procedures were repeated using another bottle for the purpose of obtaining the average. The results are giving in Table 8.

Tests on concrete

Tests were conducted on both fresh and hardened concrete. The

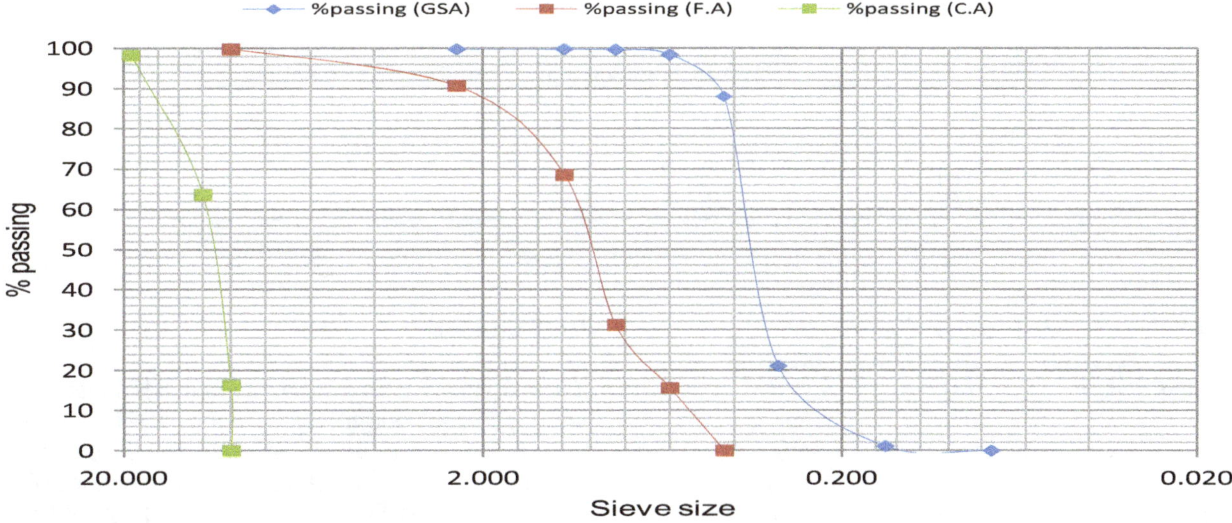

Figure 1. Particle size distribution of aggregates and GSA.

tests are slump test, weight development, compaction factor, compressive strength and splitting tensile strength.

Slump test

A slump cone, straight edge, scoops, steel rule and tamping rod were used. The slump cone was filled with freshly mixed concrete in three approximately equal layers, roding each exactly 25 times while standing the cone on a solid, flat impermeable and clean surface of concreting and bricklaying floor. The final layer slightly protruded above the cone was strike off from the cone while concrete droppings around the base were cleaned. The cone was lifted steadily, vertically and the slump was measured as the difference between the highest points on the slumped concrete and its original level in the cone by inverting the empty cone alongside the slumped concrete, placing a straight edge with a rule; results are shown in Table 6.

Weight development determination

The cubes and cylindrical specimens were weighed before testing and the densities of cubes at different time of testing were measured. Prior to testing, the specimens were brought out of the tank, left outside in the open air for about 2 h before crushing.

Splitting tensile test

Splitting tensile test was conducted at the Mechanical Laboratory of the Polytechnic, Ibadan. After the specimens had been cured for the proper length of time in the water tank, the immersed specimens were taken out from water and allowed to dry. The machine was set for the required range and diametrical lines were drawn on the two ends of the specimen to ensure that they are on the same axial place, after noting the weight and dimension of the specimen. A plywood strip was placed on the lower plate, then the specimen was placed above the lower plate and the other plywood strip was placed above the specimen. The specimen was loaded continuously without shock at uniform rates until failure occurred and the failure load was recorded. The results are given in Tables 9 and 10.

Compressive test

After the specimens had been cured for the proper length of time in the water tank, the concrete cube specimens were crushed at ages 7, 14, 21, 28 days of curing using the compression testing machine available in the Civil Engineering Laboratory of the Polytechnic, Ibadan. The cube was placed between the compressive plates parallel to the surface and then compressed at uniform rate (that is, without shock) until failure occurred. The maximum load at failure and the compressive strength were read through the screen at the top of the machine. The compressive strength was manually calculated by dividing the maximum load in Newtons (N) by the average cross sectional area of the specimen in square millimeters (mm^2) (Tables 7 and 8).

RESULTS

The chapter presents and discuses the results obtained from the preliminary test, tests carried out on both fresh and hardened concrete.

Chemical analysis of GSA

Chemical analysis of GSA is given in Table 3.

Sieve analysis of GSA, fine aggregate and coarse aggregate

The graph of the sieve analysis of the GSA, fine aggregate and coarse aggregate is given in Figure 1.

Aggregate impact value (AIV) and aggregate crushing value (ACV)

This is given in Table 4 Where M_1 = the mass of the test

Table 4. Result of AIV and ACV tests.

	Aggregate impact test	Aggregate crushing test
W1 (g)	642	759
W2 (g)	177	205
AIV (%)	28	-
ACV (%)	-	27

Table 5. Result of specific gravity of G.S.A.

Sample weight	Test A	Test B
W1	25.60	25.30
W2	60.12	59.59
W3	80.20	80.10
W4	68.24	68.00
G	1.54	1.55

specimen (in gram); M_2 = the mass of the material passing the 2.36 mm test sieve (in gram). Both AIV and ACV were calculated from the equation as follows:

$$AIV = \frac{M1}{M2} \times 100$$

Specific gravity determination

Table 5 gives result of specific gravity determination.

$$S.G = \frac{W_2 - W_1}{(W_4 - W_1) - (W_3 - W_2)}$$

Where W1 = weight of empty flask, W2 = weight of flask + cement, W3 = weight of flask + cement + water, W4 = weight of flask + water. Average = 1.54.

Slump test result

Table 6 and Figure 2 shows that the slump decreases with increasing %GSA replacement for both mixes.

Compressive test

Tables 7 and 8 and Figures 3 and 4 shows results of the compressive test.

Splitting tensile test

Tables 9 to 12 and Figures 5 to 8 shows results of the splitting tensile test.

DISCUSSION

Table 3 shows that groundnut shell ash contains the main chemical constituents of cement. ASTM C-618 (2007) specifies that any pozzolan that will be used as cement replacement in concrete requires a minimum of 70% for SiO_2, Al_2O_3 and Fe_2O_3 and that silica, of all the oxides, which is normally considered the most important, should not fall below 40% of the total. From Table 3, the total amount of SiO_2, Al_2O_3 and Fe_2O_3 was 54.79% which was less than the value specified by ASTM C-618 (2007). However, since calcination temperature has significant effect on these three oxides, GSA could still be a good and suitable pozzolan, when it is calcined at higher or lower temperature than 600°C. The particle size distribution of GSA, fine and crushed granite is shown in Figure 1. The uniformity coefficient for crushed granite is greater than 4.0, which implies that the material is suitable for concrete works and its coefficient of curvature of 1.1 lies within the required range of values that is, 1.0 and 3.0. The S-shaped curve of sand and GSA shows that it is well graded. BS 812: Part 110 and Part 112:1990 specify that ACV and AIV respectively should not be greater than 30% for construction purpose. The ACV (27%) and AIV (28%) from Table 4 fall within the acceptable limit (that is, 30% and below), which implies that the aggregate used in this research is suitable for concrete works. The specific gravity of the GSA (1.54) was less than that of the OPC (3.15) it replaced, this means that a considerable greater volume of cementitious materials will result from mass replacement. It was observed that the splitting tensile strength increases with curing age but decreases with GSA inclusion. Tensile strength was roughly about 10% of compressive strength, this agrees with the specification of BS 8110.

The compressive strengths of concrete cube specimens for different percentages of GSA are shown in the Figures 3 and 4 for concrete mixes 1: 2:4 and 1:2.3:2.6. For each mix, compressive strength decreases as GSA contents increases (that is, as percentage of cement decreases). Generally, compressive strength increases with curing age for both mixes. For 1:2:4 concrete mix, 0% ash (100% cement) that served as the control, strength increased from 10.80 N/mm^2 at 7 days to24.06 N/mm^2 at 28 days that is about 140% increment. At 5% ash, strength increased by 140% while increments

Table 6. Results of slump test on concrete with GSA partial replacement (1:2:4 and 1:2.3:2.6 mix).

OPC/GSA	0%	5%	10%	15%	20%
Slump 1:2:4 mix (mm)	30	23	18	15	-
Slump 1:2.3:2.6 mix (mm)	20	-	15	-	13

Table 7. Compressive test result for normal concrete 1:2:4 mix.

Sample	GSA%	7 days	14 days	21 days	28 days
CN_0	0	10.80	13.29	17.91	24.06
CN_5	5	9.29	12.30	18.32	21.34
CN_{10}	10	10.50	12.20	16.35	22.33
CN_{15}	15	8.23	10.23	11.33	15.34

Table 8. Compressive test result for HSC (1:2.3:2.6).

Sample	GSA%	7 days	14 days	21 days	28 days
CH_0	0	14.05	19.34	27.11	35.11
CH_{10}	10	10.11	19.11	22.34	27.33
CH_{20}	20	8.16	17.74	16.56	21.34

Table 9. Splitting tensile test result for normal concrete (1:2:4).

Sample	GSA%	7 days	14 days	21 days	28 days
TN_0	0	1.01	1.48	2.01	2.67
TN_5	5	0.54	1.15	1.86	2.56
TN_{10}	10	1.46	1.30	1.64	2.11
TN_{15}	15	0.88	0.95	1.15	1.99

Table 10. Splitting tensile test result for HSC (1:2.3:2.6).

Sample	GSA%	7 days	14 days	21 days	28 days
TH_0	0	2.36	2.61	2.72	4.21
TH_{10}	10	1.20	1.11	4.33	4.01
TH_{20}	20	0.98	1.74	1.56	2.80

Table 11. Compressive and splitting tensile strengths of concrete compared (normal concrete, 1:2:4).

Sample	GSA%	7 days	14 days	21 days	28 days
CN_0	0	10.80	13.29	17.91	24.06
TN_0	0	1.01	1.48	2.01	2.67
CN_5	5	9.29	12.30	18.32	22.33
TN_5	5	0.54	1.15	1.86	2.56
CN_{10}	10	10.50	12.20	16.35	21.34
TN_{10}	10	1.46	1.30	1.64	2.11
CN_{15}	15	8.23	10.23	11.33	15.34
TN_{15}	15	0.88	0.95	1.15	1.99

Table 12. Compressive and splitting tensile strengths of concrete compared (HSC, 1:2.3:2.6).

Samples	GSA%	7 days	14 days	21 days	28 days
CH_0	0	14.05	19.34	27.11	35.11
TH_0	0	2.36	2.61	2.72	4.21
CH_5	10	10.11	19.11	22.34	27.33
TH_{10}	10	1.20	1.11	3.33	4.01
CH_{20}	20	8.16	17.74	16.56	21.34
TH_{20}	20	0.98	1.74	1.56	2.80

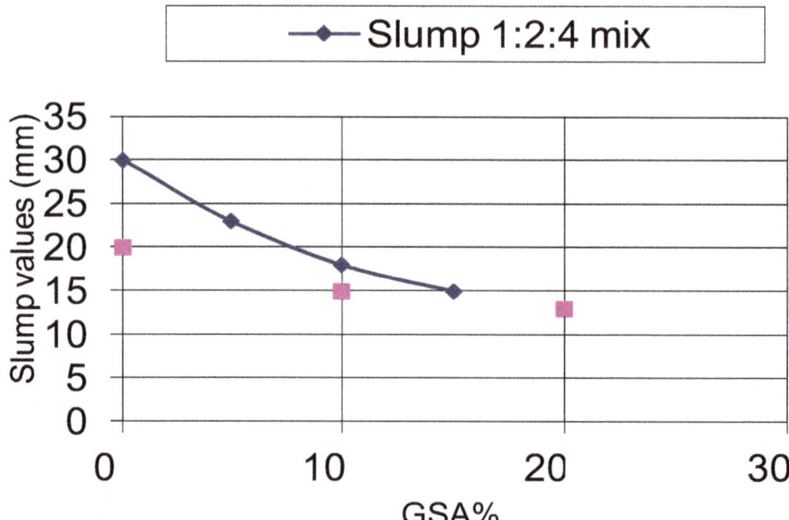

Figure 2. Slump value for various percentages of GSA in concrete 1:2:4 and 1:2.3:2.6 mixes.

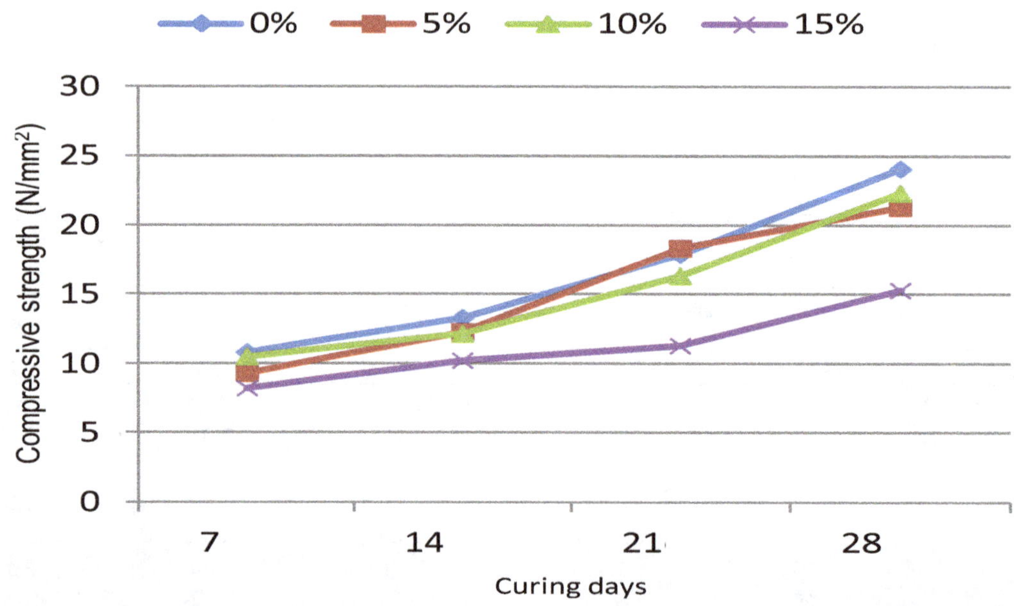

Figure 3. Compressive strength variation with different percentages of GSA in concrete with mix 1:2:4.

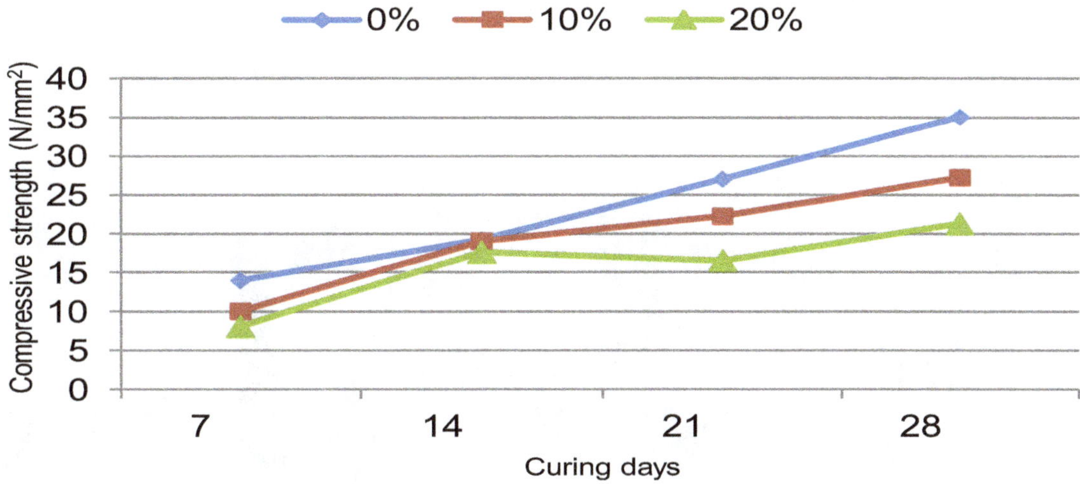

Figure 4. Compressive strengths variation with different percentages of GSA in concrete with mix 1:2.3:2.6.

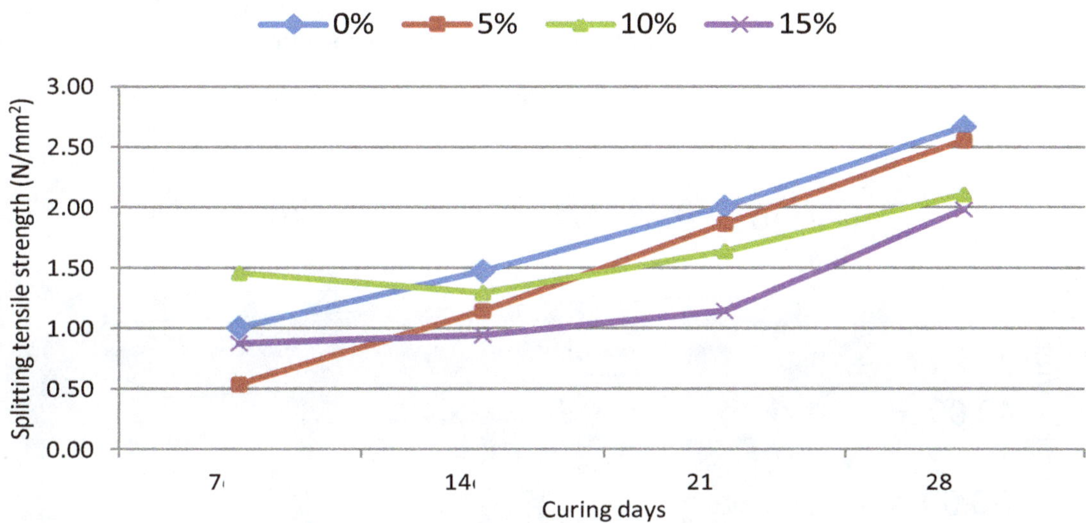

Figure 5. Splitting tensile strength variation with different percentages of GSA in concrete with mix 1:2:4.

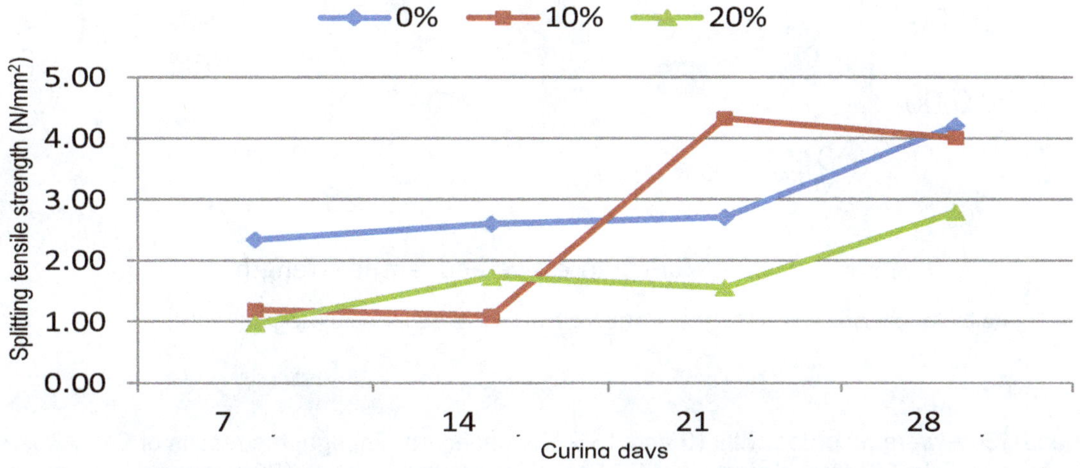

Figure 6. Splitting tensile strengths variation with different percentages of GSA in concrete with mix 1:2.3:2.6.

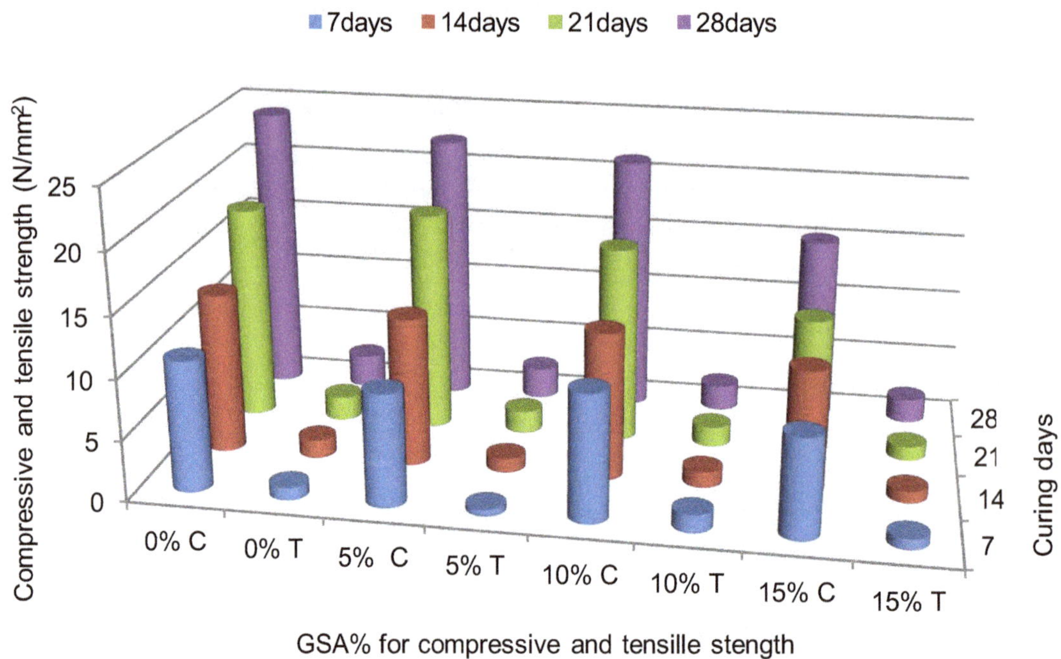

Figure 7. Compressive and splitting tensile strength of concrete compared (normal concrete, 1:2:4).

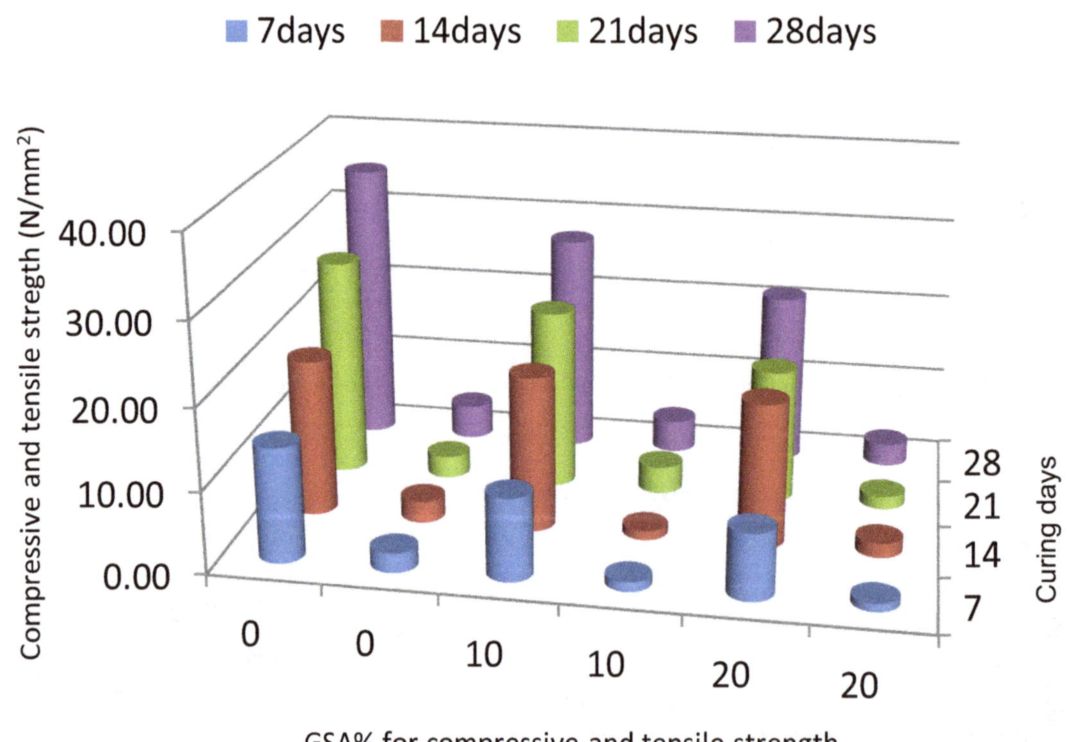

Figure 8. Compressive and splitting tensile strength of concrete compared (HSC, 1:2.3:2.4).

of about 150 and 130% were recorded with 10 and 15% ash respectively from 7 to 28 days curing period. For 1:2.6:2.3 concrete mix, the same trend of increments was observed. Though, the results of OPC/GSA concrete was lower than that of 100% cement in both cases, it can be used for light load bearing elements.

Conclusion

The following conclusions are drawn after this study which investigated the strength performance of modified concrete with groundnut shell ash (GSA). It was discovered that the groundnut shell ash contains all the main chemical constituents of cement though in different proportions compared to that of OPC. This means it will be a good replacement for cement, if the optimum calcinations temperature is established and the right proportion is used:

(1) The experimental results showed that GSA is a good pozzolanic material which reacts with calcium hydroxide forming calcium silicate hydrate. The pozzolanic activity of GSA increases with increase of time.
(2) The slump values for both concrete mixes show that the slump decreases with increasing GSA replacement.
(3) The specific gravity of the GSA gotten was less than that of the OPC it replaced, this means that a considerable greater volume of cementitious materials will result from mass replacement.
(4) Though the strength of OPC/GSA concrete was lower than that of 100% cement, it can be used for light load bearing elements.

RECOMMENDATIONS

The following recommendations are made for further investigations:

(1) Superplasticicer should be introduced so that early strength could be generated and lower water/cement ratio is maintained.
(2) Other test such as corrosion resistance, shrinkage properties, and absorption rate should be carried out on the GSA concrete.
(3) It is recommended that the concrete curing should be extended beyond 28 days to ascertain the long term strength development of ash modified concrete.
(4) GSA calcinations temperature should be varied to establish optimal temperature for pozzolanic activity of GSA.

REFERENCES

ASTM C618 (2007). Standard Specification for Portland Cement", American Society for Testing and Materials, 100 Barr Harbor Drive, P. O. Box C700, West Conshohocken, PA 19428-2959,United States.

Alabadan BA, Olutoye MA, Abolarin, MS, Zakariya M (2005). Partial Replacement of Ordinary Portland Cement (OPC) with Bambara Groundnut Shell Ash in Concrete", Leonardo Electronic J. Pract. Technol. (LEJPT) 6:43-48.

BS 8110: Part 2, (1997), Structural use of concrete. British Standards Institution, her majesty stationary office, London.

BS 812, Part 110 (1990), Methods for determination of aggregate crushing value. British Standards Institution, her majesty stationary office, London.

BS 812: Part 112 (1990). Methods for determination of aggregate Impact value. British Standards Institution, her majesty stationary office, London.

Shetty MS (2005). Concrete Technology, Theory and Practice, S. Chand and Company LTD. 7361, RAM \AGAR, NEW DELHI-1 10 055.

Malhotra VM, Mehta PK (1996). "Pozzolanic and Cementitious Materials" Gordon and Breach Publishers, 1996 SA.

Assessing the properties of freshly mixed concrete containing paper-mill residuals and class F fly ash

Bashar S. Mohammed* and Ong Chuan Fang

Department of Civil Engineering, College of Engineering, University Tenaga Nasional, Km-7, Jalan Kajang-Puchong, 43009 Kajang, Selangor, Malaysia.

The disposal of wastewater treatment-plant solids collected from the paper-mill has become a crucial problem as the landfill space is limited. The use of paper-mill residuals in concrete formulations was investigated as an alternative to landfill disposal. Currently, there is still lack of information on the effect of fresh properties of residual concrete. Furthermore, the effect of class F fly ash on fresh properties of residual concrete is not yet discovered. The fresh properties of concrete containing paper-mill residuals and class F fly ash were investigated by different workability tests, fresh concrete unit weight and air content tests. The influence of the variables such as superplasticizer dosage, water/cementitous material ratio, residual content, and fly ash content on the workability tests was revealed in this study.

Key words: Recycling, paper-mill, concrete, fly ash.

INTRODUCTION

Wastewater treatment-plant solid residues from pulp and paper industry are those solid materials collected in the process of treating water used in the mill prior to its release into the environment. These wastewater treatment plant residuals consist predominantly of primary and secondary solids derived from primary and secondary treatment.

The solid residuals from pulp and paper industry are also one of the largest solid waste streams generated from the industries. For instance, the pulp and paper industry in United State generated 5.8 million dry tons of wastewater treatment-plant residuals in 1995, and only 2.8 million dry tons (49%) were managed in beneficial use applications, with the balance being disposed in a landfill (NCASI, 1999). The continuous increasing of wastewater treatment-plant residuals generated from paper industry has increased concern over the amount and quality of future landfill space. ACI 544.1R reported that the kraff pulp fibre that used as a processed natural fibre to reinforced concretes posses an ultimate tensile strength of 101,500 psi (700 MPa) (ACI Committee, 1996). The residual solids from wastewater treatment-plant which still remains certain amount of cellulose fibres may have a potential to become an economical source for micro-fibre reinforcement of concrete.

Naik et al. (2003), mentioned that by using proper amounts of fibrous residuals in concrete formulation, water, and HRWRA, concrete mixtures containing the residuals were produced comparable to reference concrete mixtures (no residuals) in slump and compressive strength (Naik et al., 2003; Naik et al., 2005; Chun and Naik, 2005). With almost equivalent density, the concrete containing residuals can achieve higher splitting tensile and flexural strength than the reference concrete (Naik, 2005).

Fly ash, known as pulverized-fuel ash, is the ash precipitated electrostatically or mechanically from the exhaust gases of coal-fired power stations, is used in concrete for reasons including economics, improvements and reduction in temperature rise in fresh concrete, workability, and contribution to durability and strength in hardened concrete (Lane and Best, 1982; ACI Committee, 2003). It was found that adding to the paper mill residuals which are derived from recycled fibre source has shown no avail on the mechanical, durability and workability of concrete; however the including of fly ash in concrete showed a promising improvement in durability and long term strength

*Corresponding author. E-mail: bashar@uniten.edu.my, bashar_sami@hotmail.com.

Table 1. Chemical composition of PC and class F fly ash.

Oxide composition	PC, %	Class F fly Ash, %	Requirements for ASTM C 618
SiO_2	21.54	62.5	
Fe_2O_3	3.63	3.5	
Al_2O_3	5.32	23.4	
CaO	63.33	1.8	
MgO	1.08	0.34	5.0 max
SO_3	2.18	1.2	5.0 max
K_2O	-	0.95	
Na_2O	-	0.24	
Loss in ignition	2.5	5.61	6.0 max

development of paper-mill residual concrete (Mohammed and Fang, 2010).

The environmental and economic reasons are one of the important reasons for using paper mill residuals in concrete. According to Naik's study in United State (Naik et al., 2003), saving on disposal cost of residuals by using of paper mill residual in concrete is about $0.375/cubic yard of concrete. If residuals were used for microfiber reinforcement of 20% of concrete produced in the U.S., there could be economic benefits of $360 million for the concrete industry and $30 million for the paper industry per year.

Since 1997, Naik has initiated several studies on the use of paper-mill residuals in concrete production (Chun and Naik, 2005; Naik et al., 2004). However, information on the properties of freshly mixed concrete containing paper-mill residuals is not enough and furthermore there is no research yet to study the rheological effect of fly ash as partial replacement of Portland cement in producing of concrete containing paper-mill residuals.

Primary residuals which are derived from the primary wastewater treatment in the paper-mill factory and class F fly ash were used to produce concrete in this study. The use of paper-mill residuals and class F fly ash in concrete formulation exhibited different rheological properties. This paper reports some quality characteristics of freshly mixed concrete which included paper-mill residuals, and class F fly ash as partial replacement of Portland cement.

MATERIALS AND MIXTURE PROPORTIONS

Portland cement, fly ash, coarse and fine aggregates

Portland cement (PC) Type I, which conforms to the requirement of ASTM C 150 – 04 (ASTM, 2004) was used to produce the concrete in this research work. The PC used has a specific gravity of 3.1. Class F fly ash, which conforms to the requirement of ASTM C 618 – 03 (ASTM, 2004) was used as partial replacement of PC. The fly ash was produced and received from Kapar Energy Ventures power plant in Malaysia. The specific gravity of Class F fly ash used is 2.04. The chemical composition of PC and Class F fly ash is presented in Table 1. The percentage loss on ignition (LOI) of Class F fly ash, which indicates the carbon content, is within the requirement of the standard.

The coarse aggregates (CA) used were graded 9.5-mm nominal maximum-sized crushed stone. The crushed stone had a bulk density of 1,571 kg/m^3, a specific gravity of 2.61, and 0.81% absorption. 9.5-mm crushed stone was used to ensure better mechanical performance, so that the differences in the mechanical properties between mixtures containing residuals and reference mixtures can be easily detected. The sand (S) used had a bulk density 1,706 kg/m^3, a specific gravity of 2.66, 1.9% absorption, and a fineness modulus of 2.45.

The superplaticizer (SP) used in this research was an aqueous solution of a modified polycarboxylate conforming to the requirements of EN 934-2 (British Standards Institution, 2001). The manufacturer recommends a dosage rate of 0.2 to 0.8% of the cement mass for medium workability.

Paper-mill residuals

The paper-mill residue used in this study was the primary sludge recovered from the first processing stage (primary clarifier) in a paper mill. Tables 2 and 3 presented the properties of the residuals used in this research work. Figure 1 shows scanning electron micrographs of the oven-dried primary residuals.

Mixture proportions

The mixture proportions are presented in Table 4. A total of 77 concrete mixtures were produced in this study. The effects between water / cementitious material ratios (W/CM) with residual contents, and fly ash contents with residual contents on the compressive strength were first investigated by using two factors factorial experiment. The W/CM of this study was set at 0.37, 0.40 and 0.45. The residual content of 0, 0.25, 0.5, 0.75, 1.0, 1.5 or 2% (% of total cementitious material mass) was included in the concrete mixtures. The residual used in this research had undergone mechanical dewatering process in the paper mill, and assumed to be in saturated condition and would not further contribute to form the cement paste in the mixing. The w/cm is calculated from the mass of water included to react with the cementitious material to form cement paste. Superplasticizer dosage of 0, 0.2, 0.4 or 0.8% (L/100 kg of cementious material) was applied in the concrete mixtures. The Class F fly ash contents ranged from 20 to 60% of the total cementitious material mass (20, 30, 40, 50 and 60% replacement) which were used as partial replacement of PC.

Preparation of concrete specimens

Prior to the mixing of concrete mixture, deflocculating of residual fibres is necessary to ensure the residual clumps can be dispersed into individual fibre, and subsequently distributed evenly in the concrete mixtures.

Naik and Chun's method (2003) was employed for the deflocculation of residuals. A high speed mixer was used to deflocculate or repulp the residual. Mechanical repulping was performed by immersing the fibrous residuals in room-temperature water until no further clumps were observed.

Test specimens of concrete were made and cured according to the requirement of ASTM C 192M – 02 (ASTM, 2004). The mixing of concrete started with adding the coarse aggregate and some of the mixing water into the mixer. The mixer was then allowed to start and stop after it turned a few revolutions. Then the fine aggregate was added, the mixer was allowed to start and stop after it turned a few revolutions. Then the cement, the rest of the water, and SP

Table 2. Physical properties of paper-mill residuals.

Type of residual	Type of mill	Fibre origin	Moisture content (%)a	Apparent specific gravity	Loss on Ignition at 590˚C (%)a	Wood- fibre content (%)b
Primary	Paper	Recycled	225	1.661	46	36

[a]Percent of oven-dried (105 ˚C) mass of residuals; [b] Wood Fibre Content (%) = 1.083 * LOI at 590°C - 14.1 (Naik et al., 2003).

Table 3. Oxides composition of ash left after ignition of dried residuals at 1,000˚C.

Element	% by mass a
LOI at 1,000˚C	53.450
MgO	2.140
Al2O3	8.620
SiO2	15.100
K2O	0.290
CaO	17.400
TiO2	0.707
Fe2O3	1.660
Total	99.985

[a]Percent of oven-dried (105˚C) mass of residuals.

Figure 1. Scanning electron micrographs of oven-dried primary residuals.

Table 4. Mixture proportions.

Total mixtures	CM:S:CA:W	Residual content, % or kg/100 kg of cement [a]	SP, % or L/100 kg of cement	Fly ash replacement, %
28	1:0.767:1.275:0.37	0 - 2	0 - 0.8	0
35	1:0.767:1.275:0.37	0 - 2	0.4	20 - 60
7	1:0.889:1.378:0.4	0 - 2	0	0
7	1:1.093:1.55:0.45	0 - 2	0	0

[a]As-received moist residuals were used. The quantities of residuals shown are on oven-dried basis.

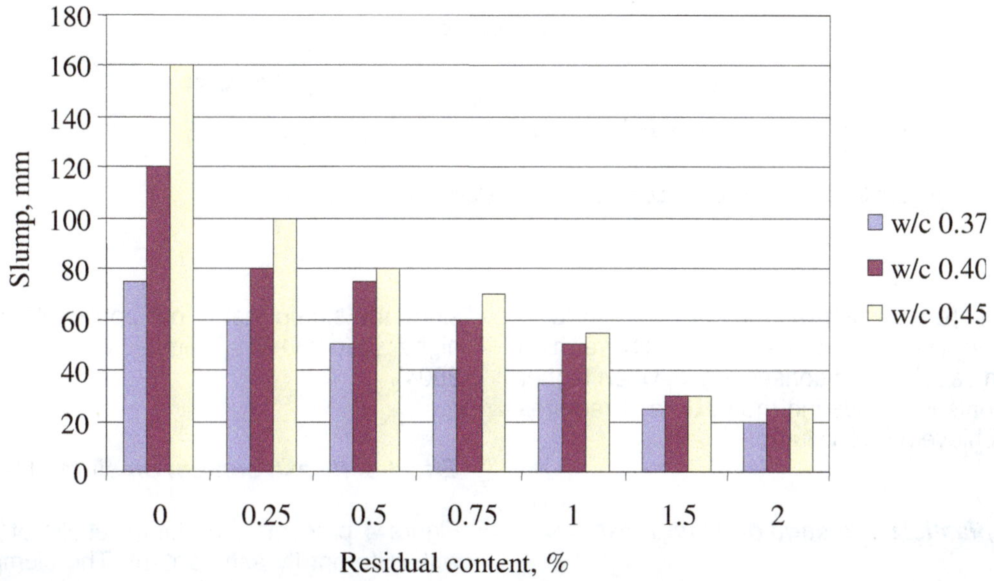

Figure 2. Effect of residual content on slump.

were added into the mixer. After all ingredients were in the mixer, the mixing continued for three minutes followed by a three minutes rest, and a two minutes final mixing.

The fresh concrete properties were further determined and 100-mm cube specimens were cast to determine the 28 days compressive strength.

Experimental program and test procedures

Immediately after the completion of mixing, the fresh concrete was tested on the workability properties, unit weight and air content. The types of workability test and its conformity standard are stated as:

1. Slump test – ASTM C 143-03 (ASTM, 2004)
2. Compacting factor test – BS 1881: Part 103:1993 (British Standard Institution, 1993).
3. Vebe test – BS EN 12350-3:2000 (British Standards Institution, 2000).

The unit weight and air content tests were performed simultaneously with the workability tests and the test procedures are conformed to ASTM C 138M – 01a (ASTM, 2004).

Compressive strength of the 100-mm cube specimens was determined at 28 days in accordance with BS 1881-116:1983 (British Standard Institution, 1983).

RESULTS AND DISCUSSIONS

Compressive strength

From the experimental results, the 28 days compressive strength of the concrete decreased when residual content and fly ash content increased in the mixtures. Due to the mixture proportion, the compressive strength decreased when W/CM ratio increased. The effect of superplasticizer dosage on the compressive strength of concrete was found to be insignificant.

Slump test

Effect of paper-mill residual on slump test

The effect of residual content on the slump is shown in Figure 2. The slump values reduced when residuals contents increased. The results shown are in good agreement with the observation of Tarun (Naik et al., 2003; Chun and Naik, 2005), where the paper-mill

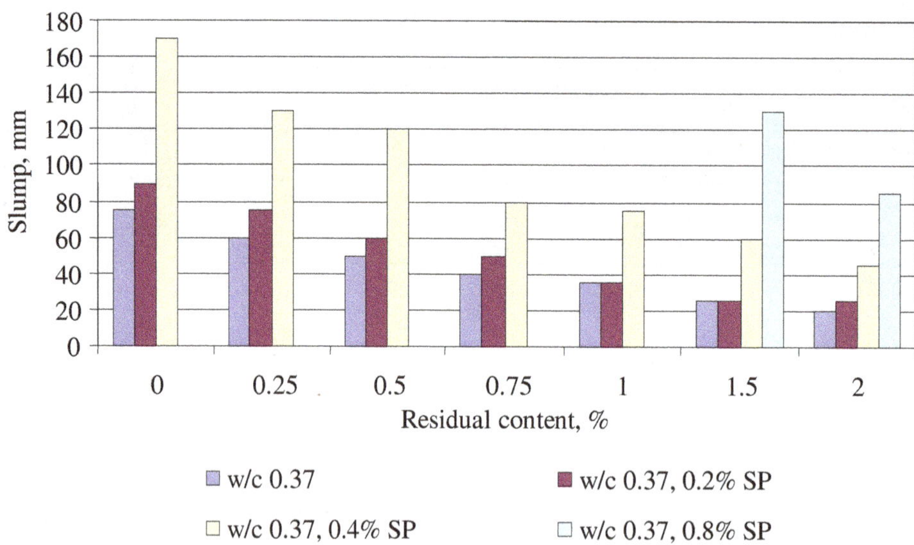

Figure 3. Effect of superplasticizer dosage on slump test.

residuals could increase water demand of concrete for a given slump. The as-received residuals exhibit a high water absorption capability. In consequence, when higher amount of residuals is included in the mixture, it requires more water to achieve a given slump.

Effect of superplasticizer dosage on slump test

Figure 3 presents the slump results of residual concrete with different superplasticizer dosage. Slight or no improvement in the slump values were observed between the mixtures with no superplasticizer and 0.2% superplasticizer at each of their respective residual content. The concrete mixture containing residuals has improved their workability through the incorporation of 0.4% superplasticizer, and has achieved workability range from medium to high depending on the residual content. The mixture with 0.8% of superplasticizer recorded collapse slump for residuals content from 0 to 1% of cement mass, and a relatively high workability could be observed in the mixtures of 1.5 and 2% residual content.

With proper combination of residuals and superplasticizer, workability of concrete can be adjusted. The workability of concrete containing paper-mill residuals was improved by using superplasticizer. Higher dosage of superplasticizer would result in higher workability. However, it should also be noticed that bleeding and segregation could be observed when 0.8% superplasticizer is introduced into the concrete mixtures containing of 0 to 0.5% of residual contents.

Apparently, slump test is not suitable to determine the workability of concrete with 0.8% superplasticizer and containing 0, 0.25, 0.5, 0.75 and 1% residual contents since the slump test is not applicable when slump value higher than 230-mm is limited in ASTM C 143–03 [ASTM, 2004].

Effect of fly ash content on slump test

Figure 4 presents the slump results of residual concrete with different fly ash content. The slump values reduced when fly ash contents increased. A very low slump, 5-mm was observed in the mixture of 60% of fly ash replacement and 2% residuals content. The slump test might not be suitable to determine the workability of concrete when the slump is less than 10-mm.

In this investigation, the workability of concrete was governed by the fly ash content and also the residuals content. Increasing of fly ash content and residuals content would lead to increase in water demand of fresh concrete to achieve a given slump. The replacement of cement with fly ash is known to contribute to a reduction in water demand and an improvement of workability due to the smaller size and the essentially spherical form of the fly ash particles (Berry and Malhotra, 1980). However, the literature does contain some contrary data. According to CUR Report 144 (Bijen and Selst, 1992), an amount of fly ash in excess of that required to cover the surface of the cement particles would confer no further benefit with respect to water demand. The reduction in water demand becomes larger with an increase in the fly ash content only to about 20%.

Brink and Halstead (1956) have reported that some fly ashes reduced the water requirement of test mortars, whereas others (generally of higher carbon content) showed increased water requirement above that of control mortars. Welsh and Burton (1958) reported loss of

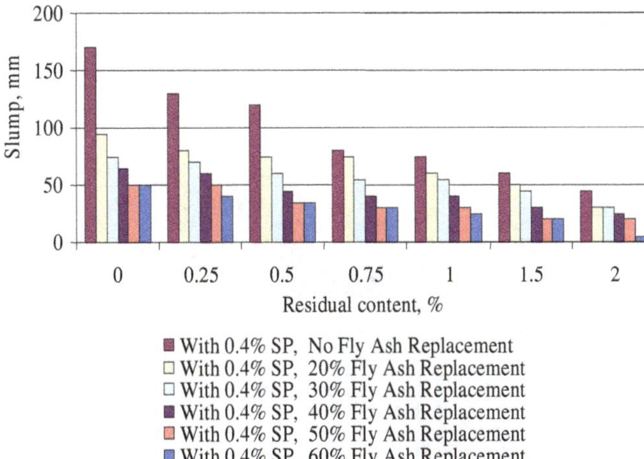

Figure 4. Effect of fly ash content on slump test.

slump and flow for concretes made with some Australian fly ashes used to partially replace cement, when water content was maintained constant. Reshi (1973) has reported that experience with a number of Indian fly ashes showed that all those examined increased the water requirement of concrete.

The carbon content of the class F fly ash used in this study was 5.61% based on LOI testing. Although the carbon content was still within the 6% allowed by ASTM, the relatively high carbon content might have led to the adverse effect on workability.

Table 5. Analysis of variance of slump tests.

	Slump tests			
Terms	W/CM	Residual	SP	Fly ash
F	8.94	31.23	35.53	23.41
P	0.0004	<0.0001	<0.0001	<0.0001
S		13.738		
R2		86.53%		

Table 6. Analysis of variance of compacting factor tests.

	Compacting factor tests			
Terms	W/CM	Residual	SP	Fly ash
F	4.07	42.28	28.04	12.68
P	0.022	<0.0001	<0.0001	<0.0001
S		0.034		
R^2		86.53%		

Table 7. Analysis of variance of Vebe tests.

Vebe tests			
W/CM	Residual	SP	Fly ash
44.48	40.27	71.26	13.04
0.022	<0.0001	<0.0001	<0.0001
	0.034		
	86.53%		

Relationship between slump, compacting factor and Vebe time

The relation between slump, compacting factor and Vebe time is presented in Figure 5. Generally, the workability was found to be reduced when residual content and fly ash content increased in the concrete mixtures, and the workability could be improved by increasing of W/CM ratio and superplasticizer dosage. Similar trend of workability could be found in the slump test results, as well as compacting factors and Vebe times. The range of slump in this study was within 5 to 170-mm. ASTM C 143–03 (ASTM, 2004) stated the limitation of slump test, which was only applicable for slump in the range of 15-mm to 230-mm. It can be seen that at low workability, most of the slump recorded the same value while Vebe test recoded a wider range of Vebe time.

Through the observation of relationship, Vebe test seems to be more sensitive in the low workability condition than slump test and compacting factor test. However, slump test and compacting factor test provide better indication of workability at high workability. It is due to the vibration time of Vebe tests of high workability mixtures are relatively close between each other since there are not much effort required in the compaction. BS EN12350-3 (British Standard Institution, 2000) also stated the Vebe time should be in the range of 5 to 30 s.

The compacting factors recorded in this study ranged around 0.7 to 1. However, BS 1881: Part 103 :1993 (British Standard Institution, 1993) stated that the normal range of compacting factor test lies between 0.8 - 0.92. The sensitivity of the compacting factor is reduced outside the normal range of workability and is generally unsatisfactory for compacting factor greater than 0.92. Thus, Vebe test provides better indication at low workability, while slump test provides better indication at high workability as compared to compacting factor test.

From the result, slump test and compacting factor test seem to be a more suitable testing for the concrete containing residuals, except for the mixture with high residual content (2%) and high fly ash portion mixture (60% fly ash replacement). On the other hand, Vebe test is merely suitable for testing with low W/CM ratio, high fly ash content, and low superplasticizer mixture.

Tables 5 - 7 show the analysis of variance for slump, compacting factor, and Vebe tests. Within the range of the test variables, the analysis of variance showed that

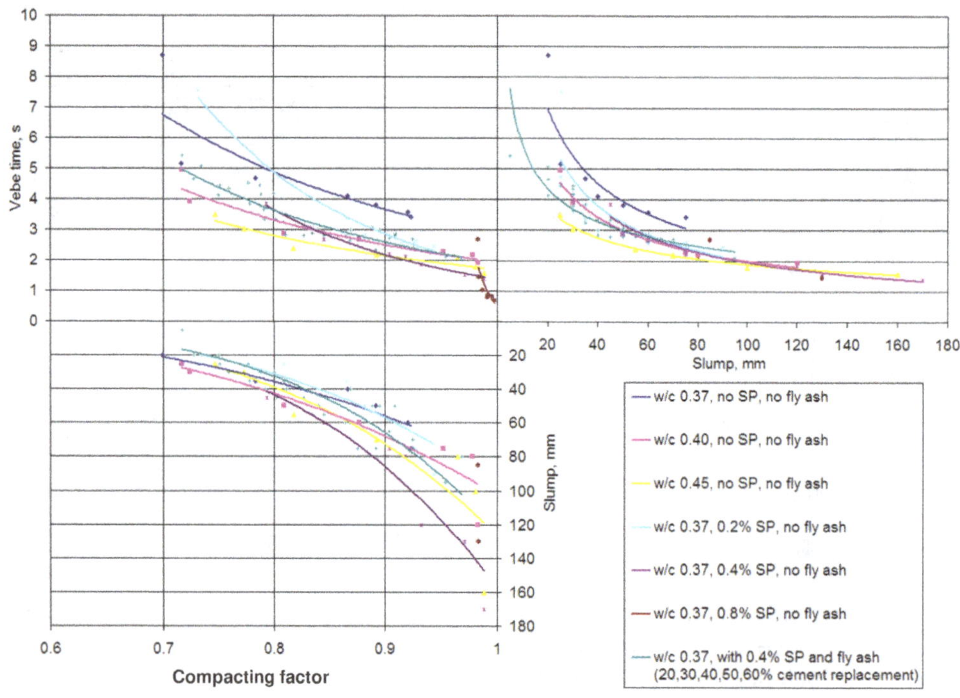

Figure 5. Relation between workability tests for different mixture category.

the factors that influence slump according to descending order are:

1. Superplasticizer dosage
2. Residual content
3. Fly ash content
4. W/CM

The result leads to the observation that slump test was more sensitive to the superplasticizer dosages while W/CM ratio has relatively less influence on the slump. When slump test was performed, the slump occurred when the concrete specimen sheared its self-weight. As the superplasticizer dosage increases, the dispersion of cement particles would reduce the friction and the shear of concrete specimen, eventually causing a great influence on the slump.

The analysis of variance showed that the factors that influence compacting factor according to descending order are:

1. Residual content.
2. Superplasticizer dosage.
3. Fly ash content.
4. W/CM ratio.

Note that the residual content plays a more important role in the compacting factor than the slump and Vebe test. When higher residual content was introduced in the mixtures, the mixtures became more cohesive, and subsequently affected the free-fall process of the compacting factor test and the mass of partially compacted concrete collected. Hence, the influence of residual content on compacting factor is more than slump test and Vebe test.

The analysis of variance showed that the factors that influence Vebe time according to descending order are:

1. Superplasticizer dosage.
2. W/CM ratio.
3. Residual content.
4. Fly ash content.

Apparently, the superplasticizer dosage is sensitive to the vibration of Vebe test. Besides, the effect of residual content and fly ash content to Vebe time is not as significant as slump test and compacting factor test. It is due to the vibration compaction of Vebe test that could compact the mixtures containing paper-mill residuals and fly ash more easily than the rod tamping method of slump test and compacting factor test.

Unit weight

The fresh concrete unit weights of concrete mixtures are presented in Figure 6. The unit weight of concrete mixtures decrease with increase of residuals content. There was no significant difference observed between the unit weights of the reference concrete mixtures with

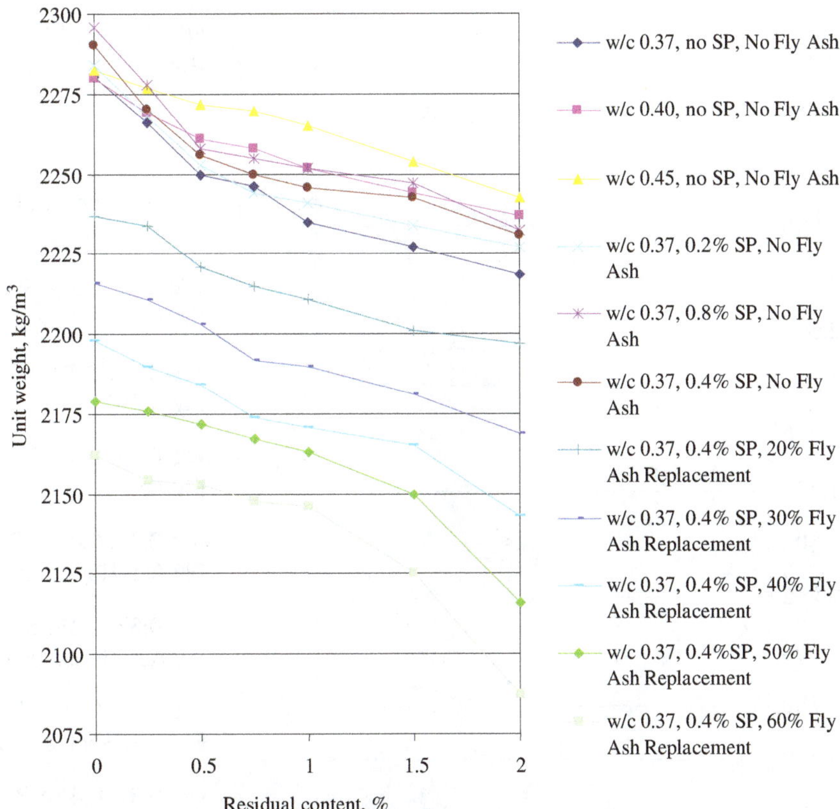

Figure 6. Unit weight versus paper-mill residual content.

different W/CM ratio. However, when higher residual content was introduced in the mixtures, the unit weight also increased as W/CM ratio increased. Higher dosage of superplasticizer could produce concrete mixtures with slightly higher unit weight. The replacements of cement with fly ash have resulted in comparative greater difference in the unit weight of concrete mixtures. The unit weights of mixtures containing fly ash decreased with higher fly ash content in the concrete mixtures.

The paper-mill residual, which has a specific gravity of 1.661, is a light weight ingredient comparing to the aggregates and cementitious material used. Thus, when more residuals are included in the mixtures, more volume in the concrete would be occupied by the residuals and it would reduce the overall unit weight.

The increase of unit weight when W/CM ratio increased in the concrete mixtures which have the same residual content was due to the decreasing of air content when higher W/CM ratio was introduced in the mixtures.

The superplasticizer improved the concrete mixture workability and orientated the paste more compactly. Hence, higher dosage of superplasticizer probably would slightly increase the unit weight of the concrete mixtures containing the same residual content.

Fly ash has a lower specific gravity (2.04 kg/m^3) than Portland cement (3.1 kg/m^3). Substitution of fly ash for an equal weight of Portland cement therefore increased the paste volume in the concrete and reduced the density of concrete. Hence increasing fly ash content in concrete mixture would reduce the overall concrete unit weight.

Air content

The air contents of concrete mixtures are presented in Figure 7. The air contents of concrete mixtures were affected by residual content, W/CM ratio, dosage of superplasticizer, and fly ash content.

Higher residuals content in the concrete mixtures resulted in higher air content. The air contents decreased as W/CM ratio increased and higher superplasticizer dosage resulted in lower air content. Generally, higher fly ash content showed lower air content in the concrete mixtures, exceptions were found in mixture of 50% cement/fly ash replacement with 2% of cement mass of residuals, and 60% cement /fly ash replacement with 1.5 and 2% of cement mass of residuals.

Since paper-mill residuals exhibited high water absorption characteristic, higher residual content included into the concrete mixture would absorb more water which is used to form cement paste. Eventually, the workability of concrete mixtures would reduce and more pores would

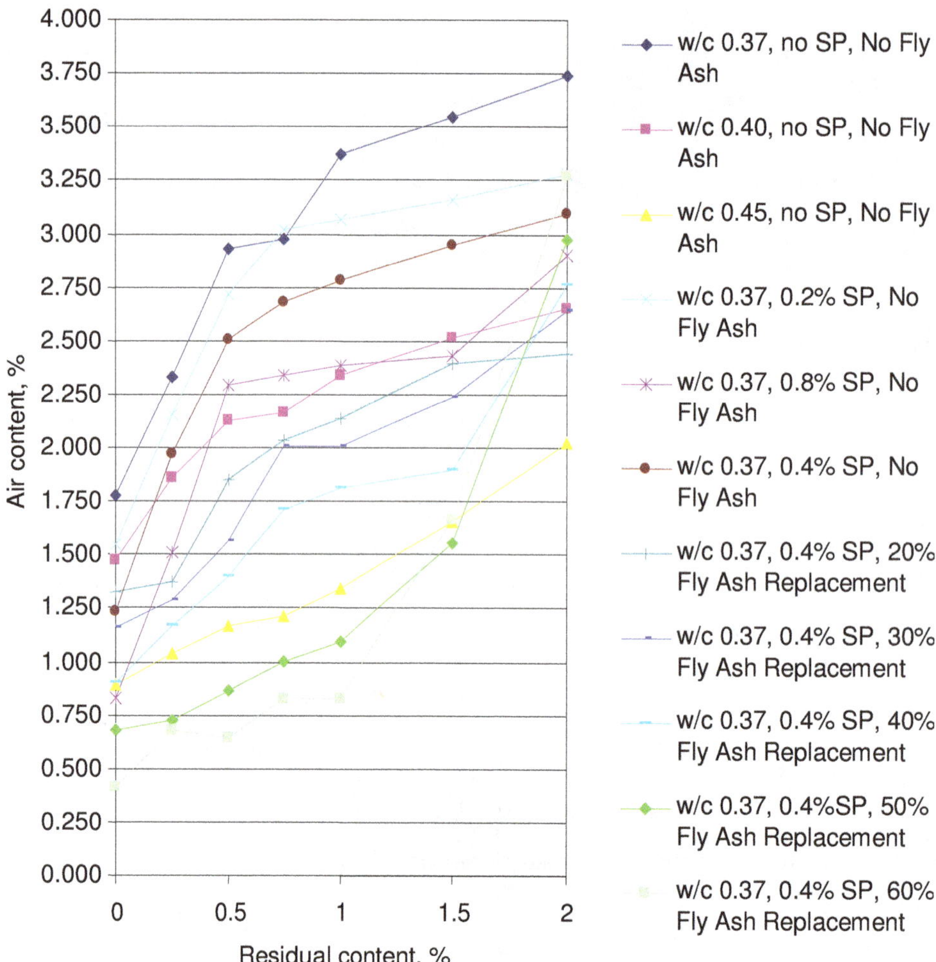

Figure 7. Air content versus paper-mill residual content.

exist in the concrete mixture which contributes to higher air content.

The concrete mixtures of higher W/CM ratio would have more water content for cementitious material to form cement paste rather than been absorbed by the fibrous residuals and eventually create pores around the residuals in the mixtures. Hence, when W/CM ratio increased, the air content of concrete mixtures decreased.The results also indicate that by adding superplasticizer in the mixtures, it improved the workability of concrete and orientated the mixture more compactly, leaving lesser space for air content.

The effect of fly ash on the air content of concrete mixtures generally agrees with the previous researchers (Lane and Best, 1982; ACI Committee, 2003; Lane, 1983). Higher fly ash content in concrete mixture could decrease the air content since fly ash is more effective to fill the voids in the concrete mixtures, which are due to its fineness and shape. The exceptions found in the high cement replacement might be caused by the reducing of workability of concrete.

Conclusions

Based on the results presented, the following conclusions can be drawn:

(i) Generally, higher residual content and fly ash content in the concrete mixtures would increase the water demand of concrete for a given slump, thus, decreasing the workability of fresh concrete. The workability of concrete containing paper-mill residuals and fly ash content could be adjusted and improved by using proper amount of superplasticizer. The Class F fly ash decreased the workability of concrete due to its high percentage of fly ash replacement in mixture proportion and high carbon content which increases the water demand.

(ii) The slump test and compacting factor test were the most suitable testing for the workability of concrete containing paper-mill residuals as compared to Vebe test. However, the influence of the variables such as superplasticizer dosage, W/CM ratio, residual content,

and fly ash content were varied for each respective type of tests.
(iii) The unit weight of fresh concrete decreased with increasing of residuals content and fly ash content. Including superplasticizer in the concrete mixtures resulted in slightly higher unit weight compared with mixtures without superplasticizer content.
(iv) Higher residuals content in the concrete mixtures resulted in higher air content. Superplasticizer and Class F fly ash could reduce the air content of the concrete mixtures containing paper-mill residuals.

REFERENCES

ACI Committee 226 (2003). "Use of Fly Ash in Concrete (ACI 226.3R-87)," American Concrete Institute, Farmington Hills, MI, p. 41.

ACI Committee 544 (1996). "Report on Fibre Reinforced Concrete (ACI 544.1R-96)," American Concrete Institute, Farmington Hills, MI, pp. 56-64.

ASTM (2004). "Annual Book of ASTM Standards, Vol. 04.01: Cement; Lime; Gypsum," American Society for Testing and Materials, Philadelphia, PA.

ASTM (2004). "Annual Book of ASTM Standards, Vol. 04.02: Concrete and Aggregates," American Society for Testing and Materials, Philadelphia, PA.

Berry EE, Malhotra VM (1980). "Fly Ash for Use in Concrete - A Critical Review," ACI Publications. J. Proc., 77(2): 59-73.

Bijen J, Selst I (1992). CUR Report 144, "Fly Ash as Addition to Concrete," Research carried out by INTRON, Institute for Material and Environmental Research B.V., A.A. Balkema, Rotterdam,.

Brink RH, Halstead WJ (1956). "Studies Relating to the Testing of Fly Ash for Use in Concrete," Proceeding, ASTM, 56: 1161-1206.

BS 1881(1983). "Method for determination of compressive strength of concrete cubes," British Standard Institution, London, Part 116: 1983.

BS 1881(1993). "Method for Determination of Compacting Factor," British Standard Institution, London. Part 103:1993.

BS EN (2000). "Testing fresh concrete – Part 3: Vebe test," London: British Standards Institution, 12350-3:2000.

BS EN 934 (2001). Admixtures for concrete, mortar and grout – Part 2: Concrete admixtures – Definitions, requirements, conformity, marking and labelling, London: British Standards Institution.

Chun Y, Naik TR (2005). "Concrete with Paper Industry Fibrous Residuals: Mixture Proportioning." ACI Mater. J., 102(4): 237-243.

Lane RO (1983). "Effects of Fly Ash on Freshly Mixed Concrete," Concrete International: Des. Constr., 5(19): 50-52.

Lane RO, Best JF (1982). "Properties and Use of Fly Ash in Portland Cement Concrete." Concrete Int. Mag., 4(7): 81-92.

Mohammed BS, Fang OC (2010). Mechanical and durability properties of concretes containing paper-mill residuals and fly ash. Constr. Build. Mater. doi:10.1016/j.conbuildmat. 2010.07.015.

Naik TR, Chun Y, Kraus RN (2003). "Use of residual solids from pulp and paper mills for enhancing strength and durability of ready-mixed concrete," Final report submitted to the U.S. Dept. of Energy for the Project DE-FC07-00ID13867, September.

Naik TR, Chun Y, Kraus RN (2005). "Paper Industry Fibrous Residuals in Concrete and CLSM," Report No. CBU-2005-10, UWM Center for By-Products Utilization, Department of Civil Engineering and Mechanics, The University of Wisconsin-Milwaukee.

Naik TR, Friberg T, Chun Y (2004). "Use of Pulp and Paper-mill Residual Solids in Production of Cellucrete." Cement Concrete Res., 34(7): 1229-1234.

National Council for Air and Stream Improvement (NCASI) (1999). "Solid Waste Management Practices in the U.S. Paper Industry-1995," Technical Bulletin 793, September.

Rehsi SS (1973). "Studies on Indian Fly Ashes and Their Use in Structural Concrete," Proceedings, Third International Ash Utilization Symposium (Pittsburgh, Mar. 1973), Information Circular IC 8640, U.S. Bureau of Mines, Washington, D.C. 1973, pp. 231-245.

Welsh GB, Burton JR (1958). "Sydney Fly Ash in Concrete." Commonw. Eng., 45: 62-67.

Permissions

All chapters in this book were first published in JCECT, by Academic Journals; hereby published with permission under the Creative Commons Attribution License or equivalent. Every chapter published in this book has been scrutinized by our experts. Their significance has been extensively debated. The topics covered herein carry significant findings which will fuel the growth of the discipline. They may even be implemented as practical applications or may be referred to as a beginning point for another development.

The contributors of this book come from diverse backgrounds, making this book a truly international effort. This book will bring forth new frontiers with its revolutionizing research information and detailed analysis of the nascent developments around the world.

We would like to thank all the contributing authors for lending their expertise to make the book truly unique. They have played a crucial role in the development of this book. Without their invaluable contributions this book wouldn't have been possible. They have made vital efforts to compile up to date information on the varied aspects of this subject to make this book a valuable addition to the collection of many professionals and students.

This book was conceptualized with the vision of imparting up-to-date information and advanced data in this field. To ensure the same, a matchless editorial board was set up. Every individual on the board went through rigorous rounds of assessment to prove their worth. After which they invested a large part of their time researching and compiling the most relevant data for our readers.

The editorial board has been involved in producing this book since its inception. They have spent rigorous hours researching and exploring the diverse topics which have resulted in the successful publishing of this book. They have passed on their knowledge of decades through this book. To expedite this challenging task, the publisher supported the team at every step. A small team of assistant editors was also appointed to further simplify the editing procedure and attain best results for the readers.

Apart from the editorial board, the designing team has also invested a significant amount of their time in understanding the subject and creating the most relevant covers. They scrutinized every image to scout for the most suitable representation of the subject and create an appropriate cover for the book.

The publishing team has been an ardent support to the editorial, designing and production team. Their endless efforts to recruit the best for this project, has resulted in the accomplishment of this book. They are a veteran in the field of academics and their pool of knowledge is as vast as their experience in printing. Their expertise and guidance has proved useful at every step. Their uncompromising quality standards have made this book an exceptional effort. Their encouragement from time to time has been an inspiration for everyone.

The publisher and the editorial board hope that this book will prove to be a valuable piece of knowledge for researchers, students, practitioners and scholars across the globe.

List of Contributors

M. A. Ezzat
Department of Mathematics, Faculty of Education, Alexandria University, Alexandria, Egypt

H. M. Atef
Department of Mathematics, Faculty of Education, Alexandria University, Alexandria, Egypt

A. Sivakumar
Department of Civil Engineering, Vellore Institute of Technology (VIT) University, Vellore - 632007, India

Ayman Ahmed Seleemah
Structural Engineering Department, Faculty of Engineering, Tanta University, Egypt

Mohamed S. Issa
Housing and Building National Research Center, Giza, Egypt

S. M. Elzeiny
Housing and Building National Research Center, Giza, Egypt

A. Sivakumar
VIT University, India

P. Gomathi
VIT University, India

Menghong Wang
School of Civil and Transportation Engineering, Beijing University of Civil Engineering and Architecture, Beijing 100044, China

Xiaodong Guo
School of Civil and Transportation Engineering, Beijing University of Civil Engineering and Architecture, Beijing 100044W, China

C Freeda Christy
School of Civil Engineering, Karunya University, Coimbatore, TamilNadu, India, 641 114

D Tensing
School of Civil Engineering, Karunya University, Coimbatore, TamilNadu, India, 641 114

R Mercy Shanthi
School of Civil Engineering, Karunya University, Coimbatore, TamilNadu, India, 641 114

B. Madhusudana Reddy
Department of Civil Engineering, Sri Venkateswara University (S. V. U.) College of Engineering, Sri Venkateswara (S. V.) University, Tirupati -517502, Andhra Pradesh, India

I. V. Ramana Reddy
Department of Civil Engineering, Sri Venkateswara University (S. V. U.) College of Engineering, Sri Venkateswara (S. V.) University, Tirupati -517502, Andhra Pradesh, India

A. Sivakumar
Structural Engineering Division, SMBS, VIT University, Vellore, Tamilnadu, India

V. M. Sounthararajan
Structural Engineering Division, SMBS, VIT University, Vellore, Tamilnadu, India

A. Sivakumar
Structural Engineering Division, Vellore Institute of Technology (VIT) University, India

A. Sivakumar
Department of civil engineering, Vellore Institute of Technology (VIT) University, India

M. Prakash
Department of civil engineering, Vellore Institute of Technology (VIT) University, India

J. I. Aguwa
Department of Civil Engineering, Federal University of Technology, Minna, Nigeria

Rohit Ghosh
Department of Construction Engineering, Jadavpur University, Kolkata, India

T. E. Omoniyi
Department of Agricultural and Environmental Engineering, Faculty of Technology, University of Ibadan Oyo State, Nigeria

B. A. Akinyemi
Department of Agricultural and Environmental Engineering, Faculty of Technology, University of Ibadan Oyo State, Nigeria

T. Karech
Université de Batna Dep de Génie-Civil Algérie

K. Lupogo
Department of Geology, College of Natural and Applied Sciences, University of Dar es Salaam, P. O. Box 35052, Dar es Salaam Tanzania

A. E. Abalaka
Department of Building, Federal University of Technology, Minna, Nigeria

O. G. Okoli
Department of Building, Ahmadu Bello University, Zaria, Nigeria

T. E Omoniyi
Department of Agricultural and Environmental Engineering, University of Ibadan, Nigeria

A.O. Olorunnisola
Department of Agricultural and Environmental Engineering, University of Ibadan, Nigeria

B.A Akinyemi
Department of Agricultural and Environmental Engineering, University of Ibadan, Nigeria

A. Sivakumar
Structural Engineering Division, School of Mechanical and Building Sciences, VIT University, India

V. M. Sounthararajan
Structural Engineering Division, School of Mechanical and Building Sciences, VIT University, India

Omer Nawaf Maaitah
Faculty of Engineering, Mutah University, Karak, Jordan

Nafeth A. Abdel Hadi
Balq'a Applied University, Jordan

Monther Abdelhadi
Department of Civil Engineering, Al-Ahliyya Amman University, Jordan

S. B. Raheem
Civil Engineering Department, the Polytechnic, Ibadan, Oyo State, Nigeria

G. F. Oladiran
Civil Engineering Department, the Polytechnic, Ibadan, Oyo State, Nigeria

F. A. Olutoge
Civil Engineering Department, University of Ibadan, Oyo State, Nigeria

T. O. Odewumi
Civil Engineering Department, the Polytechnic, Ibadan, Oyo State, Nigeria

Bashar S. Mohammed
Department of Civil Engineering, College of Engineering, University Tenaga Nasional, Km-7, Jalan Kajang-Puchong, 43009 Kajang, Selangor, Malaysia

Ong Chuan Fang
Department of Civil Engineering, College of Engineering, University Tenaga Nasional, Km-7, Jalan Kajang-Puchong, 43009 Kajang, Selangor, Malaysia

www.ingramcontent.com/pod-product-compliance
Lightning Source LLC
Chambersburg PA
CBHW080658200326
41458CB00013B/4904

* 9 7 8 1 6 8 2 8 5 0 8 2 4 *